Harappa Script Primer

--Cryptography for metalwork trade

S. Kalyanaraman
Sarasvati Research Center, 2016

ISBN-13: 978-1541274921

ISBN-10: 154127492X

Harappa Script Primer

--Cryptography for metalwork trade

Table of Contents

Harappa Script Primer

--Cryptography for metalwork trade

The thesis is organized in the following sections and posits that Meluhha language speakers followed the spiritual values of Veda cultural traditions, used Harappa Script inscriptions to create data archives of metalwork, accounting for Bronze Age trade transactions

Section 1. Pre-Sanskrit civilization of Meluhha speakers. Harappa Script is hieroglyphic in nature and decipherment can be attempted the way Egyptian hieroglyphs were decrypted

On the salient features of Harappa (Indus) Script, some remarkable observations -- related to decipherment researches -- were made by John Hubert Marshall, who was Director-General of Archaeological Survey of India from 1902 to 1908 and associated with excavations of two major sites: Mohejo-daro and Harappa. That he made these observations based on the then available corpus of 541 seal impressions should be underscored. “The proper names and names of professions on these seals do not supply sufficient material for successful decipherment. It is not possible to separate word and sign groups; the declensions and verb inflections cannot be detected here, and the pronouns are entirely absent. Until longer inscriptions of a literary and historical character are discovered, not much advance in the interpretation can be expected. A good many important facts can be determined, however, to clear the ground for satisfactory research. In the first place this script is in no way even remotely connected with either the Sumerian or Proto-Elamitic signs…The Indus inscriptions resemble the Egyptian hieroglyphs far more than they do the Sumerian linear and cuneiform system. And secondly, the presence of detached accents in the Indus Script is a feature which distinguishes it from any of these systems. Although vowels must be inherent in all the signs, nevertheless some of the signs and accents must be pure vowel signs. For this reason alone, it is necessary to resign further investigation to Sanskrit scholars. If future discoveries make it possible to transliterate the signs, and the language proves to be agglutinative, it will then be a problem for Sumerolotists…I am convinced that all attempts to derive the Brahmi alphabet from Semitic alphabets were complete failures…This study of the script of a pre-Sanskrit civilization of the Indus Valley is made from the material supplied by 541 impressions of small press seals." (Marshall, J.H., 1931, *Mohenjo-daro and the Indus Civilization, Repr. Asian Educational Services*, 1931, Vol. I, Delhi, pp. 423-424)

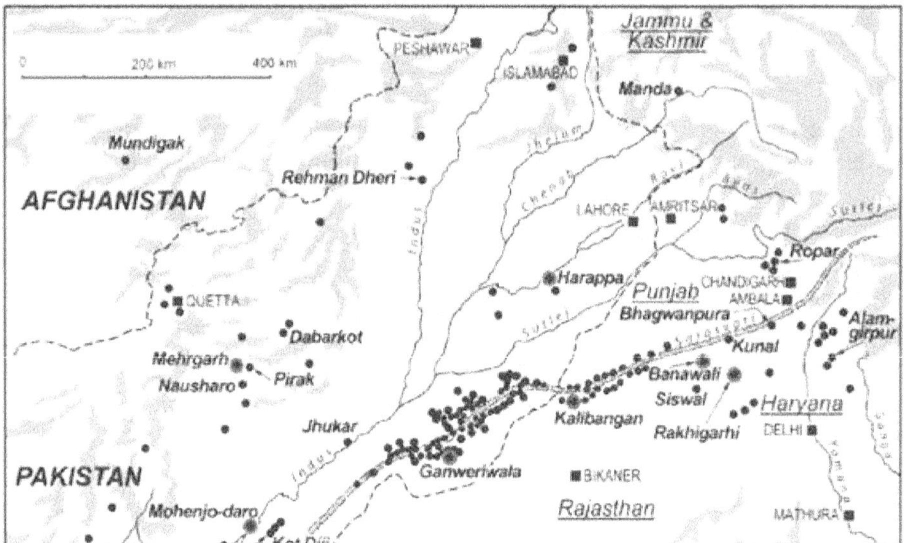

These observations provide the framework for the decipherment attempted by this researcher who started the investigations by delineating the courses of a 'Pre-Sanskrit' Vedic River called River Sarasvati in North-western Bharata. Since the days of Marshall's archaeological work of 1920's, remarkable progress has been made by the explorations identifying over 2000 archaeological sites (or 80% out of a total of over 2600) on the Sarasvati River Basin. These explorations and limited excavations in about twenty sites have now taken the Harappa Script Corpora to a substantial size of over 8000 inscriptions making them fit for cryptographic analyses or cryptanalyses. The Corpora constitute a quantum leap from 541 seal impressions studied by Marshall.

Corpora of Harappa Script inscriptions

Based on numerous resources and from the collections of inscribed objects held in many museums of the world, such as the Metropolitan Museum of Art, the Harappa Writing Corpora include Sarasvati heiroglyphs, representing many facets of glyptic art of Harappa Civilization. The corporas also transcribes many texts of inscriptions, corresponding to the epigraphs inscribed on objects. The compilation is based mostly on published photographs in archaeological reports right from the days of Alexander Cunningham who discovered a seal at Harappa in 1875, of Langdon at Mohenjodaro (1931) and of Madhu Swarup Vats at Harappa (1940). The corpus includes objects collected in Bharata, Pakistan, other countries and the finds of the excavations at Harappa by Kenoyer and Meadow during the seasons 1994-1995 and 1999-2000.

Parpola's initial corpus (CISI 1973) included a total number of 3204 texts. After compiling the pictorial corpus, Parpola notes that there are approximately 3700 legible inscriptions (including 1400 duplicate inscriptions, i.e. with repeated texts). Both the concordances of Parpola and Mahadevan complement each other because of the sort sequence adopted. Parpola's concordance is sorted according to the sign following the indexed sign. Mahadevan's concordance is sorted according to the sign preceding the indexed sign. The latter sort ordering helps in delineating signs which occur in final position.

Two subsequent volumes of the pictorial corpus include a total of inscriptions as collections from Bharata and Pakistan (e.g., Harappa 2590, Mohenjo-daro 2129, Lothal 281, Chanhudaro 50). Additional inscriptions have been discovered which are not included in the three volumes of Corpus of Inscriptions of Parpola et al.: e.g. Khirsara, Farmana, Gilund, Bhirrana, Kunal, Garo Biro, Rakhigarhi). Mahadevan concordance (1977) with only 2906 artifacts, excludes inscribed objects which do not contain 'texts; for example, this concordance excludes about 50 seals inscribed with the 'svastikā' pictorial motif and a pectoral which contains the pictorial motif of a one-horned bull with a device in front and an over-flowing pot. Parpola concordance has been used to present such objects which also contain valuable orthographic data which may assist in decoding the inscriptions. Many broken objects are also contained in Parpola concordance which are useful, in many cases, to count the number of objects with specific 'field symbolś, a count which also provides some valuable clues to support the decoding of the messages conveyed by the 'field symbolś which dominate the object space. Along the Persian Gulf, in sites such as Failaka, Bahrain, Saar, nearly 2000 Harappa script inscriptions (so-called Dilmun or Persian Gulf seals) have been found on seals and sealings. These are in addition to the Gadd seals of Ancient Near East (Gadd, CJ, Seals of Ancient Bharata Style found in Ur in: Possehl, GL, ed. 1979, *Ancient Cities of the Indus*, Delhi, Vikas Publishing House, p.119). In many sites of Ancient Near East such as Shahdad, Susa (lady spinner artifact with Harappa script hieroglyphs, pot containing metal implements with Harappa script hieroglyphs of fish, quail, flowing water), Tepe Hissar, Haifa (three tin ingots with Harappa script found in a shipwreck), Anau, Altyn Depe (also spelt as Altin Tepe), caravan routes from Ashur and Mari to Kish, Anatolia. Many cylinder seals with cuneiform inscriptions also contain uniquely characteristic Harappa script hieroglyphs such as 'overflowing pot', sun's rays, safflower, pine-cone, fish, scorpion, zebu, buffalo, hair-styles with six curls.

With the publication of CISI Vol. 3, Part 1, the total number of inscriptions from Mohenjo-daro totals 2134 and from Harappa totals 2589; thus, these two sites alone account for 4,723; bring the overall total number of inscriptions to over 8,000 from all sites (including comparable inscriptions on 'Persian Gulf type' circular seals from the total count).

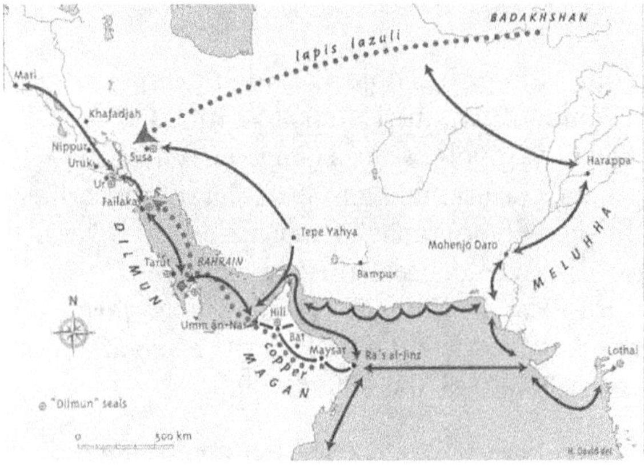

Meluhha-Magan-Dilmun=Mesopotamia[1]

Harappa Script is a corpus of symbols constituting a logo-phonetic writing system which was used in Harappa (also known as Sarasvati-Sindhu) Civilization, during the Bronze Age from ca. 3300 BCE 1 to 600 BCE[1] in the doab of Vedic Rivers Sarasvati and Sindhu. The people and their language was called Meluhha in ancient cuneiform texts. Harappa Script inscriptions occur also in the Persian Gulf (also called Magan and Dilmun in ancient cuneiform texts), in contact areas of Mesopotamia (Ancient Near East) and in Dong Son/Karen Bronze Drums Culture areas of the Ancient Far East (*L'Extrême-Orient* pace George Coedes[2])

Researches have also progressed in delineating the 'Pre-Sanskrit' language bases of the civilization. The received wisdom of 1920's of the framework of languages such as Sanskrit, Egyptian (Coptic), Sumerian, Elamite, Greek, Indo-European has been significantly expanded thanks to Proto-Indo-European language researches. "Proto-Indo-Europen (PIE) is the linguistic reconstruction of the common ancestor of the Indo-European languages. Since there is no written record of this language, the knowledge of it that we have has been reconstructed using historical linguistics, by comparing and extrapolating back from the properties fo the languages descended from it. It is thought that Proto-Indo-European may have been spoken as a single language (before divergence began) around 3500 BCE, though estimates vary by more than a thousand years...Today, there are about 445 living Indo-European (IE) languages descended from PIE, with Spanish, English, Hindi, Portuguese, Bengali, Russian, Punjabi, German, French and Marathi being the ten Indo-European languages with the most native speakers in descending order. PIE is thought to have had a complex system of morphology that included inflectional suffixes as well as ablaut (vowel alterations, for example, sing, sang, sung). Nouns and verbs had complex systems of declension and conjugation respectively... In 1816 Franz Bopp published *On the System of Conjugation in Sanskrit* in which he investigated a common origin of Sanskrit, Persian, Greek, Latin, and German. In 1833 he began publishing the *Comparative Grammar of Sanskrit, Zend, Greek, Latin, Lithuanian, Old Slavic, Gothic,and German*...JuliusPokorny's *Indogermanisches etymologisches Wörterbuch* ("Indo-European Etymological Dictionary", 1959) gave a detailed, though conservative, overview of the lexical knowledge accumulated up until that time. " https://en.wikipedia.org/wiki/Proto-Indo-European_language 'The original speakers may have originated in the Pontic-Caspian steppe of Eastern Europe north of the Black Sea' or from the Ganga River Basin or Mekong-Irrawaddy-Salween River deltas in the Ancient Far East (with evidence of Austro-Asiatic or Proto-Munda languages). A major resource has been compiled called the *Indian Lexicon* with over 8000 semantic clusters from over 25 Ancient Languages of Bharata. Link:
https://www.dropbox.com/s/ykm93xf4unhordu/IndianLexicon.pdf?dl=0

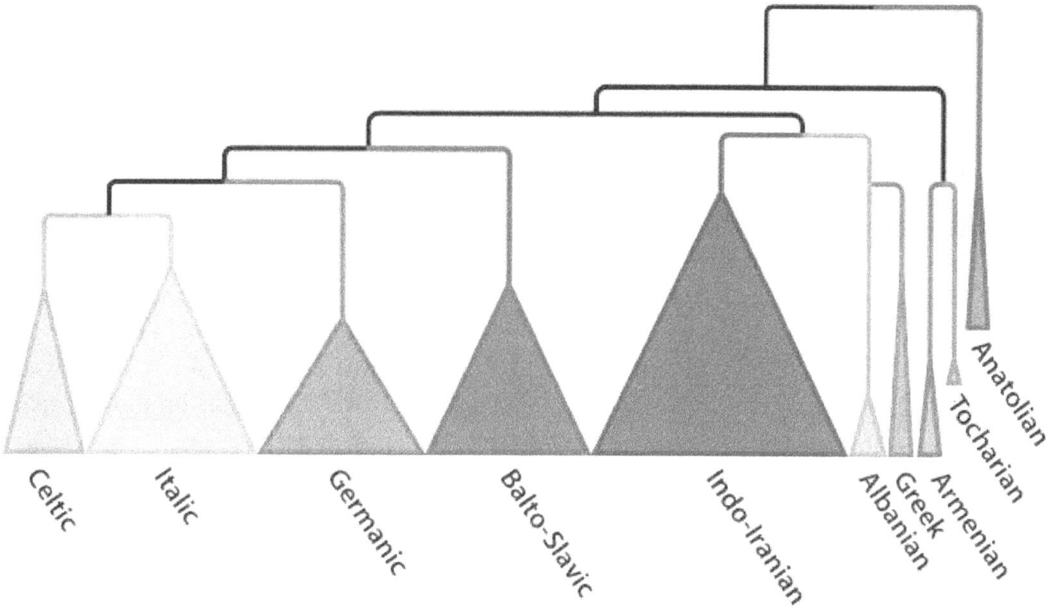

Adapted from R. Bouckaert et al, Science, 2012. More than 400 Indo-European languages diverged from a common ancestral tongue; the earliest ones (top right), Anatolian and Tocharian, arose in today's Turkey and China, respectively.
http://www.sciencemag.org/news/2015/02/mysterious-indo-european-homeland-may-have-been-steppes-ukraine-and
russia?utm_source=facebook&utm_medium=social&utm_campaign=facebook

Georges Pinault pointed to the concordance between Vedic and Tocharian: *amśu ~~ ancu*, 'iron' (Tocharian). *Amśu* (Vedic Sanskrit or Chandas) is a synonym for *Soma*. This concordance makes *Soma* a metaphor for metal of th Bronze Age. See *Harosheth hagoyim – Smithy of nations.* https://www.scribd.com/document/200092919/Harosheth-hagoyim-smithy-of-nations Harosheth hagoyim is a fortress described in the Book of Judges of the Old Bible, as the fortress or cavalry base of Sisera, commander of the army of 'Jabin, king of Canaan'. *Harosheth hagoyim* 'smithy of nations' is cognate with *kharoṣṭī* **खरोष्टी** A kind of alphabet; Lv.1.29 PLUS *gōya* (Prākṛt) '*gotra* or guild or lineage'. These two anecdotal evidences point to the Harappa Script inscriptions as repositories of metalwork catalogues. Harappa writing in Ancient Near East is a tribute to the Meluhha artisans who have established an expansive contact area in Eurasia and left for posterity the bronze-age *harosheth hagoyim*, 'the smithy of nations', an expression used in the Old Bible. This expression is cognate with Meluhha *kharoṣṭī* खरोष्टी *goya,* 'blacksmith lip + guild'. The association of Meluhha with metalwork is discovery of a knowledge system detailing contributions of Bharata artisans to the Bronze Age Revolution, traceable to ancient times of 5th millennium BCE.

It is interesting that Dr. Moti Shemtov refers to Nahal Mishmar as Nachal Mishmar. It is similar to the identification of Meluhha and Mleccha, as cognates ! – both terms derivable from *mliṣṭa* म्लिष्ट [p= 837,3] *mfn.* spoken indistinctly or barbarously Pāṇ. 7-2, 18, indistinct speech.

The hypothesis of this monograph is premised on the definition of *mleccha (meluhha)* as the spoken forms of 'Pre-Sanskrit' or 'Proto-Indo-European' (PIE). Borrowings have occurred among Dravidian, Munda and IE language-families since the present-day consensus among linguists is that Bharata was a *sprachbund* (language union or linguistic area) of the Bronze Age, since many language speakers absorbed language features from one another and made them their own.

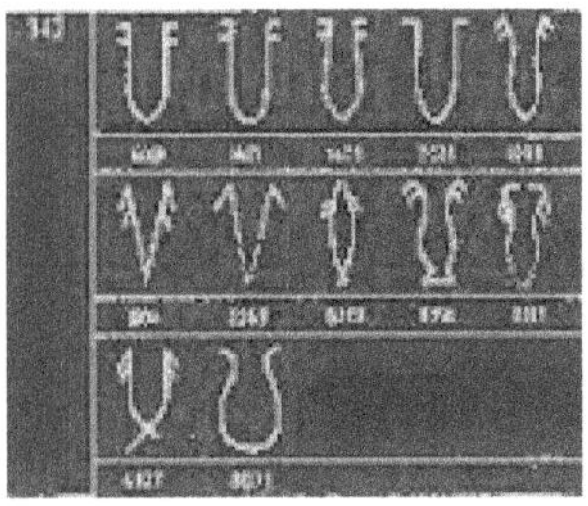

Variants of hieroglyph 'rim of jar' in Harappa Script. This characteristic feature of Bharata *sprachbund* explains the sheer variety of phonetic speech forms as exemplified by the example of Santali word, *kankha* 'rim of jar', a hieroglyph which is of the most frequent occurrence in Harappa Script corpora.

The cognates of this word and concomitant semantic expansions of the Pre-Sanskrit forms such as *kankha* (Austro-Asiatic), *ka<u>nn</u>am* (Tamil) can be seen in the vocables of Indo-Aryan languages: *kárṇaka* m. ' projection on the side of a vessel, handle ' ŚBr. [*kárṇa* --] Pa. *kaṇṇaka* -- ' having ears or corners '; Wg. *kaṇə* ' ear -- ring ' NTS xvii 266; S. *kano* m. ' rim, border '; P. *kannā* m. ' obtuse angle of a kite ' (→ H. *kannā* m. ' edge, rim, handle '); N. *kānu* ' end of a rope for supporting a burden '; B. *kāṇā* ' brim of a cup ', G. *kānɔ* m.; M. *kānā* m. ' touch -- hole of a gun' (CDIAL 2831). கன்னம்[2] *ka<u>nn</u>am, n. < karṇa.* 1. Ear; காது. கன்னமுற நாராசங் காய்ச்சிக் சொருகியபோல் (பிரமோத். 13, 16). 2. Elephant's ear; யானைச் செவி. (திவா.) 3. [K. *kanna*, M. *kannam*.] Cheek *கன்னல்[1] ka<u>nn</u>al, n. perh. கன்[1]. 1. Earthen vessel, water-pot; கரகம். தொகுவாய்க் கன்னற் றண்ணீ ருண்ணார் (நெடுநல். 65*). 2. Perforated hour- glass that fills and sinks at the expiration of a nā<u>l</u>ikai; நாழிகைவட்டில். கன்னலின் யாமங் கொள் பவர் (மணி. 7, 65). 3. Measure of time = 24 minutes; நாழிகை. காவத மோரொரு கன்னலி னாக (கந்தபு. மார்க். 142). கன்[2] ka<u>n</u> , *n.* < கல். 1. Stone; கல். (சூடா.) கன்[1] ka<u>n</u>, *n.* perh. கன்மம். 1. Workmanship; வேலைப்பாடு. கன்னார் மதில்சூழ் குடந்தை (திவ். திருவாய். 5, 8, 3). 2. Copper work; கன்னார் தொழில். (W.) 3. Copper; செம்பு. (ஈடு, 5, 8, 3.) 4. See கன்னத்தட்டு. (நன். 217, விருத்.)

Kaṇḍa. A waterpot of a certain size and shape.

Kaṅkha. / Kankha. / Khaṅkha. } Rim of a vessel, brow of a hill.

Kaṇḍa kankha. / Kaṇḍa kaṅkha. } The rim of a waterpot.

Santali glosses.

म्लेच्छित [p= <u>838</u>,1] *mfn.* = म्लिष्ट <u>Pa1n2.</u> 7-2 , 18 Sch. *mlecchitaka* 'speaking jargon unintelligible to others; म्लेच्छ [p= <u>837</u>,3] any person who does not speak Sanskrit and does not conform to the usual Hindu institutions <u>S3Br.</u> &c (*f*(ई).); a person who lives by agriculture or by making weapons <u>L.</u>; *n.* copper <u>L.</u> (*mleccha-mukha, mlecchAsya, mlecchākya* called copper, copper – so named because the complexion of the Greek and Muhammedan invaders of India was supposed to be copper-coloured); *mlecchayati* 'a person with indistinct speech' (Dhātup. xxxii, I 20) *mleccha-jāti* 'a man belonging to the Mlecchas, a mountaineer (as a Kirāta, S'abara or Pulinda)(MBh.); *mlecchana* 'the act of speaking confusedly or barbarously, Dhātup.

Mleccha, as cognate of *Meluhha* speech is attested in Mahabharata, Jatugrihaparva which records that conversation took place between Yudhishthira and Khanaka in 'mleccha' tongue. It is hypothesized that *Mleccha* aka *Meluhha* was the vernacular, spoken version of what John Maynard Marshall refers to as 'Pre-Sanskrit civilization'.

The Mekong-Irrawaddy-Salween River deltas formed by the three Himalayan glacial rivers are crucial to an understanding of the Bronze Age civilizations, because these deltas are coterminous with the largest Tin Belt of the globe. Tin was a crucial mineral to substitute for the scarcity of naturally occurring arsenical bronzes. Tin or cassiterite mineral added to copper mineral yielded Tin-Bronze alloy which created the Bronze Age revolution, from ca. 5th millennium BCE.

Let us continue to use Marshall's expression, 'Pre-Sanskrit' civilization, but call it populated by

Meluhha speakers, a hypothesis to be tested by further researches. This is suggested because this particular language *Meluhha* is attested by a seal with cuneiform text which refers to the translator of the *Meluhha* language. The cuneiform text reads in Akkadian: Shu-Ilishu EME.BAL.ME.LUH.HA.KI (i.e., interpreter of Meluhha language). Roll out of Mesopotamian Cylinder seal of Shu-Ilishu. Late Akkadian ca. 2200-2113 BCE. Musee du Louvre, Paris.
https://www.penn.museum/documents/publications/expedition/PDFs/48-1/What%20in%20the%20World.pdf

Another Meluhhan holding an antelope on his hands is shown on another cylinder seal. Cylinder seal described as Akkadian circa 2334-2154 BC, cf. figure 428, p. 30. "The Surena Collection of Ancient Near Eastern Cylinder Seals." *Christies Auction Catalogue*. New York City. Sale of 11 June 2001).

Hieroglyphic nature of Harappa (Indus) Script writing system justifies the use of rebus method to decipher and read out the plain text Meluhha inscriptions.

Examples of inscriptions reviewed are:

- Water carrier as a hieroglyph
- Scorpion as a hieroglyph
- Human face as a hieroglyph
- Feeding trough as a hieroglyph

A classic paper by Cyril John Gadd F.B.A. who was a Professor Emeritus of Ancient Semitic Languages and Civilizations, School of Oriental and African Studies, University of London, opened up a new series of archaeological studies related to the trade contacts between Ancient Far East and what is now called Sarasvati-Sindhu (Hindu) civilization.

Gadd's paper was published in the *Proceedings of the British Academy*, XVIII, 1932.

Gadd presents examples of seals from Ur and considers them 'Indian style'. Two such examples Seal impression No. 12 (Water-carrier) and Seal No.11 (Scorpion and Ellipse) are presented in this monograph.

GR Hunter commented on Gadd's insight in *Journal of Royal Asiatic Society*, 1932 pointing out that the enclosure of the pictorial moif (of water-carrier PLUS two stars on either side of the head) by 'parenthesis' marks is perhaps a way of splitting of the ellipse and that this is an unmistakable example of an 'hieroglyph' seal.

Water carrier as a hieroglyph

Seal impression. Ur. C.J. Gadd, Seals of ancient Indian style found at Ur, *Proceedings of the British Academy*, XVIII, 1932, pp. 11-12, Plate II, No. 12; Description: water carrier with a skin (or pot?) hung on each end of the yoke across his shoulders and another one below the crook of his left arm; the vessel on the right end of his yoke is over a receptacle for the water; a star on either side of the head (denoting supernatural?) The whole object is enclosed by 'parenthesis' marks. The parenthesis is perhaps a way of splitting of the ellipse (Hunter, G.R., *JRAS*, 1932, 476). An unmistakable example of an 'hieroglyphic' seal.

In the context of the water-carrier hieroglyph PLUS stars enclosed within parenthesis, CJ Gadd had rightly noted that these are determinatives that the writing system was hieroglyphic. I suggest that the hieroglyphs are read as Meluhha words and substituted by similar-sounding words (i.e., rebus) in Meluhha speech form. Rebus substitution method converts the cipher text into plain text..

kuṭi 'water-carrier' (Telugu) rebus: *kuṭhi* 'smelter' *kuṛī* f. 'fireplace' (H.); *krvṛi* f. 'granary (WPah.); *kuṛī, kuṛo* house, building'(Ku.)(CDIAL 3232) *kuṭi* 'hut made of boughs' (Skt.) *guḍi* temple (Telugu) *meḍha* 'polar star' rebus: *mẽṛhẽt, meḍ* 'iron' (Santali.Mu.Ho.) *dula* 'pair' rebus: *dul* 'metal casting'.

The most frequently occurring glyph -- rim of jar -- ligatured to Glyph 12 becomes Glyph 15 and is thus explained as a *kankha, karṇaka:* 'furnace scribe' and is consistent with the readings of glyphs which occur together in contextual sequence. *Kan-kha* may denote an artisan working with copper, *kaṉ* (Ta.) *kaṉṉār* 'coppersmiths, blacksmiths' (Ta.) Thus, the phrase *kaṇḍ karṇaka* may be decoded rebus as a brassworker, scribe. *karṇaka, karNIka* 'scribe, accountant' *karNi* 'supercargo'

Semantic expansion occurs when this hieroglyph is superscripted with 'rim-of-jar' hieroglyph:

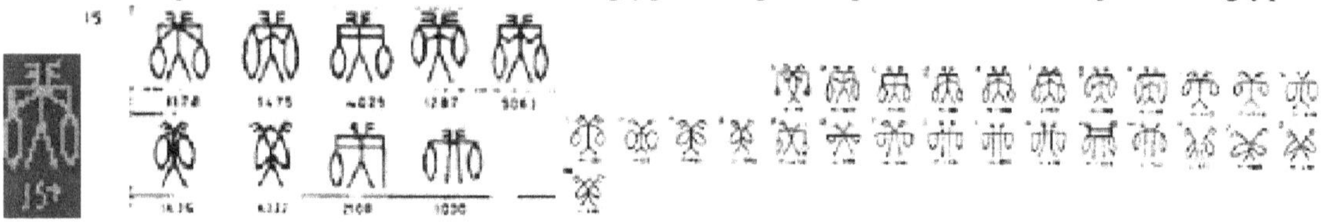

Sign 15 & orthographic variants: *kuṭhi kaṇḍa kankha* 'smelting furnace account (scribe)'. *kuTi 'water-carrier' rebus: kuThi 'smelter' kanda 'pot' rebus: kanda 'fire-altar' kanka, karanika 'rim of jar' rebus: kāraṇika* 'smelter producer'. Thus, the hieroglyph-multiplex is an expression: *kuThi kāraṇika 'smelter-maker.' kuTi karaṇī* 'Supercargo smelter' (i.e. Supercargo responsible for trading produce from smelter and carried by seafaring vessel).

Scorpion as a hieroglyph

Seal; UPenn; a scorpion and an elipse [an eye (?)]; U. 16397; Gadd, PBA 18 (1932), pp. 10-11, pl. II, no. 11 Rectangular stamp seal of dark steatite; U. 11181; B.IM. 7854; ht. 1.4, width 1.1 cm.; Woolley, *Ur Excavations*, IV (1956), p. 50, n.3.

bicha, 'scorpion' rebus: *bica* 'haematite, ferrite ore'.

Hieroglyph: oval shaped bun ingot: *mũhe* 'ingot' (Santali) *mũhã̄* = the quantity of iron produced from a furnace (Santali). Thus, the two hieroglyphs signify haematite ore ingot.

Since the insights of Gadd and Hunter in 1932, the Harappa Script Corpora has grown to include over 8000 inscriptions.

Hieroglyphic nature of the Harappa (Indus) Script writing system may also be seen in the following examples.

Human face as a hieroglyph

mũh 'a face' in Indus Script Cipher signifies *mũh, muhã* 'ingot' or *muhã* 'quantity of metal produced at one time in a native smelting furnace.'

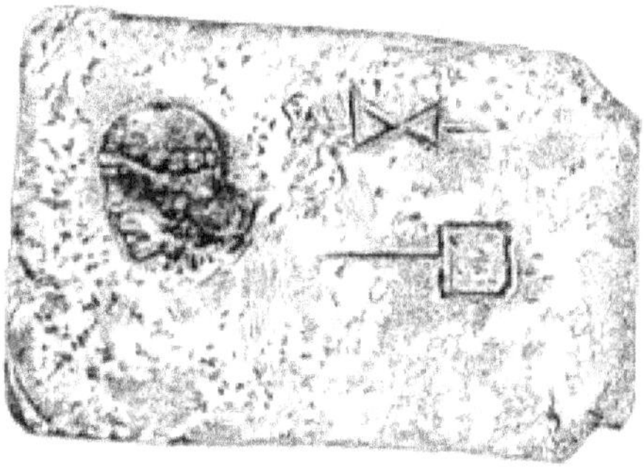

Inscribed tin ingot with a moulded head, from Haifa (Artzy, 1983: 53). (Michal Artzy, 1983, Arethusa of the Tin Ingot, Bulletin of the American Schools of Oriental Research, BASOR 250, pp. 51-55) https://www.academia.edu/5476188/Artzy-1983-Tin-Ignot Face on this tin ingot: *mũhe* 'face' (Santali) Rebus: *mũh* 'ingot' (Santali).

The three hieroglyphs are: *ranku* 'antelope' Rebus: *ranku* 'tin' (Santali) *ranku* 'liquid measure' Rebus: *ranku* 'tin' (Santali). *dāṭu* = cross (Te.); *dhatu* = mineral (Santali)
Hindi. *dhāṭnā* 'to send out, pour out, cast (metal)' (CDIAL 6771). [The 'cross' or X hieroglyph is incised on all three tin ingots found in a shipweck in Haifa.]

Feeding trough as a hieroglyph

Feeding trough signified in front of wild (non-domesticated) animals such as tiger, rhinoceros. *Pattar* 'trough'; rebus *pattar, vartaka* 'merchant, *பத்தர்*[2] *pattar , n.* < *T. battuḍu.* A caste title of goldsmiths; தட்டார் பட்டப்பெயருள் ஒன்று.*Ta.* pātti bathing tub, watering trough or basin, spout, drain; pattal wooden bucket; pattar id., wooden trough for feeding animals. *Ka.* pāti basin for water round the foot of a tree. *Tu.* pāti trough or bathing tub, spout, drain. *Te.* pādi, pādu basin for water round the foot of a tree (DEDR 4079)

prastha2 m.n. ' a measure of weight or capacity = 32 palas ' MBh.Pa. *pattha* -- m. ' a measure = 1/4 āḷhaka, cooking vessel containing 1 pattha '; NiDoc. *prasta* ' a measure '; Pk. *pattha* -- , °*aya* -- m. ' a measure of grain '; K. *path* m. ' a measure of land requiring 1 trakh (= 9 1/2 lb.) of seed '; L. *patth*, (Ju.) *path* m. ' a measure of capacity = 4 boras '; Ku. *pātho* ' a measure = 2 seers '; N. *pāthi* ' a measure of capacity = 1/10 man '; Bi. *pathiyā* ' basket used by sower or for feeding cattle '; Mth. *pāthā* ' large milk pail ', *pathiyā* ' basket used as feeding **trough** for animals '; H. *pāthī* f. ' measure of corn for a year '; Si. *pata* ' a measure of grain and liquids = 1/4 näliya '. *prasthapattra -- .Addenda: prastha -- 2: WPah.poet. *patho* m. ' a grain measure about 2 seers ' (prob. ← Ku. Mth.

pā´tra n. ' drinking vessel, dish ' RV., °*aka* -- n., *pātrī´*- ' vessel ' GṛŚrS. [√pā1] Pa. *patta* -- n. ' bowl ', °*aka* -- n. ' little bowl ', *pātī̆* -- f.; Pk. *patta* -- n., °*tī* -- f., amg. *pāda* -- , *pāya* -- n., *pāī* -- f. ' vessel '; Sh. *păṭi̯* f. ' large long dish ' (← Ind.?); K. *pāthar*, dat. °*tras*m. ' vessel, dish ', *pôturu* m. ' pan of a pair of scales ' (*gahana* -- *pāth*, dat. *pöčü* f. ' jewels and dishes as part of dowry ' ← Ind.); S. *pāṭri* f. ' large earth or wooden dish ', *pāṭroṛo* m. ' wooden **trough** '; L. *pātrī* f. ' earthen kneading dish ', *parāt* f. ' large open vessel in which bread is kneaded ', awāṇ. *pātrī* ' plate '; P. *pātar* m. ' vessel ', *parāt* f., *parāṭrā* m. ' large wooden kneading vessel ', ḍog. *pāttar* m. ' brass or wooden do. '; Ku.gng. *pāi* ' wooden pot '; B. *pātil* ' earthern cooking pot ', °*li* ' small do. ' Or. *pātiḷa*, °*tuḷi* ' earthen pot ', (Sambhalpur) *sil* -- *pā* ' stone mortar and pestle '; Bi. *patīlā* ' earthen cooking vessel ', *patlā* ' milking vessel ', *pailā* ' small wooden dish for scraps '; H. *patīlā* m. ' copper pot ', *patukī* f. ' small pan '; G. *pātrũ* n. ' wooden bowl ', *pātelũ* n. ' brass cooking pot ', *parāt* f. ' circular dish ' (→ M. *parāt* f. ' circular edged metal dish '); Si. *paya* ' vessel ', *päya* (< *pātrī´* --). *kācapātra -- , khaḍgapātra -- , tāmrapātra -- .**pāthá --** m. ' way, path ' Pāṇ.gaṇa. [pánthā --]śabdapātha -- .Addenda: **pā´tra --** : S.kcch. *pātar* f. ' round shallow wooden vessel for kneading flour '; WPah.kṭg. (kc.) *pərāt* f. (obl. -- *i*) ' large plate for kneading dough ' ← P.; Md. *tilafat* ' scales ' (+ *tila* < tulā´ --)(CDIAL 8055).

Mth. *pāthā* ' large milk pail ', *pathiyā* ' basket used as feeding **trough** for animals '***Tu.*** **pāti** trough or bathing tub. These variant pronunciations in Maithili and Tulu indicate the possibility that the early word which signified a feeding trough was *pattha, patthaya* 'measure of grain' (Prakrtam). The suffix -mar in Pattimar which signifies a dhow, seafaring vessel is related to the word.

Procession of animals as a hieroglyph set

Harappa (Indus) Script inscription on a Mohenjo-daro tablet (m1405) including 'rim-of-jar' glyph as component of a ligatured glyph (Sign 15 Mahadevan)

This inscribed object is decoded as a professional calling card: a blacksmith-precious-stone-merchant with the professional role of copper-miner-smelter-furnace-scribe-Supecargo

m1405At Pict-97: Person standing at the centre points with his right hand at a bison facing a trough, and with his left-hand points to the ligatured glyph.

The inscription on the tablet juxtaposes – through the hand gestures of a person - a 'trough' gestured with the right hand; a ligatured glyph composed of 'rim-of-jar' glyph and 'water-carrier' glyph (Sign 15) gestured with the left hand.

A characteristic feature of Indus writing system unravels from this example: what is orthographically constructed as a pictorial motif can also be deployed as a 'sign' on texts of inscriptions. This is achieved by a stylized reconstruction of the pictorial motif as a 'sign' which occurs with notable frequency on Indus Script Corpora -- with orthographic variants (Signs 12, 13, 14).

Signs 12 to 15. Indus script: 

 Indus inscription on a Mohenjo-daro tablet (m1405) including 'rim-of-jar' glyph as component of a ligatured glyph (Sign 15 Mahadevan) This tablet is a clear and unambiguous example of the fundamental orthographic style of Indus Script inscriptions that both signs and pictorial motifs are integral components of the message conveyed by the inscriptions. Attempts at 'deciphering' only what is called a 'sign' in Parpola or Mahadevan concordances will result in an incomplete decoding of the complete message of the inscribed object.

This inscribed object is decoded as a professional catalogue calling card: a blacksmith-precious-stone-merchant with the professional role of copper-miner-smelter-furnace-scribe-Supercargo.

The inscription on the tablet juxtaposes – through the hand gestures of a person - a 'trough' gestured with the right hand; a ligatured glyph composed of 'rim-of-jar' glyph and 'water-carrier' glyph (Glyph 15) gestured with the left hand.

The Pali expression *usu -- kāraṇika* -- m. ' arrow -- maker ' provides the semantics of the word *kāraṇika* as relatable to a 'maker' of a product. *usu-kāraṇika* is an arrow-maker. Thus, *kuThi kāraṇika* can be explained as a smelter-maker. Supercargo is a representative of the ship's owner on board a merchant ship, responsible for overseeing the cargo and its sale. The Marathi word for Supercargo is: *kārṇī*. Thus, it can be suggested that kuTi kāraṇika was an ovrseer of the cargo (from smelter) on a merchantship. In the historical periods, the Supercargo has specific duties "The duties of a supercargo are defined by admiralty law and include managing the cargo owner's trade, selling the merchandise imports to which the vessel is sailing, and buying and receiving goods to be carried on the return voyage...A new supercargo was always appointed for each journey who also had to keep books, notes and ledgers about everything that happened during the voyage and trade matters abroad. He was to present these immediately to the directors of the Company on the ship's return to its headquarters."
https://en.wikipedia.org/wiki/Supercargo While a captain was in charge of navigation, Supercargo was in charge of trade.

कारण 1[p= 274,2] a number of scribes or कायस्थs W. instrument , means;that on which an opinion or judgment is founded (a sin, mark; a proof; a legal instrument, document), Mn. MBh.

कारणिक [p= 274,3] *mfn.* (g. काश्य्-आदि) " investigating, ascertaining the cause " , a judge Pan5cat. a teacher MBh. ii , 167.

B. *kerā* ' clerk ' (*kerāni* ' id. ' < **kīraka* -- ***karaṇika***<-> ODBL 540): very doubtful. -- Poss. ← Ar. *qārī*, pl. *qurrā*' ' reader, esp. of Qur'ān '.(CDIAL 3110) कर्णक kárṇaka, kannā **'legs spread'**, 'rim of jar', 'pericarp of lotus' karaṇī 'scribe, supercargo', kañi-āra 'helmsman'. kāraṇika m. ' teacher ' MBh., ' judge ' Pañcat. [*kā- raṇa* --] Pa. *usu* -- *kāraṇika* -- m. ' arrow -- maker '; Pk. *kāraṇiya* -- m. ' teacher of Nyāya '; S. *kāriṇī* m. ' guardian, heir '; N. *kārani* ' abettor in crime '; M. *kārṇī* m. ' prime minister, supercargo of a ship ', *kul -- karṇī* m. ' village accountant '.(CDIAL 3058) kāraṇa n. ' cause ' KātyŚr. [√**kr̥1**] Pa. *kāraṇa* -- n. ' deed, cause '; Aś. shah. *karaṇa* --, kāl. top. *kālana* -- , gir. *kāraṇa* -- ' purpose '; Pk. *kāraṇa* -- n. ' cause, means '; Wg. (Lumsden) "*kurren*" ' retaliation ', Paš. *kāran* IIFL iii 3, 97 with (?); S. *kāraṇu* m. ' cause '; L. *kārnā* m. ' quarrel '; P. *kāraṇ* m. ' cause ', N. A. B. *kāran*, Or. *kāraṇa;* Mth. *kāran* ' reason ', OAw. *kārana*, H. *kāran* m., G. *kāraṇ* n.; Si. *karuṇa* ' cause, object, thing '; -- postpositions from oblique cases: inst.: S. *kāraṇi*, *kāṇe*, °*ṇi* ' on account of ', L. awāṇ. *kāṇ*, Addenda: kāraṇa -- : Brj. *kāran* ' on account of '.(CDIAL 3057) *kiraka* m. ' scribe ' lex.

eraka 'raised arm' Rebus: *eraka* 'metal infusion' (Kannada. Tulu)

Sign 15: *kuṭhi kaṇḍa kankha* 'smelting furnace account (scribe)'.

Thus, the hieroglyph multiplex on m1405 is read rebus from r.: *kuṭhi kaṇḍa kankha eraka bharata pattar*'goldsmith-merchant guild -- helmsman, smelting furnace account (scribe), molten cast metal infusion, alloy of copper, pewter, tin.'

m290 tiger PLUS trough

m276

Mohenjo-Daro rhinoceros

M-1910 a

h088 Rhinoceros PLUS trough

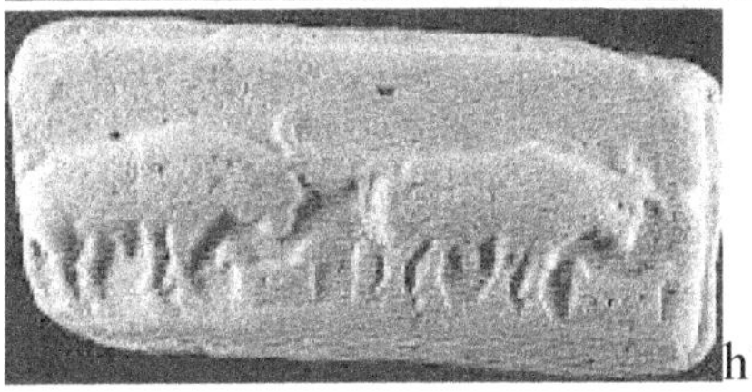

h1966A h1966B

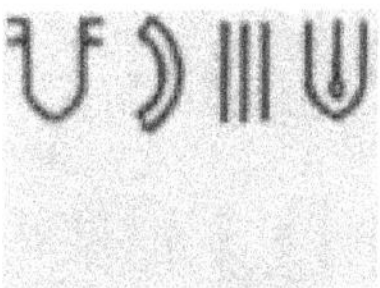

m1486B Text 1711

Obverse: *karibha* 'trunk of elephant' *ibha* 'elephant' rebus: *karba* 'iron' *ib* 'iron' *khAr* 'blacksmith'. Thus, ironsmith.

Reverse: Inscription of hypertext:

baTa 'rimless pot' Rebus: *bhaTa* 'furnace' PLUS *muka* 'ladle' rebus; *mũ̃h* 'ingot', quantity of metal got out of a smelter furnace (Santali)

kolom 'three' Rebus: *kolimi* 'smithy, forge'

 Doubling of this signifies *dula* 'pair' rebus: *dul* 'cast metal'. Thus doubling of the right parenthesis results in a hieroglyph-multiplex (hypertext) as shown on the elephant copper plate inscription m1486 text

 This hieroglyph-multiplex is thus read as: *kuṭilika* 'bent, curved' *dula* 'pair' rebus: *dul* 'metal casting'; *kuṭila, katthīl* = bronze (8 parts copper and 2 parts tin)

 The 'curve' hieroglyph is a splitting of the ellipse. *kuṭila* 'bent' CDIAL 3230 kuṭi— in cmpd. 'curve', *kuṭika*— 'bent' MBh.

Rebus: *kuṭila, katthīl* = bronze (8 parts copper and 2 parts tin) cf. *āra-kūṭa,* 'brass' Old English *ār* 'brass, copper, bronze' Old Norse *eir* 'brass, copper', German *ehern* 'brassy, bronzen'. kastīra n. ' tin ' lex. 2. *kastilla -- .1. H. *kathīr* m. ' tin, pewter '; G. *kathīr* n. ' pewter '.2. H. (Bhoj.?) *kathīl*, °*lā* m. ' tin, pewter '; M. *kathīl* n. ' tin ', *kathlẽ* n. ' large tin vessel '.(CDIAL 2984)

Hieroglyphs: कौटिलिकः kauṭilikḥ कौटिलिकः 1 A hunter.-2 A blacksmith. कौटिलिक [p= 315,2] *m.* (fr. कुटिलिका Pa1n2. 4-4 , 18) " deceiving the hunter [or the deer Sch.] by particular movements " , a deer [" a hunter " Sch.] Ka1s3. *f.* (Pa1n2. 4-4 , 18) कुटिलिका crouching , coming stealthily (like a hunter on his prey ; a particular movement on the stage) Vikr. कुटिलिक " using the tool called कुटिलिका " , a blacksmith ib. कुटिलक [p= 288,2] *f.* a tool used by a blacksmith Pa1n2. 4-4 , 18 Ka1s3.*mfn.* bent, curved , crisped Pan5cat.

The hieroglyph-multiplex may be a variant of split ellipse curves paired: *dula* 'pair' rebus: *dul* 'cast metal' PLUS *mũ̃h* 'ingot' (Paired split ellipse or a pair of right parentheses) -- made of -- *kuṭila, katthīl* = bronze (8 parts copper and 2 parts tin) *karNika* 'rim of jar' rebus: *karNI* 'supercargo'; *karNaka* 'account'; Alternative: *kankha* 'rim of jar' rebus: *kanga* 'brazier'.

Thus, the entire inscription is a metalwork catalogue: supercargo of iron, cast bronze metal ingots, out of smithy furnace and forge.

 Kalibangan033  8025  m0527Bt  3336

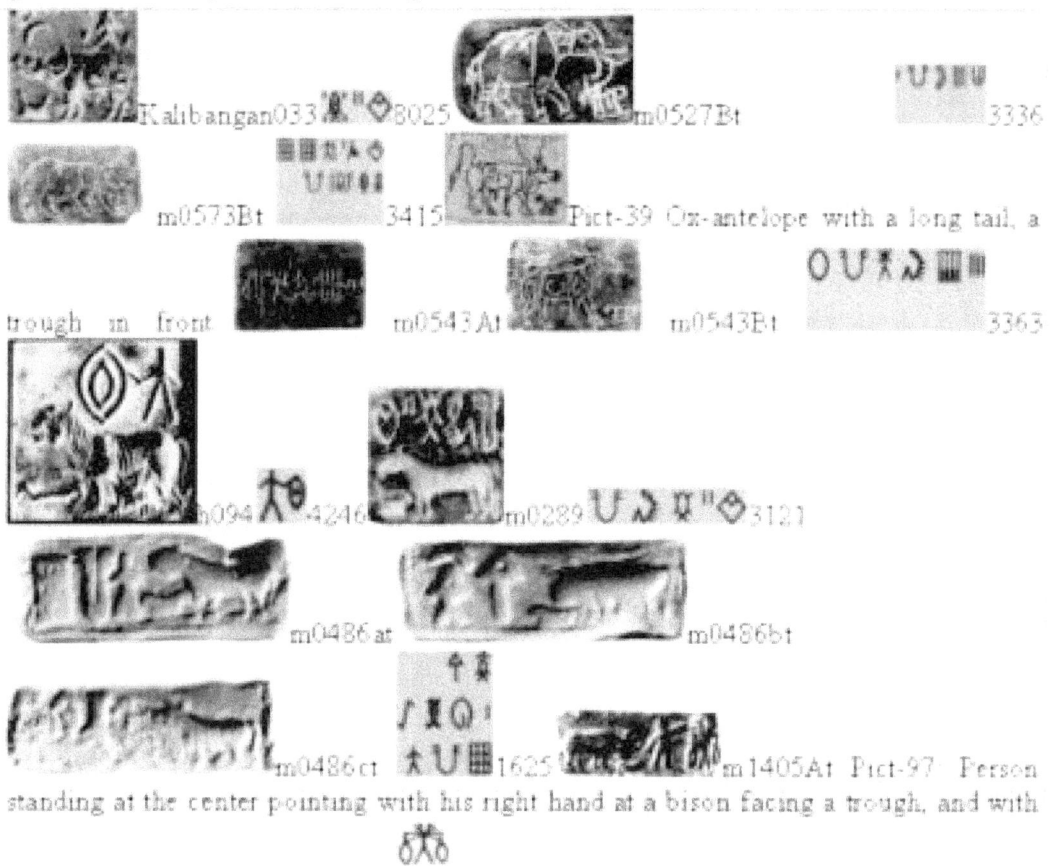

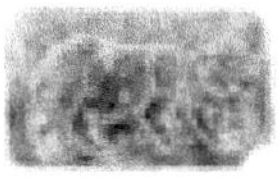 m0573Bt  3415  Pict-39 Ox-antelope with a long tail, a trough in front  m0543At 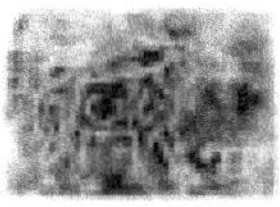 m0543Bt  3363

 m094  4246 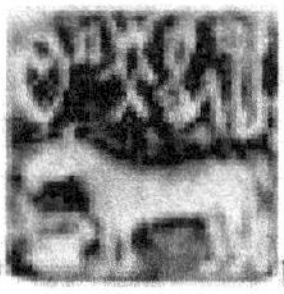 m0289  3121

 m0486at  m0486bt

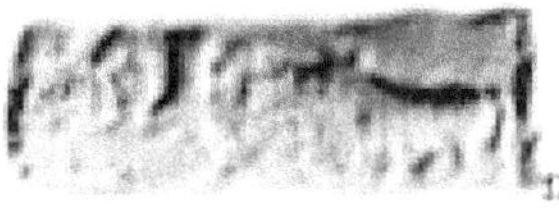 m0486ct 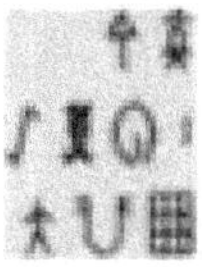 1625  m1405At Pict-97 Person standing at the center pointing with his right hand at a bison facing a trough, and with his left hand pointing to the sign 

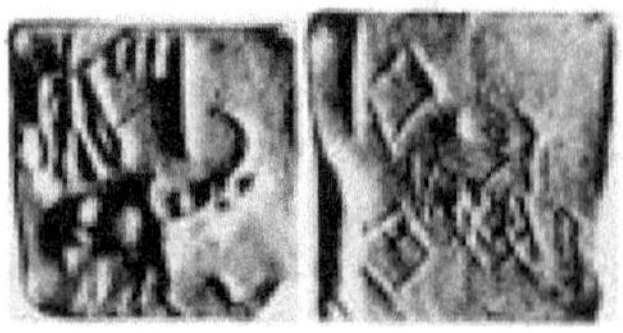 m1928 a,b  h0087  m0266

 1306  m0267 Water-buffalo  2257

 m0268 Water-buffalo  2445  1804 

Pict-39 Ox-antelope with a long tail; a trough in front.

Trough PLUS buffalo/bull

Other examples of trough as a hieroglyph on Indus writing seals shown in front of animals.

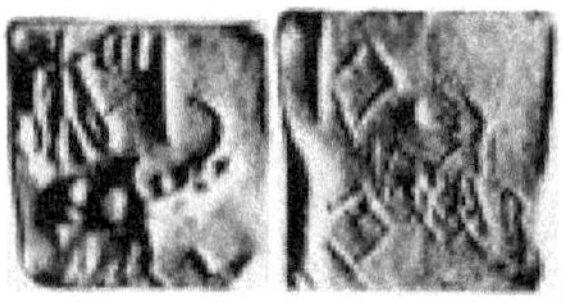
m1928 a,b

h0087

m0266

1306

m0267 Water-buffalo

2257

m0268 Water-buffalo

2445

1804

Pict-39 Ox-antelope with a long tail; a trough in front.

A trough is shown in front of some domesticated animals and wild animals like rhinoceros, tiger, elephant. The trough glyph is clearly a hieroglyph, in fact, a category classifier. Trough as a glyph occurs on about one hundred inscriptions, though not identified as a distinct pictorial motif in the corpus of inscriptions. Why is a trough shown in front of a rhinoceros which was not a domesticated animal? A reasonable deduction is that 'trough' is a hieroglyph intended to classify the animal 'rhinoceros' in a category.

ḍhangar 'trough'; *ḍhangar* 'bull'; rebus: *ḍhangar* 'blacksmith'

Chanhudaro22a *ḍhangar* ‘bull’. Rebus: *ḍhangar* ‘blacksmith’ *pattar* ‘trough’. Rebus: *pattar* (Ta.), *battuḍu* (Te.) goldsmith guild (Tamil.Telugu) *khōṭ* ‘alloyed ingot’; *kolmo* ‘rice plant’. Rebus: *kolami* ‘smithy’. *koḍi* ‘flag’ (Ta.)(DEDR 2049). Rebus: *koḍ* ‘workshop’ (Kuwi) Vikalpa: *baddī* = ox (Nahali); *baḍhi* = worker in wood and metal (Santali) *ḍāngrā* = a wooden trough just enough to feed one animal. cf. *iḍankaṛi* = a measure of capacity, 20 *iḍankaṛi* make a par-r-a (Ma.lex.) ḍangā = small country boat, dug-out canoe (Or.); *ḍõgā* trough, canoe, ladle (H.)(CDIAL 5568). Rebus: *ḍānro* term of contempt for a blacksmith (N.) (CDIAL 5524)

Stamp seal with a water-buffalo, Mohenjo-daro. “As is usual on Indus Valley seals that show a water buffalo, this animal is standing with upraised head and both hornsclearly visible. (Mackay, 1938b, p. 391). A feeding trough is placed in front of it, and a double row of undecipherable script fills the entire space above. The horns are incised to show the natural growth lines. During the Akkadian period, cylinder seals in Mesopotamia depict water buffaloes in a similar pose that may have been copied from Indus seals (see cat. No.135) (For a Mesopotamian seal with water buffalo, see Parpola1994, p. 252 and Collon 1987, no.529 – Fig. 11).”(JMK –Jonathan Mark Kenoyer, Professor of Anthropology, University of Wisconsin, Madison) (p.405). பத்தர்[1] pattar, n. 1. See பத்தல், 1, 4, 5. 2. Wooden trough for feeding animals; தொட்டி. பன்றிக் கூழ்ப்பத்தரில் (நாலடி, 257).

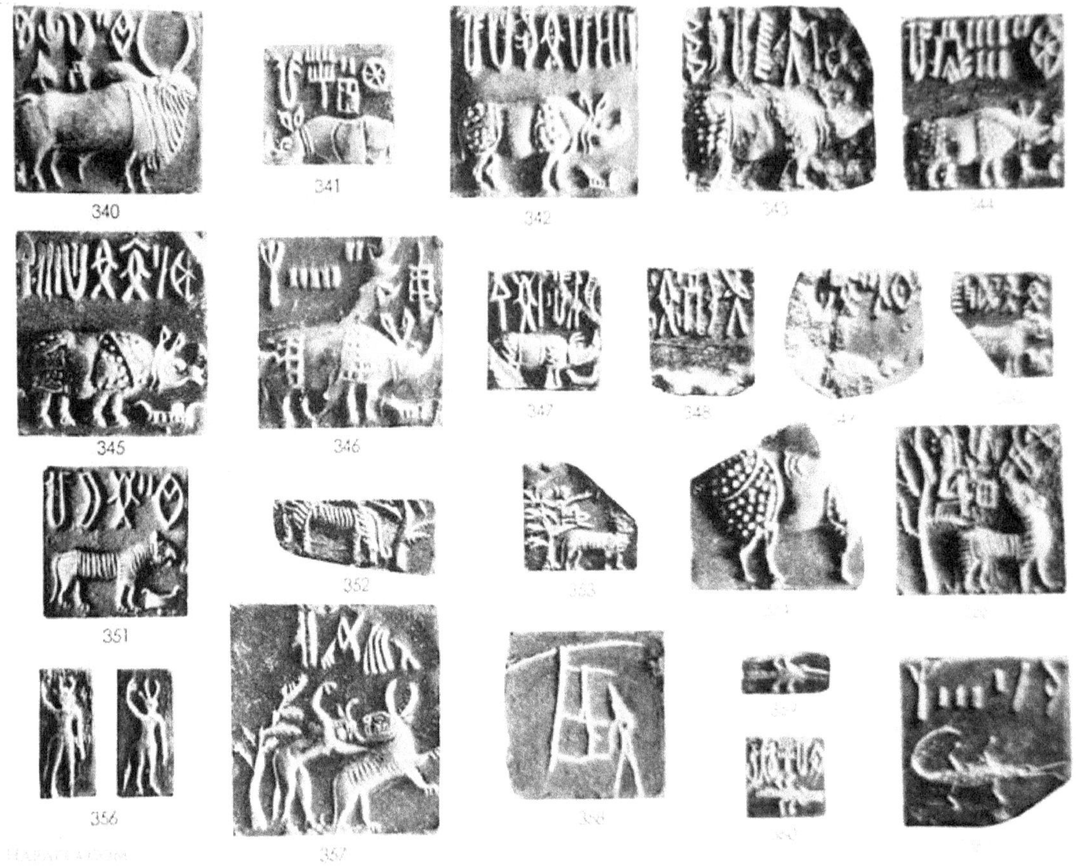

Hieroglyph: *pattar* 'trough' Rebus: *pattharaka* 'merchant' *pattar* ‘guild, goldsmith’.

Section 2. Preparation for the decipherment attempt

Preparation for the decipherment effort has been exerted in two parallel stages: Stage 1. Compilation of the Harappa Script Corpora; and Stage 2. Compilation of the Bharata *sprachbund* lexicon called *Indian Lexicon*[3].as a comparative dictionary for over 25 ancient languages of Bharata in semantic clusters to reconstruct the phonetic forms of the language spoken by the creators of the Corpora. Since the preliminary indicators point to the Corpora to be metalwork catalogues, these two parallel stages are complemented by the compilation of data archives related to archaeo-metallurgy in respect of 1. Metal alloys and 2. *Cire Perdue* (lost-wax) techniques of metalcastings.

A tribute has to be paid to many savants who have researched into Harappa Script ever since 1875 when Alexander Cunningham published a surface find of Harappa seal with Harappa script. Many concordance lists have been compiled. Many inscriptions are not restricted to the River Basins of Rivers Sarasvati and Sindhu (Indus) but extend over an expansive contact area which stretches from Hanoi (Vietnam, e.g. Dong Son Bronze Drum) to Haifa (Israel, three pure tin ingots with Harappa Script).

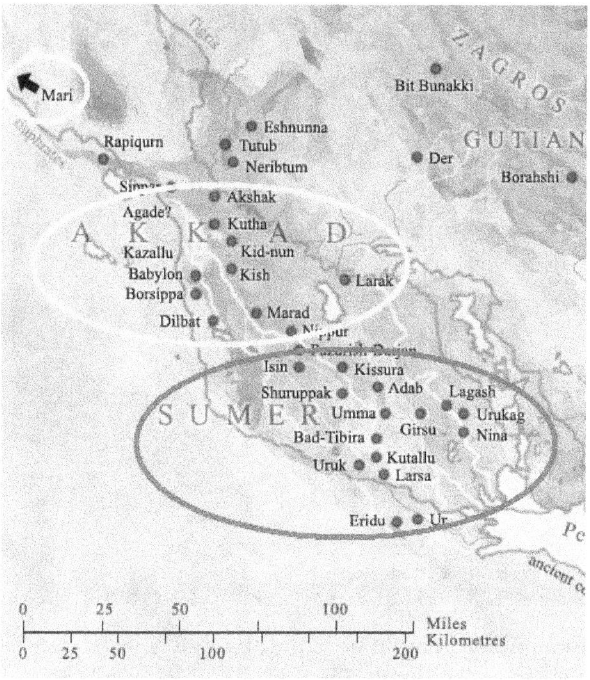

Mari in relation to Persian Gulf. The concordance lists constitute the Harappa Script Corpora which now number more than 8000, thanks to the discoveries of Persian Gulf or Dilmun seals and also artifacts such

as the sculptural frieze of Spinner lady with Harappa Script hieroglyphs, now in Louvre Museum or Standard of Mari held in a procession. The standard upheld by the central figure is itself an emphatic hieroglyph because the post which holds the standard of 'one-horned young bull' a very frequently used hypertext in Harappa Script Corpora, is neither a stick nor a rod, but 'culm of millet'. *karb* 'culm of millet' rebus: *karba* 'iron'. *kaḍambá, kalamba* -- 1, m. ' end, point, stalk of a pot- herb ' lex. [See kadambá --] B. *kaṛamba* ' stalk of greens '; Or. *kaṛambā*, °*mā* stalks and plants among stubble of a reaped field '; H. *kaṛbī, karbī* f. ' tubular stalk or culm of a plant, esp. of millet ' (→ P. ***karb*** m.); M. *kaḍbā* m. ' the culm of millet '. -- Or. *kaḷama* ' a kind of firm -- stemmed reed from which pens are made ' infl. by H. *kalam* ' pen ' ← Ar.? (CDIAL 2653) Rebus: *Ta. ayil iron. Ma. ayir, ayiram any ore. Ka. aduru* native metal. Tu. ajirda karba very hard iron. (DEDR 192) (a) Ta. karu black; To. kary- (kars-) to be singed, scorched, fried too much; (karc-) to heat (new pot, etc., to purify it); kary charcoal; ka, kax, kaxt black; kabïn iron; Ka. kabbiṇa iron; Te. krãgu to be burned; Go. (Tr. W.) karw-, (SR. Ph.) karv-, (Mu.) kar-, kaṛ- to burn (intr.); (G. Ma. Ko.) karv-, (Ph.) karsahtānā, (Mu.) kaṛih- id. (tr.); (Mu.) kaṛha field for burning cultivation (Voc. 563); (Grigson) kare area set apart for penda cultivation when left fallow for a term (Voc. 536); (Ma.) gaŋga darkness, mist (Voc.1016); (Mu.) kark- rice to burn while cooking; (Ko.) karr- to be charred or burnt (Voc. 539) (DEDR 1278)

खोंड (p. 122) *khōṇḍa* m A young bull, a bullcalf) one-horned young bull and *karb* 'culm of millet' (Punjabi), respectively. (NOTE: कोंद *kōnda* 'engraver, lapidary setting or infixing gems' is a phonetic variant of a worker with gold and lathe: *kunda* 'fine gold, lathe.

Panel standard of Mari shell limestone and bitumen, Mari Temple of Ishtar, Early Dynastic period, w: 72 cm from the Early Dynastic - Northern Mesopotamian Period, 2900 BCE - 2350 BCE.

Copper rein ring ca. 2750 BCE Tomb of Queen Puabi of Ur.

https://67.media.tumblr.com/b8e5d9edad7568e304eb2d5c4004bc05/tumblr_inline_o50zc0kXEo1t79fgm_540.jpg

"The central figure has been reconstructed as carrying an animal standard which is in fact a rein ring similar to the electrum one found in the death pit of Puabi in the Royal Cemetery at Ur. Both the Mari and Ur standards were constructed using a shell-inlay technique and bitumen." http://www.newsnfo.co.uk/pages/sheeple%20war%20sheep%20emblem%20carrier%20.htm

bāga 'bridle, rein' (Oriya)Valgā वल्गा A bridle, rein; आलाने गृह्यते हस्ती वाजी वल्गासु गृह्यते Mk.1.5; also वल्गः (in the same sense); वङ्कावलग्नैक- सवल्गापाणयः Śi.12.6. వల్గ (p. 1140) valga valga. [Skt.] n. A bridle, a rein. గుర్రపుకళ్లెము తగిలించేవాగె. వాగె. valgā f. ' bridle ' Mr̥cch. [S. L. P. poss. indicate earlier *vālgā --] Pk. *vaggā* -- f. ' bridle '; K. *wag* f. ' rein, tether '; S. *vāga* f. ' rein, halter ' (whence *vāgaṇu* ' to tie up a horse '); L. (Ju.) *vāg* f. ' rein ', awāṇ. *vāg*; P. *vāg*, *bāg* f. ' bridle ', N. *bāg -- ḍori*; B. *bāg* ' rein '; Or. *bāga* ' bridle, rein ', *gaja -- bāga* ' elephant goad '; Bi. *bāg -- ḍor* ' tether for horses '; Mth. Bhoj. *bāg* ' rein, bridle ', OAw. *bāga* f., H. *bāg* f. (→ K. *bāg -- ḍora* m., M. *bāg -- dor*); G. *vāg* f. ' rein, guiding rope of a bullock '; M. *vāg -- dor* m. ' bridle '. Addenda: valgā -- and *vālgā -- [Dial. *a ~ ā* < IE. *o* (Lett. *valgs* ' rope, cord ') T. Burrow BSOAS xxxviii 72] (CDIAL 11420)

bā̃g f. ' tin, lead, calx of tin ' (Hindi) vaṅga1 n. ' tin, lead ' lex. [Cf. raṅga -- 3, nāga -- 2] S. *vaṅga* f. ' calx of tin (used as an aphrodisiac) ', P. *vaṅg*, *baṅg* f.; H. *bā̃g* f. ' tin, lead, calx of tin ', *bā̃gā* ' having a metallic or brackish taste (of water) '(CDIAL 11195) vaṅgḥ वङ्गम् 1 Lead. -2 Tin; ताम्रं लोहं च वङ्गं च काचं च स्वर्णमाक्षिकम् Śiva B.3.11. -Comp. वङ्गः -अरिः yellow orpiment. -ज 1 brass. -2 red lead. -जीवनम् silver. -शुल्यजम् bell- metal (कांस्य).

Thus, the *karb* 'culm of millet' rebus: *karba* 'iron' PLUS *bāga* 'bridle, rein' rebus: *bā̃g* f. ' tin, lead, calx of tin ' (Hindi) PLUS खोंड (p. 122) *khōṇḍa* m A young bull, a bullcalf) one-horned young bull rebus: कोंद *kōnda* 'engraver, lapidary setting or infixing gems' is a phonetic variant of a worker with gold and lathe: *kunda* 'fine gold, lathe.

Thus, the Mari procession of the Mari standard is a celebration of three metals: *kunda*, 'fine gold' PLUS *bā̃g* 'tin, lead, calx of tin' PLUS *karba* 'iron'. The lapidary turner, *kōnda* had gained the competence to 'turn' fine gold, calx of tin and iron. The calx of tin *is Vajra smghAta* 'adamantine metal glue', a remarkable metallurgical advance in tinning and gold-plating using *bhasma.*

Section 3. Methodology developed

There are enormous documented resources available from ancient texts of Bharata for trying out various research methods. This resource relates to the method of documenting a research and is called *tantra yukti.* This method of research documentation is in over 32 categories. The method is also called, in the context of mathematical and astronomical researches as *Yuktibhāṣā* (Malayalam: യുക്തിഭാഷ; "Rationale in the Malayalam/Sanskrit language"[(K V Sarma; S Hariharan (1991). "Yuktibhāṣā of Jyeṣṭhadeva: A book on rationales in Indian Mathematics and Astronomy: An analytic appraisal" in: *Indian Journal of History of Science*. 26 (2), pp. 185-207.)
http://web.archive.org/web/20060928203221/http://www.new.dli.ernet.in/insa/INSA_1/20005ac0_185.pdf

Yukti or rationale is evaluated in terms of the meanings and many categories such as 32 devices listed in *Arthaśāstra:*

अधिकरण (topic), विधानं (statement of contents), योगः (employment of sentences),पदार्थः (meaning of the word), हेत्वर्थः (reason), उद्देशः (mention), निर्देशः (explanation), उपदेशः (advice), अपदेशः (reference),अतिदेशः (application), प्रदेशः (indication), उपमानं (analogy), अर्थापत्ति (implication), संशयः (doubt), प्रसंगः (situation), विपर्ययः (contrary), वाक्यशेषः (completion of a sentence), अनुमतं (agreement), व्याख्यानं (emphasis), निर्वचनं (derivation), निदर्शनं (illustration),अपवर्ग (exception), स्वसंज्ञा (technical term), पूर्वपक्षः (prima facie view), उत्तरपक्षः (correct view), एकान्तः (invariable rule), अनागतावेक्षणं (reference to a future statement), अतिक्क्रंत्तातवेक्षणं (reference to a past statement), नियोगः (restriction), विकल्प (option, alternative), समुच्चय (combination), ऊह्यं (what is understood). (Kautilya *Arthas'Astra,* Vol. II, 1972, Delhi, MLBD).

तत्रायुर्वेदः शाखा विद्या सूत्रं ज्ञानं शास्त्रं लक्षणं तन्त्रमित्यनर्थान्तरम्

Tantra is synonymously used with äyurveda, *a branch of Veda, education, aphorism, knowledge,* s'Astra *and definition.* (*Carakasamhitä*, siddhisthänam, uttaravastisiddhih ,12 adhyäyaù, s'lokaù 29, 30 païcamakhANDah, Edited and Revised by Kaviraja Narendranath Sengupta, Kaviraja Balaichandra Sengupta, Caukhambä Orientalia Varanasi–1, 1991 (Reprinted) https://www.academia.edu/12132105/Tantrayukti

समुच्चय (combination) method is apparent from the orthographic device of combining animal parts to सांगडणें (p. 495) sāṅgaḍaṇēṃ v c (*sangada*), that is, 'To link, join, or unite together (boats, fruits, animals).' Similar combination occurs on other hieroglyphs, the so-called signs, using the 'ligatures', for e.g. by infixing short linear strokes within 'rim of jar' or by ligaturing the 'water-carrier' hieroglyph with a 'rim of jar'. सांगड (p. 495) sāṅgaḍa is thus 'f A body formed of two or more (fruits, animals, men) linked or joined together.' It is also a lathe or a joined canoe called 'catamaran', सांगड (p. 495) *sāṅgaḍa* m f (संघट्ट S) A float composed of two canoes or boats bound together.

Such a समुच्चय (combination) method is referred to as 'hypertexting' in modern computer jargon to signify cross-referencing between sections of text and associated graphic material.

Harappa Script inscriptions are very short. The average number of 'text signs' is 5 with or without pictorial motifs or field symbols on seals and tablets. The short texts and the 400+ symbols on 'sign list' are pointers to the logo-phonetic nature of the Script. These 'signs' are NOT pictograms but logograms because they represent words in Harappa language.

Unfortunately, many attempts at decipherment were premised on a conjecture of alphabetical or syllabic nature of the Script and searching for 'names or titles or sentences composed of morphemes and grammatical units such as prepositions' on the assumption that the seals and tablets should have recorded names or titles or even poems of prayer. Such assumptions ignored

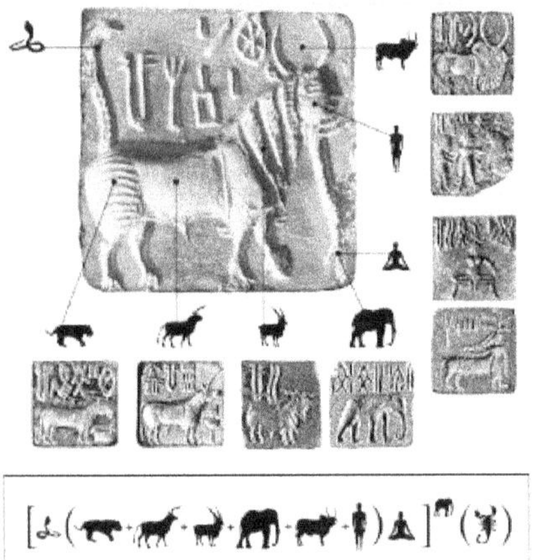

the essentially pictorial nature of the Script composed of hieroglyphs both as 'text signs' and as 'pictorial motifś and pictorial narratives such as a tiger looking back and a person sitting on a tree branch. An approach framed on the hypothesis that Harappa Script was necessitated for trade transactions to cope with the Bronze Age Revolution, provides a framework for the use of rebus method evidenced on Egyptian hieroglyphs which are rebus renderings of consonants in words (See discussions on rebus readings of hieroglyphs which are complete messages on Narmer palette and on Ivory tags with one to four hieroglyphs, from U-J of Abydos tomb dated to ca. 32nd cent. BCE). Thus, it will be a futile and speculative exercise to look for long texts in Harappa Script which has achieved comprehensive messaging to render utterances in Harappa language, using rebus principle to compose hypertexts in Meluhha cipher (pairs of homonymous words, one set signifying logograms and another set signifying similar sounding metalwork catalogue items).

The hypertext principle of Harappa Script is demonstrated effectively by Dennys Frenez and Massimo Vidale[4] for Harappa Script, identifying pictorial components which constitute a composite orthographic construction as in the example of a composite animal orthographed by various parts of various animals on m0300 Mohenjo-daro seal. Orthographic components are identified and explained by Dennys Frenez and Massimo Vidale.

The evidence is remarkable that many single glyphs or glyptic elements of the Harappa writing can be read rebus using the repertoire of artisans (lapidaries working with precious shell, ivory, stones and terracotta, mine-workers, stone-masons, metal-smiths working with a variety of minerals, furnaces and other tools) who created the inscribed objects and used many of them to authenticate their trade transactions. Many of the inscribed objects are seen to be calling cards of the professional artisans, listing their professional skills and repertoire. Many are veritable mining- and metal-work catalogs.

Continuing legacies of glyptic art noted by Huntington: "There is a continuity of composite creatures demonstrable in Indic culture since Kot Diji ca. 4000 BCE."[5]

The identification of glosses from the present-day languages of Bharata on Sarasvati river basin is justified by the continuation of culture evidenced by many artifacts evidencing civilization continuum from the Vedic Sarasvati River basin, since language and culture are intertwined, resulting in a unique, logo-semantics related to a logo-phonetic writing system. .

Over 8000 inscriptions of Harappa Script are metalwork documents. Data mining techniques are applied to this Corpora for knowledge discovery about aspects of the Bronze Age Revolution such a: 1. *Cire perdue* lost-wax technique for creating metal sculptures of exquisite beauty and artistry (e.g. dancing girl bronze statue of Mohenjodaro); and 2. Techniques of alloying of minerals to create hard alloys necessary to forge useful tools, implements, weapons, pots and pans; 3. Organization of guilds of artisans and merchants as Corporate forms to control the works-in-process (e.g. circular platforms of Harappa) and shipment of cargo of metalwork merchandise using dhows, seafaring vessels.

Tin-Bronze Revolution as a trans-Eurasian phenomenon is attested by archaeological discoveries of artifacts and ancient cuneiform text records, which validate Harappa Script inscriptions, datable to ca. 5th-4th millennium BCE.

Harappa Script is mlecchita vikalpa cryptography, uses rebus method of substitution cipher

Mlecchita vikalpa is an expression used as the 52nd item in the list of 64 arts learnt by youth (Chapter 3, Part 1, Vatsyayana's *Kamasutra* dated to ca. 1st cent. CE). Translators Richard Burton, Bhagavanlal Indrajit, Shivaram Parashuram Bhide explain the expression *mlecchita vikalpa* as 'the art of understanding writing in cipher' or cryptography or, simply, encryption.

Definition of terms

The intended message is plain text and the encrypted *vikalpa* (alternative representation) is ciphertext. Decrypting the ciphertext using *Meluhha (mleccha)* language of Bharatiya *sprachbund* (language union), is decipherment of Harappa Script. Mleccha of Meluhha is the reconstructed spoken form characterized by mispronunciations and variant spellings identified from the comparative lexicon of over 25 ancient languages of Bharata (called *Indian Lexicon*[6])

Potsherd. Harappa. Ca. 3300 BCE (HARP Harvard Archaeology Project) Three hieroglyphs. *tagaraka 'tabernae montana'* rebus (substitution of similar sounding word): *tagara* 'tin'. Note: The discovery of tin as an alloying mineral with copper to substitute for arsenical bronze, to create Tin-Bronzes created the Bronze Age revolution. This

writing system could be one of the earliest writing systems in the story of civilization. The date of this potsherd compares with the date of bone/ivory tags with Egyptian hieroglyphs discovered in Abydos U-j tomb ca. 3200 BCE. It is remarkable that the writing systems of Egyptian hieroglyphs, cuneiform texts on cylinder seals with pictorial motifs which include Harappa Script hieroglyphs were invented almost simultaneously with the invention of Harappa Script ca. 3300 BCE.

Yasodhara's *Jayamangala* (13th cent.) is a commentary on *Kamasutra* and describes that the *mlecchita vikalpa* method of writing is 'substitution cipher' and that -- as in the methods called *Kautiliyam* and *Muladeviya*--, the letter substitutions are based on phonetic relations.

Kautiliyam is a *mlecchita* named after Kautilya, the author of the ancient Indian political treatise, the *Arthaśāstra.* In this system, the short and long vowels, the *anusvara* and the spirants are interchanged for the consonants and the conjuncts. The following table shows the substitutions used in the *Kautiliyam* cipher. The characters not listed in the table are left unchanged.

a	ā	i	ī	u	ū	ṛ	ṝ	ḷ	ḹ	e	ai	o	au	ṃ	ḥ	ñ	ś	ṣ	s	i	r	l	u
kh	g	gh	ṅ	ch	j	jh	ñ	ṭh	ḍ	ḍh	ṇ	th	d	dh	n	ph	b	bh	m	y	r	l	v

Muladeviya is also called *Gudhalekhya.* The cipher alphabet of *Muladeviya* consists of the reciprocal one specified in the table below.

a	kh	gh	c	t	ñ	n	r	l	y
k	g	ṅ	ṭ	p	ṇ	m	ṣ	s	ś

See: Friedrich L Brauer (2007). *Decrypted Secrets: Methods and Maxims of Cryptology.* Springer. p. 47. Source: https://en.wikipedia.org/wiki/Mlecchita_vikalpa

Cryptanalysis or breaking of codes and ciphers is the method to decipher Harappa Script writing system. Atbash (name derived from first, last, second and second to last Hebrew letters, Aleph-Tav-Beth-Shrin) is a method of monoalphabetic substitution cipher dated to ca. 600 BCE. In this cipher, the modern Hebrew alphabet would be:

Plain	אבגדהוזחטיכלמנסעפצקרשת
Cipher	תשרקצפעסנמלכיטחזוהדגבא

Source: https://en.wikipedia.org/wiki/Atbash

The reference to *kautiliyam* is significant because the text of Kautilya is titled *Kauṭilīyaṃ Arthaśāstram,* 'the science of wealth creation'.

The glyphs of the Harappa script or Harappa Writing include both pictorial motifs and signs. Both categories of glyphs are read rebus. As a first step in delineating the Harappa language, an *Indian lexicon* provides a resource, compiled semantically in clusters of over 1240 groups of words/expressions from ancient Bharata languages as a Proto-Indic substrate dictionary.

Rebus method to signify phonetic sounds of words, is a logo-phonetic writing system

The writing system seems to use the rebus principle to phonetize its hieroglyphs. Such a rebus principle is seen in ca. 3100 BCE Egyptian hieroglyphs of Narmer and Abydos tomb. At Tomb UJ at Abydos in Upper Egypt (dated to ca. 3250 BCE), Dreyer found place names written in Egyptian hieroglyphs (up to four in number) recognizable as hieroglyphs which persisted and were employed during later periods and which are written and read phonetically. The rebus principle may explain the pictorial motifs of some cylinders of Ancient Near East of ca. 3100 BCE.Ivory tags from tomb UJ of Umm El Qa'ab at Abydos. The hieroglyphs possibly signified specific goods and localities. Dryer notes: "As most of the signs manifest themselves as hieroglyphics in the dynastic period (i.e. after 3170 BCE or so), and since their later arrangement can already be observed in the beginning, it makes sense to take them, at least in part, not simply as symbols/markers, but to read them like hieroglyphics…Also other groups of signs can be read with the same phonetic values…The stork beside the chair (No. 103)…b..st = Basta. The fact that names of places occur among the signs, can be proven on a non-decipherable (*nicht lesbaren*) sign, the wrestlers, which are (also) inscribed as a hieroglyphic, identifying a place on the pallet of cities in one of the city-rings. A series of tags with the combination of tree + animal can be read, similarly to inscriptions on vessels, as designations of commodities that are named after their originator…Although it is often difficult to decide whether a sign is an ideogram or a phonogram. In some cases only one definitive interpretation is possible." (loc.cit. Richard Mattessich, 2002, The oldest writings, and inventory tags of Egypt, A review essay of Gunter Dreyer's Umm El-Quaab I-Das pradynastische Konigsgrab U-j und seine fruhen Schriftzeugnisse, in: *The Accounting Historians Journal* 29 (June 2002), pp. 195-208; Contaduria No. 41, Medellin, September 2002). https://aprendeenlinea.udea.edu.co/revistas/index.php/cont/article/viewFile/25609/21149

Clearly, the ivory and bone tags of royal tomb U-j Abydos were inventory tags ('vouchers' or 'inventory labels') with some accounting information. Early token and token-envelope accounting systems of Mesopotamia were also precursors of the syllabic cuneiform writing.

The hieroglyphs on inventory tags not merely show but say. Thus a remarkable advance has occurred over pictographic writing as it moved into the phase of logographic writing. Hieroglyphs: N'r M'r (N'r 'cuttle fish' M'r 'awl') rebus: Narmer (king's name of 33rd cent. BCE in Egypt). "Linguistic terminology makes it possible to identify the various units of language that helped to transform communication in early Egypt from merely pictorial expression to speech writing, which is important in identifying the nature of early graphic material:

"1) Logograms: symbols representing specific words
"2) Phonograms: symbols representing specific sounds
"3) Determinatives: symbols used for classifying words
"Moreover, writing on the tags shows that the Egyptian writing system had adopted the rebus principle, which broadened the meaning of symbols to include their homophones—words with the same sound but different definitions..." (Elise V. Macarthur, "The Concept and Development of the Egyptian Writing System" IN: Woods (ed), *Visible Language. Inventions of Writing in the Middle East and Beyond* [2010] 120).

Early writing from Abydos was used to label containers. (Courtesy Günter Dreyer)

Bone/ivory tags measuring 2 by 1 ½ cms. With one to four glyphs, clay seal impressions bearing hieroglyphs were unearthed by Germn Archaeological Institute from the tomb of predynastic ruler Scorpion at Abydos, south of Cairo. These are dated to ca. 3400-3200 BCE and constitute the earliest examples of Egyptian hieroglyph writing. "According to Jim Allen of the Metropolitan Museum of Art in New York, such early hieroglyphs represent a rebus system, akin to modern Japanese, in which pictures are used according to the way they sound. In early phonetic systems phrases such as "I believe," for example, might be rendered with an eye, a bee, and a leaf. The Abydos hieroglyphs are simple precursors to the complex hieroglyphic forms discovered at later sites such as Metjen and Turin."

http://archive.archaeology.org/9903/newsbriefs/egypt.html

Mlecchita vikalpa exemplifies *tantra yukti*, the Bharatiya method of composing a writing system. *Tantra* can be termed as that which discusses and details subjects and concepts; *yukti* is "... that which removes blemishes like impropriety, contradiction, etc., from the intended meaning and thoroughly joins the meanings together." The expression, *Tantra-yukti* denotes those devices that

aid the composition of a text in a systematic manner to convey intended ideas clearly. Cakrapāṇi lists 40 distinct devices of *tantra yukti*; 32 or more of these devices are exemplified in the treatises of Suśruta, Caraka, Vāgbhaṭa, Kautilya, Panini on knowledge domains of Ayurveda, Arthas'Astra and grammar of Samskrtam. Caraka notes:
तंत्रेसमासव्यासोक्ते भवन्त्येता हि कृत्स्नशः एकदेशने दृश्यन्ते समासाभिहिते तथा 'all these *tantrayukti*-s occur in a scientific work in brief and in detail. But only some of them occur in a work written in brief."

A characteristic feature of the structure and form of Harappa script is crispness of expression. This is governed by the cardinal principle of *tantra yukti* :
स्फुटता न पदैरपाकृता न च न स्वीकृत मर्थगौरवम् 'Crispness (of an expression) is not obliterated by verbosity, nor is the depth of meaning that is intended to be conveyed compromised (to attain crispness).

Precision of speech expression is achieved by unambiguously orthographed devices of hieroglyphs signified by images of wild and domestic animals (tiger, elephant, rhinoceros, boar, buffalo, zebu, ox, goat, markhor, ram, serpent hood), narratives such as a tiger or an antelope with head turned backwards, an archer or person seated in penance, a worshipper, tumblers, drummer, *ficus religiosa* leaf, claws of crab, rice-plant sprout, pincers, harrow, comb, scarf, lid of jar, rim of jar, rimless pot, ladle, stool, hayrick, platform, crocodile, frog, turtle, fish, fish-eye, quail, duck or aquatic birds, black ant, svastika, fire-altar, hillock, mountain-range, twig or sprig, numerical markers – one, two, three, four to signify numeral words. Semantic expansion to signify speech expressions is achieved by the 'crispness' feature of ligaturing combining animal heads, animal parts, infixed or circumscribed or superscripted hieroglyph multiplexes creating hypertexts. Such hypertexts have been matched with words and expressions of Harappa language (Meluhha) using the lexis of *Indian Lexicon*. Such words and expressions of animals, etc. hieroglyph-muliplexes have homonyms (rebus, similar sounding words) in the *parole*(speech forms) of Meluhha (again using the lexis of *Indian Lexicon*). All the homonyms for words and expressions so discovered relate to metalwork catalogues.

Some typical *tantra yukti* devices in the narratives of Harappa Script Corpora are: *upamānam* (or दृष्टान्त *dRṣṭānta* or analogy), *vākyaviśeṣa* (completion of a sentence meaningfully even in the absence of a word which is understood), *pūrvapakṣa* (objections, *prima facie* or provisional view), *uttarapakṣa* (correct view or answers). These devices are among 32 devices in*Arthaiśāstra* list of *Tantra yukti*. *Nidarśana (illustration)* The illustration of the devices of *tantra yukti* occurs on a cylinder seal from Ancient Near East. Cylinder Seal of Ibni-Sharrum Agade period, reign of Sharkali-Sharri (c. 2217-2193 BCE) Mesopotamia Serpentine H. 3.9 cm; Diam. 2.6 cm Formerly in the De Clercq collection; gift of H. de Boisgelin, 1967 AO 22303 The signifiers: *rango* 'buffalo' rebus: *rango* 'pewter' *lo*'overflow' *kanda* 'pot' rebus: *lokhanda* 'metal implements' *baTa* 'six' rebus: *bhaTa* 'furnace' *meD* 'curl' rebus:*meD* 'iron'. Thus, Ibni-Sharrum is a smelter working with pewter and metal implements.

Another example is a Mohenjodaro pectoral. m1656 Mohenjodro The message of the inscription is a *Dharma saṁjñā* ‘responsibility indicator’: arranger, manager of metal implements.

Hieroglyph: *sãghāṛɔ* 'lathe, brazier'.(Gujarati). Rebus: सं-ग्रह *saṁgraha, saṁgaha* ‘a guardian , ruler , manager , arranger’ R. BhP. *Vajra Sanghāta* 'binding together' (Varahamihira) **saṁgaḍha* ‘ collection of forts ’. [*gaḍha --]L. *sãgaṛh* m. ‘line of entrenchments, stone walls for defence’.(CDIAL 12845). Hieroglyph: खोंड (p. 216) [*khōṇḍa*] m A young bull, a bullcalf; खोंडा [*khōṇḍā*] m A कांबळा of which one end is formed into a cowl or hood. खोंडरूं [*khōṇḍarūṃ*] n A contemptuous form of खोंडा in the sense of कांबळा-cowl (Marathi. Molesworth[17]); *kōḍe dūḍa* bull calf (Telugu); *kōṛe* 'young bullock' (Konda)Rebus: *kõdā* ‘to turn in a lathe’ (Bengali) Rebus 2: koTiya ‘dhow, seafaring vessel’. *kāṇḍam* காண்டம்² *kāṇṭam,* n. < *kāṇḍa.* 1. Water; sacred water Rebus: *khāṇḍā* ‘metal tools, pots and pans’ (Marathi) (B) {V} *lo* (‘pot’, etc.) to ^overflow". See `to be left over'. @B24310. #20851. Re(B) {V} See `to be left over'. (Munda) Rebus: *loh* ‘copper’ (Hindi) The hieroglyph clearly refers to the metal tools, pots and pans of copper.

Section 4. Steps of the Decipherment with illustrations

Mĕrhĕt. Iron.
Mĕṛhĕt icena. The iron is rusty.
Ispat mĕṛhĕt. Steel.
Dul mĕṛhĕt. Cast iron.
Mĕṛhĕt khanḍa. Iron implements.

When hieroglyphs or hypertexts are ‘organized’ on an object, for e.g. a seal, there is an integral unity in the message (plain text) intended for communication. The Bogazkoy seal is intended to create a seal impression, say, on a cargo consignment by a metal merchant. All the hieroglyph components of the inscription have to be read together to generate a sentence. In this case, the sentence reads: three mineral ores metal casting.

Bogazkoy seal impression with 'twisted rope' hieroglyph (ca. 18th cent. BCE) मेढा [mēḍhā]twist (rope) rebus: *mẽṛhẽt, meḍ* ‘iron (metal)’ and a cognate word,मृदु mṛdu 'iron' (Samskritam)

_*skambha2 ‘ shoulder -- blade, wing, plumage ’. [Cf. **skapa* -- s.v. *khavaka --]S. *khambhu*, *°bho* m. ‘ plumage ’, *khambhuṛi* f. ‘ wing ’; L. *khabbh* m., mult. *khambh* m. ‘ shoulder -- blade, wing, feather ’, khet. *khamb* ‘ wing ’, mult. *khambhaṛā* m. ‘ **fin** ’; P. *khambh* m. ‘ wing, feather ’; G. *khā̆m* f., *khabhɔ* m. ‘ shoulder ’. (CDIAL 13640) rebus: *kammaTa* ‘mint, coiner, coinage’ eruvai 'kite' rebus: eruvai 'copper' *dhAv* ‘strand’ rebus1: *dhAv* ‘mineral, element’ rebus2. *dhAvaD* ‘smelter’ *dula* ‘two’ rebus: *dul* ‘metal casting’

m1406 Hieroglyphs: thread of three stands + drummer + tumblers

dhollu ‘drummer’ (Western Pahari) *dolutsu* 'tumble' Rebus: *dul* ‘cast metal’

karaḍa 'double-drum' Rebus: *karaḍa* 'hard alloy'.

dhAtu, dhAv 'strands of rope' Rebus: *dhAtu* 'mineral, metal, ore'

Kalibangan seal. k020 Hieroglyphs: thread of three strands + water-carrier + one-horned young bull. *kuTi* 'water-carrier' Rebus: *kuThi* 'smelter'. *dhAv* 'strands of rope' rebus: *dhAv* 'element, ore'; *dhAtu* id.
The following three examples from Mehrgarh, Mohenjo-daro and Banawali show acrobats in bull-jumping or buffalo-leaping.

Mehrgarh. Terracotta circular button seal. (Shah, SGM & Parpola, A., 1991, Corpus of Indus Seals and Inscriptions 2: Collections in Pakistan, Helsinki: Suomalainen Tiedeakatemia, MR-17. A humped bull (water buffalo?) and abstract forms (one of which is like a human body) around the bull. The human body is tossed from the horns of the bovine.

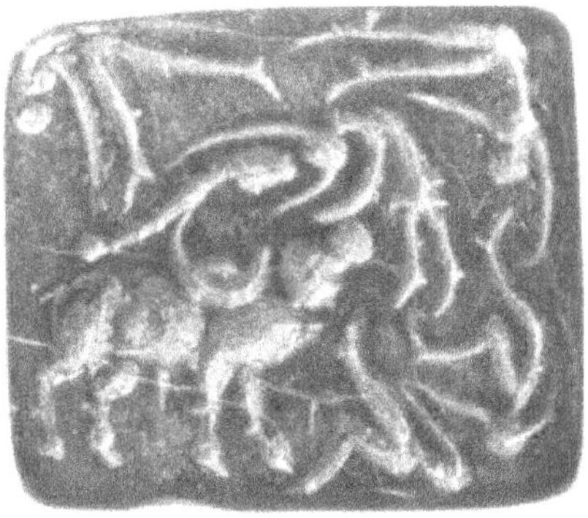

m0312 Persons vaulting over a water buffalo. The water buffalo tosses a person on its horns. Four or five bodies surround the animal. Rounded edges indicate frequent use to create clay seal impressions.

Impression of a steatite stamp seal (2300-1700 BCE) with a water-buffalo and acrobats. Buffalo attack or bull-leaping scene, Banawali (after UMESAO 2000:88, cat. no. 335). A figure is impaled on the horns of the buffalo; a woman acrobat wearing bangles on both arms and a long braid flowing from the head, leaps over the buffalo bull. The action narrative is presented in five frames of the acrobat getting tossed by the horns, jumping and falling down. Two Indus script glyphs are written in front of the buffalo. (ASI BNL 5683).
Rebus readings of hieroglyphs: '1. arrow, 2. jag/notch, 3. buffalo, 4.acrobatics':
1. kaṇḍa 'arrow' (Skt.) H. kā̃ḍerā m. ' a caste of bow -- and arrow -- makers (CDIAL 3024). Or. kāṇḍa, kā̃ṛ 'stalk, arrow '(CDIAL 3023). ayaskāṇḍa 'a quantity of iron, excellent iron' (Pāṇ.gaṇ)
2. खांडा [khāṇḍā] m A jag, notch, or indentation (as upon the edge of a tool or weapon). (Marathi) Rebus: khāṇḍā 'tools, pots and pans, metal-ware'.
3. rāngo 'water buffalo bull' (Ku.N.)(CDIAL 10559) Rebus: rango 'pewter'. ranga, rang pewter is an alloy of tin, lead, and antimony (anjana) (Santali).
4. ḍullu to fall off; ḍollu to roll over (DEDR 2698) Te. ḍul(u)cu, ḍulupu to cause to fall; ḍollu to fall; ḍolligillu to fall or tumble over (DEDR 2988) డొలుచు [ḍolucu] or ḍoluṭsu. [Tel.] v. n. To tumble head over heels as dancing girls do (Telugu) Rebus 1: dul 'to cast in a mould'; dul mẽṛhẽt, dul meṛeḍ, 'cast iron'; koṭe meṛeḍ 'forged iron' (Santali) Bshk. ḍōl ' brass pot (CDIAL 6583).
Rebus 2: WPah. ḍhō`ḷ m. 'stone', ḍhòḷṭɔ m. 'big stone or boulder', ḍhòḷṭu 'small id.' Him.I 87(CDIAL 5536). Rebus: K. ḍula m. ' rolling stone'(CDIAL 6582).

Decipherment of free-hand, painted inscription on a gold pendant-needle

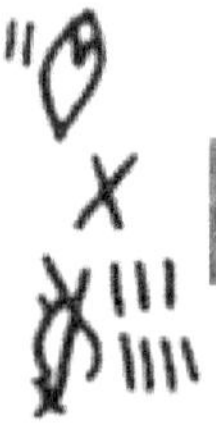

2.5 inch long Mohenjo-daro gold pendant has a 0.3 inch nib; its ending is shaped like a nib of a stylus pen or sewing or netting needle. It bears an inscription painted (perhaps

with ferric oxide pigment) in Harappa Script. This inscription is deciphered as a proclamation of metalwork competence.

Hieroglyph, needle: *ṭáṅkati1, ṭaṅkáyati* ' ties ' Dhātup. 2. **ṭañcati.* 1. S. *ṭā̃kaṇu* ' to stitch ', *ṭā̃ko* m. ' a stitch ' Rebus:*ṭaṅkaśālā* **--** , *ṭaṅkakaś°* f. ' mint ' lex. [ṭaṅka -- 1, śā´lā --]N. *ṭaksāl*, *°ār*, B. *ṭāksāl*, *ṭā̃k°*, *ṭek°*, Bhoj. *ṭaksār*, H. *ṭaksāl*, *°ār* f., G. *ṭā̃ksāḷ* f., M. *ṭā̃ksāl*, *ṭāk°*, *ṭā̃k°*, *ṭak°*. -- Deriv. G. *ṭaksāḷī* m. ' mint -- master ', M. *ṭāksāḷyā* m.
Addenda: *ṭaṅkaśālā* **-- :** Brj. *ṭaksāḷī*, *°sārī* m. ' mint -- master ' (CDIAL 5434)
kanac 'corner' Rebus: *kancu* 'bronze'; *sal* 'splinter'
Rebus: *sal* 'workshop'; *dāṭu* 'crosś(Telugu) *bāṭa* 'cross road'
Rebus: *dhatu* 'mineral'; *bhaṭa* 'furnace'; *gaṇḍa* 'four' Rebus: *khaṇḍa* 'implements; *kolmo* 'three' Rebus:*kolimi* 'smithy, forge'; Vikalpa: *?ea* 'seven' (Santali); rebus: ?*eh-ku* 'steel' (Te.)

aya, ayo 'fish' Rebus: *aya* 'iron'(Gujarati) *ayas* 'metal' (Rigveda) *khambhaṛā* 'fish fin' rebus: *kammaṭa* 'mint, coiner, coinage'

The inscription is a professional calling card -- describing professional competence of *ṭaksāḷī* m. *ṭāksāḷyā* m. 'mint master' (Gujarati.Marathi) and ownership of specified items of property -- of the wearer of the pendant.
Thus, the inscription is: *ṭaksāḷī* m. *ṭāksāḷyā* 'mint master' PLUS *kancu sal* (bronze workshop), *dhatu aya kaṇḍkolimi kammaṭa* 'mineral, metal, fire-altar, workshop, mint, coiner, coinage' mineral, metal, furnace/fire-altar smithy.

Three such needles are identified by John Marshall, in the excavation report.[7]

Section 5. Decipherment. Instances of the decipherment covering all aspects of the matter deciphered.

Harappa Script is a *mlecchita vikalpa* which employs the method of word or expression substitution cipher, i.e. substituting the word or expression signified by a picture with a similar-sounding word or expression (called homonym).

Harappa Script uses the method of word substitution cipher which can be called rebus, i.e. substituting the word signifying a hieroglyph with a similar-sounding word to convey the intended meaning, plain message. Thus, original pictorial message in cipher is substituted by the plain message.

For example, the picture of 'rim of jar' signifies the expression *kanda kankha* (variant pronunciation *karṇika*). This pictorial expression is substituted by a similar-sounding expression to signify the plain text: *khāṇḍā* 'tools, pots and pans, metal-ware' PLUS *karṇī* 'supercargo, a representative of the ship's owner on board a merchant ship, responsible for overseeing the cargo and its sale."; *karṇika* 'helmsman'. Thus, the picture 'rim of jar' signifies rebus: *khāṇḍā karṇika* '(metal) equipment, (metal) ware helmsman, supercargo, clerk (accountant)'.

loa 'ficus glomerata' rebus: *loh* 'copper' PLUS

kaṇḍa kankha 'rim of jar' rebus: *khāṇḍā karṇī, karṇika* 'metal equipment account scribe, supercargo'.

khōṇḍa m A young bull, a bullcalf rebus: *kunda* 'fine gold'.

Thus, the inscription with three hieroglyphs on the seal conveys the inventory of: copper, gold of supercargo, a representative of the ship's owner on board a merchant ship, responsible for overseeing the cargo and its sale.

The young bull is a hypertext with a number of hieroglyph components: one-horned young bull + rings on neck + pannier

खोंड (p. 122) *khōṇḍa* m A young bull, a bullcalf) one-horned young bull and *karb* 'culm of millet' (Punjabi), respectively. (NOTE: कोंद *kōnda* 'engraver, lapidary setting or infixing gems' is a phonetic variant of a worker with gold and lathe: *kunda* 'fine gold, lathe.' खोंड [*khōṇḍa*] m A young bull, a bullcalf. (Marathi) खोंडा [khōṇḍā] m A कां बळा of which one end is formed into a cowl or hood; खोंडरूं (p. 216) [khōṇḍarūṃ] n A contemptuous form of खोंडा in the sense of कांबळा -cowl. (Marathi) *khōṇḍa* A tree of which the head and branches are broken off, a stock or stump Rebus: *kõdār* 'turner' (Bengali); *kõdā* 'to turn in a lathe' (Bengali). *koḍiya* 'rings on neck', *koḍ* 'horn' rebus: *koḍ* 'workshop'. కోడియ (p. 326) *kōḍiya* కోడె (p. 326) *kōḍe* [Tel.] n. A bullcalf. *kodeduda*. A young bull (Telugu) (NOTE: the hieroglyph is a hypertext composed of young bull, one horn, pannier (a कां बळा 'sack' of which one end is formed into a cowl or hood), rings on neck.)

The importance of the 'jar' khāṇḍā can be seen from the use of the jar as cargo container, exemplified by the discovery of Susa pots as storage devices. A particular storage pot which contained metal implements from Meluhha is the defining archaeological evidence for the semantics of the 'rim-of-jar': 'metal equipment account scribe, supercargo'. This is a rivetting evidence for the purport of the entire Harappa Script corpora of over 8000 inscriptions since the 'rim-of-jar' hieroglyph occurs in about 70% of the inscriptions.

Below the rim of the storage pot, the contents are described in Harappa Script hieroglyphs/hypertexts: 1. Flowing water; 2. fish with fin; 3. aquatic bird tied to a rope Rebus readings of these hieroglyphs/hypertexts signify metal implements from the Meluhha mint.

Clay storage pot discovered in Susa (Acropole mound), ca. 2500-2400 BCE (h. 20 ¼ in. or 51 cm). Musee du Louvre. Sb 2723 bis (vers 2450 avant J.C.)

The hieroglyphs and Meluhha rebus readings on this pot from Meluhha are: 1. *kāṇḍa 'water' rebus: khāṇḍā* 'metal equipment'; 2. *aya, ayo* 'fish' rebus: *aya* 'iron' *ayas* 'metal alloy'; khambhaṛā *'fish fin' rebus: kammaṭ a* 'mint, coiner, coinage' 3. करड m. a sort of duck -- f. a partic. kind of bird ; S. karaṛa -ḍhī˜gu m. a very large aquatic bird (CDIAL 2787) *karaṇḍa* 'duck' (Samskrtam) rebus: *karaḍā* 'hard alloy'; PLUS 4. *meṛh* 'rope tying to post, pillar' rebus *meḍ* 'iron' med 'copper' (Slavic)

Susa pot is a 'Rosetta stone' for Harappa Script

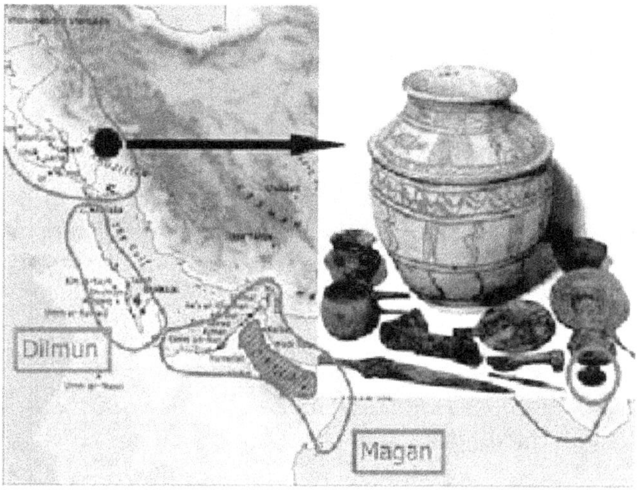

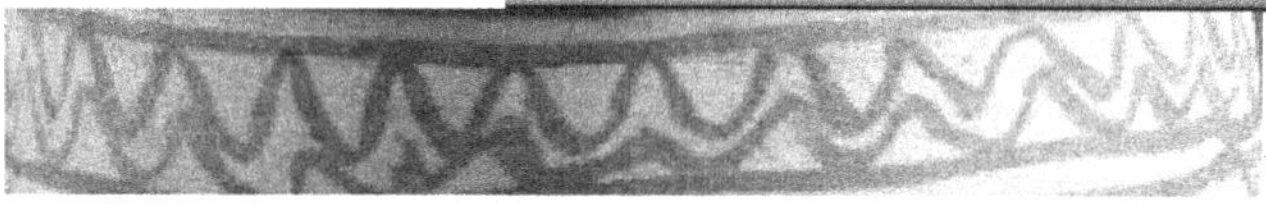Water (flow)

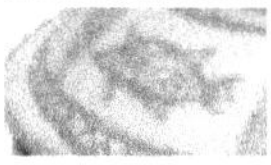Fish fish-fin

aquatic bird on wave (indicating aquatic nature of the bird), tied to rope, water
kāṇḍa 'water' rebus: *kāṇḍa* 'implements

The vase a la cachette, shown with its contents. Acropole mound, Susa.[20]
It is a remarkable 'rosetta stone' because it validates the expression used by Panini: *ayaskāṇḍa* अयस्--काण्ड [p= 85,1] m. n. " a quantity of iron " or " excellent iron " , (g. कस्का*दि q.v.). The early semantics of this expression is likely to be 'metal implements compared with the Santali expression to signify iron implements: *med'* 'copper' (Slovāk), *mẽṛhẽt, khaṇḍa* (Santali) मृदु *mṛdu*,'soft iron' (Samskrtam).

Mĕrhĕt. Iron.
Mĕṛhĕt icena. The iron is rusty.
Ispat mĕṛhĕt. Steel.
Dul mĕṛhĕt. Cast iron.
Mĕṛhĕt khanda. Iron implements. Santali glosses.

Harappa Script hieroglyphs painted on the jar are: fish, quail and streams of water;
aya 'fish' (Munda) rebus: *aya* 'iron' (Gujarati) *ayas* 'metal' (Rigveda) *khambhaṛā* 'fin' rebus: *kammaṭa* 'mint' Thus, together *ayo kammaṭa*, 'metals mint'
baṭa 'quail' Rebus: *bhaṭa* 'furnace'.
karaṇḍa 'duck' (Sanskrit) *karaṛa* 'a very large aquatic bird' (Sindhi) Rebus: करडा karaḍā 'Hard from alloy--iron, silver &c'. (Marathi) PLUS *meRh* 'tied rope' *meṛh* f. ' rope tying oxen to each other and to post on threshing floor ' (Lahnda)(CDIAL 10317) Rebus: *mūhā mẽṛhẽt* = iron smelted by the Kolhes and formeḍinto an equilateral lump a little pointed at each end; *mẽṛhẽt, meḍ* 'iron' (Mu.Ho.)
Thus, read together, the proclamation on the jar by the painted hieroglyphs is: *baṭa meṛh karaḍā ayas kāṇḍa* 'hard alloy iron metal implements out of the furnace (smithy)'.

This is a jar closed with a ducted bowl. The treasure called "vase in hiding" was initially grouped in two containers with lids. The second ceramic vessel was covered with a copper lid. It no longer exists leaving only one. Both pottery contained a variety of small objects form a treasure six seals, which range from Proto-Elamite period (3100-2750 BCE) to the oldest, the most recent being dated to 2450 BCE (First Dynasty of Ur).

Therefore, it is possible to date these objects, this treasure. Everything included 29 vessels including 11 banded alabaster, mirror, tools and weapons made of copper and bronze, 5 pellets crucibles copper, 4 rings with three gold and a silver, a small figurine of a frog lapis lazuli, gold beads 9, 13 small stones and glazed shard.
"In the third millenium Sumerian texts list copper among the raw materials reaching Uruk from Aratta and all three of the regions Magan, Meluhha and Dilmun are associated with copper, but the latter only as an emporium. Gudea refers obliquely to receiving copper from Dilmun: 'He (Gudea) conferred with the divine Ninzaga (= Enzak of Dilmun), who transported copper like grain deliveries to the temple builder Gudea...' (Cylinder A: XV, 11-18, Englund 1983, 88, n.6). Magan was certainly a land producing the metal, since it is occasionally referred to as the 'mountain of copper'. It may also have been the source of finished bronze objects."

Writing instruments/devices and pigments

240 copper tablets with inscriptions and scores of metal impements, tools, weapons inscribed with Harappa Script are evidence of metalwork and competence to write on metal. Many Harappa Script inscriptions are incised using a sharp stylus. Some inscriptions are hypertexts composed as raised script on metal. Some inscriptions on metal (for e.g. on a gold pendant) are written in some form of ink of ferric oxide or carbon black as pigment (perhaps using a writing brush). Some inscriptions are created with dotted orthography as on a gold fillet signifying the standard device which normally occurs in front of a one-horned young bull on many inscriptions. Hard metal styluses have been used to create over 240 copper plate Harappa Script inscriptions. Many copper inscriptions in bas-relief, raised script may have been used to created printed copies using ferric oxide ink. Evidence for writing in paint is provided by an inscription on a gold pendant and on pots with painted with Harappa Script hieroglyphs, for e.g. Susa pot with 'fish' hieroglyph and Nausharo pot showing a tied to a post.

Animals on Harappa Script are Meluhha metalwork hieroglyphic hypertexts, NOT totems

Animal pictographs in Harappa Script narratives are NOT totems. A totem is a symbol that serves as an emblem of a group of people, such as a family, clan, lineage, or tribe. The animal pictographs are NOT totems but hieroglyphs sacred symbols of *kole.l* 'smithy' which is *kole.l* 'temple; of *pasara* 'animals' rebus: *pasara* 'forge, smithy'.

meḍh 'ram' Rebus: *meḍho* 'one who helps a merchant' This is the explanation for the recurrence of a ram as a hieroglyph on hundreds of Dilmun, Persian Gulf seals.

Animals as hieroglyph signifiers is evident from these two examples. A 'joined animal' is created as a hypertext, composed of hieroglyph components: human face, horns of zebu, trunk of elephant, forefeet of a bovine, hindfeet of a feline, serpent-hood as upraised tail, scarves on neck.

सांगडणी (p. 495) *sāṅgaḍaṇī* f (Verbal of सांगडणें) Linking or joining together. सांगड (p. 495) sāṅgaḍa; f A body formed of two or more (fruits, animals, men) linked or joined together. सांगडणी (p. 495) *sāṅgaḍaṇī* f (Verbal of सांगडणें) Linking or joining together. That member of a turner's apparatus by which the piece to be turned is confined and steadied. सांगड or canoe-float.

mũh 'a *face*' rebus: *mũh*, muhã 'ingot' or muhã 'quantity of metal produced at one time in a native smelting furnace', 'ingot' (Santali)

karabhá m. ' camel ' MBh., ' young camel ' Pañcat., ' young elephant ' BhP. 2. *kalabhá*-- ' young elephant or camel ' Pañcat. (CDIAL 2797) rebus: *karba* 'iron' *ajirda karba* 'excellent iron' (Tulu) *ibha* 'elephant' rebus: *ib* 'iron' *ibbo* 'merchant'

kola 'tiger' कोलहा [kōlhā] कोलहे [kōlhēṃ] A jackal (Marathi) Rebus: *kol, kolhe*, 'smelter'

pasra 'animals' rebus: *pasra* 'smithy, forge'; hence, *pasra meṛed*, *pasāra meṛed* = syn. of *koṭe meṛed* = forged iron (Santali.Mundari)

pōḷā 'zebu, sacred animal set at liberty' rebus: *pōḷā* 'magnetite, ferrite ore'; cf. *pōḷā* 'animal festival held annually'.

xolā 'tail' of antelope and *kulā* '*hood* of *snake*' as tail. rebus: *kol, kolhe*, 'smelter'

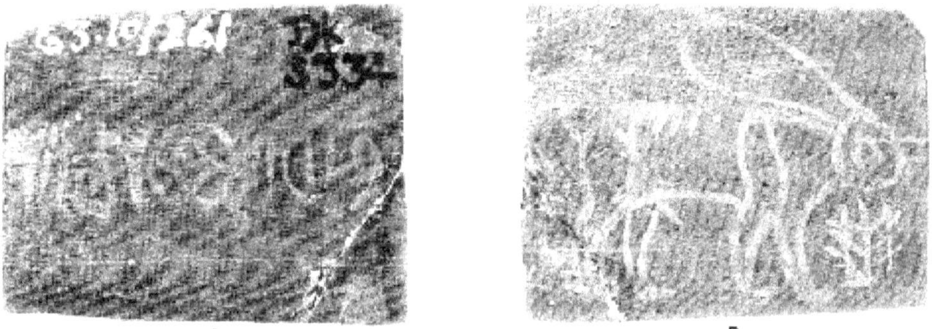

A M536 B

M319

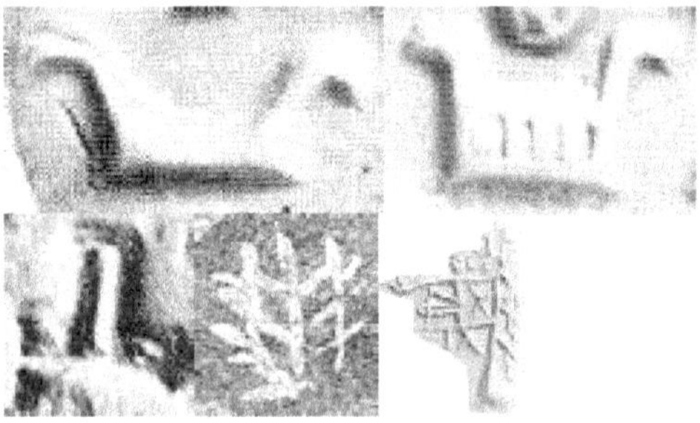

Copper tablet

m536. Hare +. Thorns. Seal impression m379 tiger + feeding trough. kharā '*hare*' (Oriya)N. *kharāyo* '*hare* ' *karA* 'crocodile' rebus: *khAr* 'blacksmith'

Both trough and tiger are hieroglyphs. It is inconceivable that tiger was a domesticated animal fed with a feeding trough. *kola* 'tiger' कोलहा [kōlhā] कोलहे [kōlhēṃ] A jackal (Marathi) Rebus: *kol, kolhe,* 'smelter' *pattar* 'feeding trough' rebus: *pattharika* 'merchant' *pattar* 'goldsmith guild' Mohenjo-daro seal impression. *bicha* 'scorpion' signifies *bica* 'haematite, ferrite ore'

barad, barat 'ox' Rebus: *bharata* 'alloy of pewter, copper, tin'

kuṭhāru = a *monkey* (Skt.lex.) Rebus: *kuṭhāru 'armourer* or weapons maker'(metal-worker)

karabhá m. ' camel ' MBh., ' young camel ' Pañcat., ' young elephant ' BhP. 2. *kalabhá*-- ' young elephant or camel ' Pañcat.(CDIAL 2797) rebus: *karba* 'iron' *ajirda karba* 'excellent iron' (Tulu) *ibha* 'elephant' rebus: *ib* 'iron' *ibbo* 'merchant'

कर्णक *kárṇaka, kannā 'legs spread*', 'rim of jar', 'pericarp of lotus' *karaṇī* 'scribe, supercargo', *kañi-āra* 'helmsman'. *karṇadhāra* m. ' helmsman ' Suśr. [kárṇa -- , dhāra -- 1] Pa. *kaṇṇadhāra* -- m. ' helmsman '; Pk. *kaṇṇahāra* -- m. ' helmsman, sailor '; H. *kanahār* m. ' **helmsman**, fisherman '.(CDIAL 2836)

mrēka, mēḻh 'goat' (Telugu. Brahui) rebus: *milakkhu*, 'copper' (Pali) *mleccha-mukha* 'copper' (Samskrtam)

gaṇḍá m. ' rhinoceros ' lex., °*aka* -- m. lex. 2. *ga- yaṇḍa -- **.** [Prob. of same non -- Aryan origin as khaḍgá -- 1: cf. *gaṇōtsāha* -- m. lex. as a Sanskritized form ← Mu. PMWS 138]
1. Pa. *gaṇḍaka* -- m., Pk. *gaṁḍaya* -- m., A. *gãr*, Or. *gaṇḍā.*
2. K. *gō̃ḍ* m., S. *geṇḍo* m. (lw. with *g* --), P. *gaĩḍā* m., °*ḍī* f., N. *gaĩṛo*, H. *gaĩṛā* m., G. *gẽḍɔ* m., °*ḍī* f., M. *gẽḍā* m. Addenda: gaṇḍa -- 4. 2. *gayaṇḍa **--:** WPah.kṭg. *geṇḍɔ mirg* m. ' rhinoceros ', Md. *genḍā* ← H. Rebus: *khāṇḍā* 'tools, pots and pans, implements'

No.642. Failaka cylinder seal.

karA 'crocodile'

rebus: *khAr* 'blacksmith'

gaṇḍá m. ' rhinoceros ' lex., °*aka* -- m. lex. 2. *ga- yaṇḍa -- **.** [Prob. of same non -- Aryan origin as khaḍgá -- 1: cf. *gaṇōtsāha* -- m. lex. as a Sanskritized form ← Mu. PMWS 138]

1. Pa. *gaṇḍaka* -- m., Pk. *gaṁḍaya* -- m., A. *gā̃r*, Or. *gaṇḍā*.
2. K. *gŏ̃ḍ* m., S. *geṇḍo* m. (lw. with *g* --), P. *gaĩḍā* m., °*ḍī* f., N. *gaĩṛo*, H. *gaĩṛā* m., G. *gẽḍɔ* m., °*ḍī* f., M. *gẽḍā* m. Addenda: gaṇḍa -- 4. 2. *gayaṇḍa -- : WPah.kṭg. *geṇḍɔ mirg* m. ' rhinoceros ', Md. *genḍā* ← H. Rebus: *khāṇḍā* 'tools, pots and pans, implements'

karabhá m. ' camel ' MBh., ' young camel ' Pañcat., ' young elephant ' BhP. 2. *kalabhá*-- ' young elephant or camel ' Pañcat.(CDIAL 2797) rebus: *karba* 'iron' *ajirda karba* 'excellent iron' (Tulu) *ibha* 'elephant' rebus: *ib* 'iron' *ibbo* 'merchant'

M-227 A

gaṇḍa four' rebus: *khāṇḍā* 'tools, pots and pans, implements'

pōḷā 'zebu, sacred animal set at liberty' rebus: *pōḷā* 'magnetite, ferrite ore'; cf. *pōḷā* 'animal festival held annually'.

खोंड (p. 122) *khōṇḍa* m A young bull, a bullcalf) one-horned young bull and *karb* 'culm of millet' (Punjabi), respectively. (NOTE: कोंद *kōnda* 'engraver, lapidary setting or infixing gems' is a phonetic variant of a worker with gold and lathe: *kunda* 'fine gold, lathe.' खोंड [*khōṇḍa*] m A young bull, a bullcalf. (Marathi) खोंडा [khōṇḍā] m A कां बळा of which one end is formed into a cowl or hood; खोंडरूं (p. 216) [khōṇḍarūṃ] n A contemptuous form of खोंडा in the sense of कांबळा -cowl. (Marathi) *khōṇḍa* A tree of which the head and branches are broken off, a stock or stump Rebus: *kõdār* 'turner' (Bengali); *kõdā* 'to turn in a lathe' (Bengali). *koḍiya 'rings on neck'*, *koḍ* 'horn' rebus: *koḍ* 'workshop'. కోడియ (p. 326) *kōḍiya* కోడె (p. 326) *kōḍe* [Tel.] n. A bullcalf. *kodeduda*. A young bull (Telugu) (NOTE: the hieroglyph is a hypertext composed of young bull, one horn, pannier (a कां बळा 'sack' of which one end is formed into a cowl or hood), rings on neck.)

aya, ayo 'fish' aya 'iron' (Gujarati) ayas 'metal alloy' (Rigveda.Samskrtam)

खाडा [khāṇḍā] m A jag, *notch*, or indentation (as upon the implement' rebus: *khāṇḍā* 'tools, pots and pans, implements'

Kuwait Museum. Gold disc.

miṇḍ 'ram', *miṇḍāl 'markhor'* (CDIAL 10310) Rebus:*mẽṛhẽt*, meḍ 'iron' (Santali.Ho.Mu.)

pasara 'domestic animals' Rebus: pasara 'smithy, forge'

rango 'buffalo' rebus: *rango* 'pewter' **raṅga**3 n. ' tin ' lex. [Cf. nāga -- 2, vaṅga -- 1]
Pk. *raṁga* -- n. ' tin '; P. *rã̄g* f., *rã̄gā* m. ' pewter, tin ' (← H.); Ku. *rāṅ* ' tin, solder ', gng. *rã̄k*; N. *rāṅ*, *rāṅo* ' tin, solder ', A. B. *rāṅ*; Or. *rāṅga* ' tin ', *rāṅgā* ' solder, spelter ', Bi. Mth. *rã̄gā*, OAw. *rāṁga*; H. *rã̄g* f., *rã̄gā* m. ' tin, pewter '; Si. *ran̆ga* ' tin '(CDIAL 10562)

ranku 'antelope' (Santal*i)* ***kuraṅgá1*** m. ' antelope ' MBh., *kulaṅgá* -- MaitrS., *kuluṅgá* -- TS.
Pa. *kuraṅga* -- , *kuruṅga* -- m., Pk. *kuraṁga* -- m., P. *kuraṅg* m., OG. *karaṁgī* f.,
G. *kurãg* m., °*gī*, °*gṇī* f.; Si. *kurun̆ga* ' antelope ', *kiran̆gu* ' the elk Rusa aristotelis '.(CDIAL 3320) Rebus: *ranku* 'tin'.

Dwaraka. Turbinella pyrum seal.

ranku 'antelope' (Santal*i) kuraṅgá1* m. ' antelope ' MBh., *kulaṅgá* -- MaitrS., *kuluṅgá* -- TS.
Pa. *kuraṅga* -- , *kuruṅga* -- m., Pk. *kuraṁga* -- m., P. *kuraṅg* m., OG. *karaṁgī* f.,
G. *kurãg* m., °*gī*, °*gṇī* f.; Si. *kurun̆ga* ' antelope ', *kiran̆gu* ' the elk Rusa aristotelis '.(CDIAL 3320) Rebus: *ranku* 'tin'.

barad, *barat* 'ox' Rebus: *bharata* 'alloy of pewter, copper, tin'

खोंड (p. 122) *khōṇḍa* m A young bull, a bullcalf) one-horned young bull and *karb* 'culm of millet' (Punjabi), respectively. (NOTE: कोंद *kōnda* 'engraver, lapidary setting or infixing gems' is a phonetic variant of a worker with gold and lathe: *kunda* 'fine gold, lathe.' खोंड [*khōṇḍa*] m A young bull, a bullcalf. (Marathi) खोंडा [khōṇḍā] m A कां बळा of which one end is formed into a cowl or hood; खोंडरूं (p. 216) [khōṇḍarūṃ] n A contemptuous form of खोंडा in the sense of कांबळा -cowl. (Marathi) *khōṇḍa* A tree of which the head and branches are broken off, a stock or stump Rebus: *kõdār* 'turner' (Bengali); *kõdā* 'to turn in a lathe' (Bengali). *koḍiya 'rings on neck'*, *koḍ* 'horn' rebus: *koḍ* 'workshop'. కోడియ (p. 326) *kōḍiya* కోడె (p. 326) *kōḍe* [Tel.] n. A bullcalf. *kodeduda*. A young bull (Telugu) (NOTE: the hieroglyph is a hypertext composed of young bull, one horn, pannier (a कां बळा 'sack' of which one end is formed into a cowl or hood), rings on neck.)

See: https://wordpress.com/post/sarasvati97.wordpress.com/769

Harappa Seal

pasra meṛed, pasāra meṛed = syn. of *koṭe meṛed* = forged iron, in contrast to *dul meṛed*, cast iron (Mundari) *dul mẽṛhẽt* 'cast iron'; mẽṛhẽt khaṇḍa 'iron implements' (Santali) *i meṛed* rusty iron, also the iron of which weights are cast (Mundari. Santali)

bica 'stone ore' as in *meṛed-bica* = iron stone ore, in contrast to *bali-bica* , iron sand ore (Mundari) *sambr.o bica* = gold ore (Mundarica)

mrēka, mē<u>lh</u> 'goat' (Telugu. Brahui) rebus: *milakkhu*, 'copper' (Pali) *mleccha-mukha* 'copper' (Samskrtam)

*PIE *melH-i, *mel-iyo-. mealie, miliary, milium, millet; gromwell, from Latin milium, millet.*

aya, ayo 'fish' aya 'iron' (Gujarati) ayas 'metal alloy' (Rigveda.Samskrtam)

Section 6. Harappa Script Decipherment in the context of wealth creation, evidenced by Archaeometallurgy

"Benoit Mille has drawn attention to copper alloy 'amulets' discovered in the early Chalcolithic (late 5th millennium) levels of Mehrgarh in Baluchistan, Pakistan. He reported that metallographic examination established that the ornaments were cast by the lost-wax method (Mille, B., 2006, 'On the origin of lost-wax casting and alloying in the Indo-Iranian world', in *Metallurgy and Civilisation: 6th international conference on the beginnings of the use of metals and alloys*, University of Science and Technology, Beijing, BUMA VI). The amulets were made from copper alloyed with lead. Mehrgarh is well recognised as a centre for early pyrotechnologies. The wax models of the amulets would have been solid and may have had a simple core inserted. This is understandably the first stage in the technology. Mille also draws attention to the 'Leopards weights' from Baluchistan, dating to about 3000 BCE which were made using a complex core keyed into the investment mould."(Davey, Christopher J., The early history of lost-wax casting, in: J. Mei and Th. Rehren, eds., Metallurgy and Civilisation: Eurasia and Beyond Archetype, London, 2009, ISBN 1234 5678 9 1011, pp. 147-154; p. 151).

Remarkable evidences of the excellence achieved in *cire perdue* metal castings are provided by bronze or copper alloy artifacts kept in the British Museum, said to have been acquired from Begram, and dated to ca. 2000 to 1500 BCE. These are also referred to as compartmented seals.

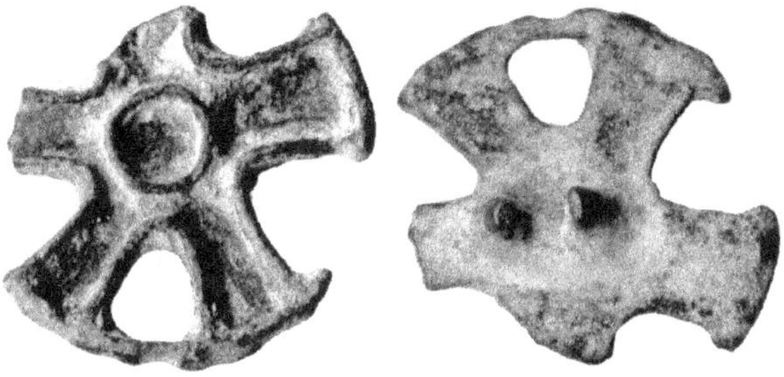

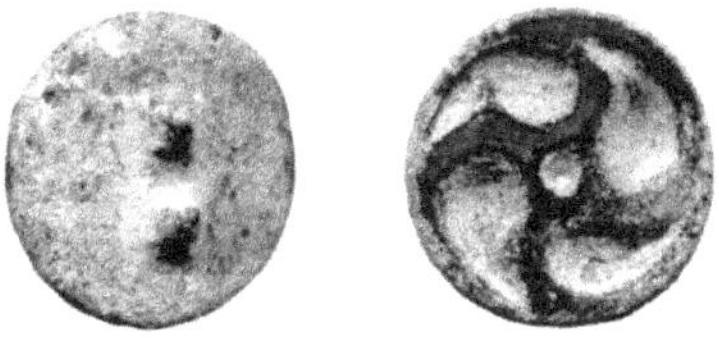

IM 154
1880-3710 (a-f)

IM 154
1880-3710 (a-f)

IM 154
1880-3710 (a-f)

IM 154
1880-3710 (a-f)

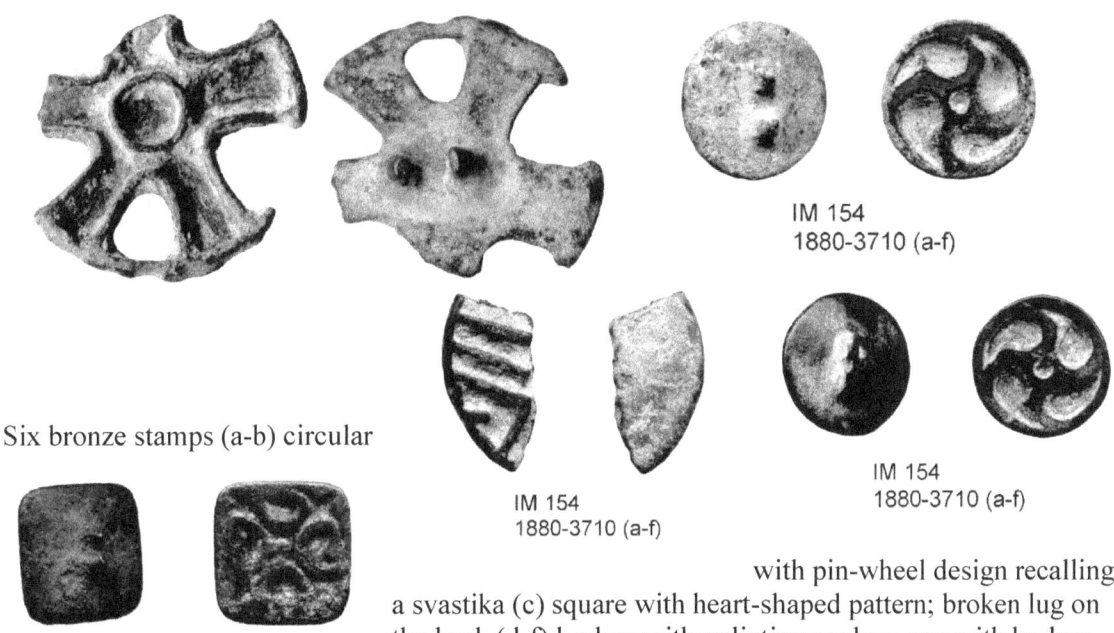

Six bronze stamps (a-b) circular with pin-wheel design recalling a svastika (c) square with heart-shaped pattern; broken lug on the back (d-f) broken with radiating spokes; one with broken lug.

Cast, copper alloy, circular, openwork seal or stamp, comprising five wide spokes with projecting rims, radiating from a circular hub also encircled by a flange. The outer rim is mostly missing and two spokes are broken. The back is flat, with the remains of a broken attachment loop in the centre.

2000BC-1500BC (circa) Copper alloy. Pierced. cast.

Made in: Afghanistan(Asia, Afghanistan)
Found/Acquired: Begram (Asia, Afghanistan,Kabul (province),Begram)

The earliest lost-wax cast object is by Bharatam Janam. Metallurgy explained -- M. Thoury et al (March 2016). Harappa Script & Language explained.

Harappa (Indus) script hieroglyph: *eraka* 'knave of wheel' rebus: eraka 'moltencast, metal infusion'; *era* 'copper'. *āra* 'spokes' arā 'brass' *erako* molten cast (Tulu) *Ka. e̲re* to pour any liquids, cast (as metal); *n.* pouring; e̲racu, ercu to scoop, sprinkle, scatter, strew, sow; *e̲raka, eraka* any metal infusion; molten state, fusion.*Tu. eraka* molten, cast (as metal); *eraguni* to melt (DEDR 866) *agasāle, agasāli, agasālevāḍu* <arka sAle= a goldsmith (Telugu) अर्क [p= 89,1]*m.* (√ अर्च्) , Ved. a ray , flash of lightning RV. &cthe sun RV. &c Rebus: copper L.அருக்கம்[1] *arukkam*, n. < arka. (நாநார்த்த.) 1. Copper; செம்பு. 2. Crystal; பளிங்கு. அக்கம்&sup4; *akkam*

, n. < arka. An ancient coin = 1/12 காசு; ஒரு பழைய நாணயம். (S. I. I. ii. 123.) అగసాలి (p. 23) *agasāli* or అగసాలెవాడు *agasāli.* [Tel.] n. A goldsmith. కంసాలివాడు.

ಅಕ್ಕಸಾಲೆಗ. = ಅಕ್ಕಸಾಲಿಗ, q. v.

ಅಕ್ಕಸಾಲೆ f. n. The workshop of a goldsmith. 2, a goldsmith.

Kannada Glosses

erka = *ekke* (Tbh. of arka) aka (Tbh. of *arka*) copper (metal); crystal (Ka.lex.) cf. *eruvai* = copper (Tamil)

The cire perdue spoked wheel of copper+lead alloy was NOT an amulet, it was a metal artifact, a metal coin, *akkam;* it was a compartmental Harappa seal with Harappa (Indus) Script hieroglyph. May or may not have been used as a coin to value and exchange goods but a proclamation of the metallurgical excellence achieved by Bharatam Janam of 4th millennium BCE.

Artisans at work in Burma making Karen drum

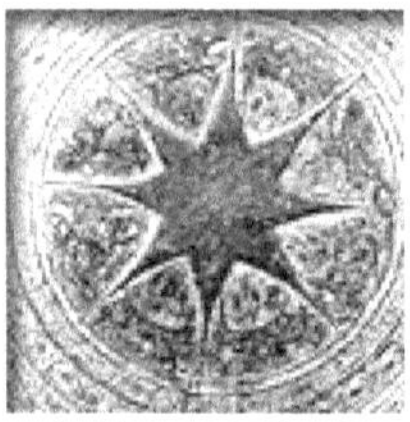

Sun motif in the centre of the tympanum, Karen drum.

arká1 m. ‘ flash, ray, sun ’ RV. [√arc] Pa. Pk. *akka* -- m. ‘ sun ’, Mth. *āk;* Si. *aka* ‘ lightning ’, inscr. *vid* -- *äki* ‘ lightning flash ’.(CDIAL 624) rebus: *erako* 'moltencast' *arka, eraka* 'gold, copper'.

Detail of the tympanum of Karen drum.

ayo 'fish' rebus; *aya* 'iron' *ayas* 'metal alloy'

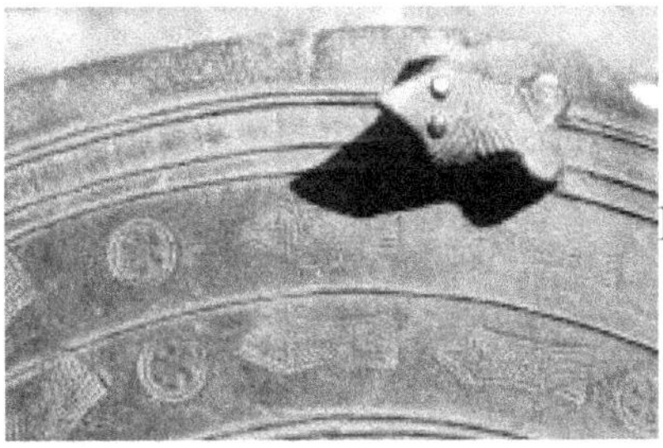Frog on the Karen bronze *pancaloha* 'five metal alloys' drum.

Kur. mūxā frog. *Malt. múqe* id. / Cf. Skt. *mūkaka-* id. (DEDR 5023) Rebus: *mū̃h* 'ingot' mũhe 'ingot' mũhā̃ = the quantity of iron produced at one time in a native furnace.

Elephant motif. karba, ibha 'elephant' rebus: karba, ib

'iron'.*Ta. ayil* iron. *Ma. ayir, ayiram* any ore. *Ka. aduru* native metal. *Tu. ajirda karba* very hard iron. (DEDR 192)

"The town of Nwe Daung, 15 km south of Loikaw, capital of Kayah (formerly Karenni) State, is the only recorded casting site in Burma. Shan craftsmen made drums there for the Karens from approximately 1820 until the town burned in 1889. Karen drums were cast by the lost wax technique; a characteristic that sets them apart from the other bronze drum types that were made with moulds. A five metal formula was used to create the alloy consisting of copper, tin, zinc, silver and gold. Most of the material in the drums is tin and copper with only traces of silver and gold. The Karen made several attempts in the first quarter of the twentieth century to revive the casting of drums but none were successful."

In ancient Indian texts, such as Manasollasa, Silparatna, Manasara,the *cire perdue* technique is referred to as *madhucchiṣṭa vidhānam*. मधु madhu -उच्छिष्टम्,-उत्थभ्,-उत्थितभ् 1 bees'-wax; शस्त्रासवमधूच्छिष्टं मधु लाक्षा च बर्हिषः Y.3.37; मधूच्छिष्टेन केचिच्च जध्नुरन्योन्यमुत्कटाः Rām.5.62.11.-2 the casting of an image in wax; Mānasāra; the name of 68th chapter. This technique was clearly attested in the Epic *Rāmāyaṇa*. मधुशिष्ट madhuśiṣṭa 'wax' (Monier-Williams, p. 780).

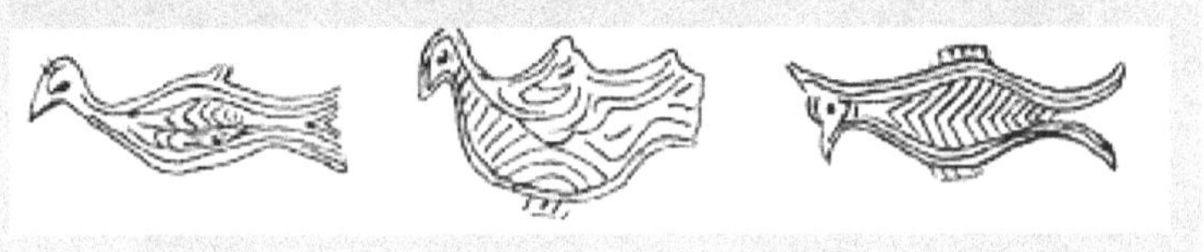

karaṇḍa 'duck' (Sanskrit) karaṛa 'a very large aquatic bird' (Sindhi) *karaDa* 'safflower' rebus:*karaḍa* 'double-drum'
Rebus: करडा [*karaḍā*] Hard from alloy--iron, silver &c kharādī = turner (Gujarati)
कारण्डवः, पुं, स्त्री, (ञमन्ताडड इति रमेर्ड । रण्डः । ईषत् रण्डः । "ईषदर्थे" ६ । ७ । १०५ । इति कोः कादेशः । कारण्डं वाति । वा गतौ + "आतोनुपेति" । ३ । २ । ३ । कः । करण्डस्येदं कारण्डं तदाकारं वाति वा ।) हंसविशेषः इत्यमरः । २ । ५ । ३४ ॥ खडहाँस इति भाषा (यथा ऋतुसंहारे । शरद्वर्णणे ८ । "कारण्डवाननविघट्टितवीचिमालाः कादम्बसारसकुलाकुलतीरदेशाः" ॥)

https://sa.wikisource.org/wiki/शब्दकल्पद्रुमः

कारण्डव पुंस्त्री रम--ड तस्य नेत्त्वम् रण्डः ईषत् रण्डः कारण्डः तं वाति वा--क करण्डस्येदं कारण्डं तदाकारं वाति वा--क वा । हंसभेदे "हंसकाण्डवोद्गीताः सारसाभिरुतास्तथा" भा० व० ३८ अ० । स्त्रियां जाति- त्वात् ङीष् । अस्य अजिरादि० पाष्ठात् मतौ संज्ञायामपि न दीर्घः कारण्डववती नदीविशेषः । "हससारसक्रौञ्च-चक्रवाककुररकादम्बकारण्डवेत्युपक्रमे" प्लवाः सधचारिणश्च" इति सुश्रुते तस्य प्लवत्वं स घचारित्वञ्चोक्तम् ।

https://sa.wikisource.org/wiki/वाचस्पत्यम्

kuṭhi 'tree' rebus *kuṭhi* 'a furnace for smelting iron ore, to smelt iron') tALa 'palm trees' rebus: DhALa 'large ingot (oxhide)'

The First Recorded Use of Lost Wax Casting

The earliest known written reference to lost wax casting comes from the Babylonian city of Sippar and is dated 1 789 B.C., during the reign of the great King Hammurabi. Written in cuneiform on a clay tablet, this is a receipt for a small quantity of wax issued to a metal worker and is composed in the typically bureaucratic manner of the period.

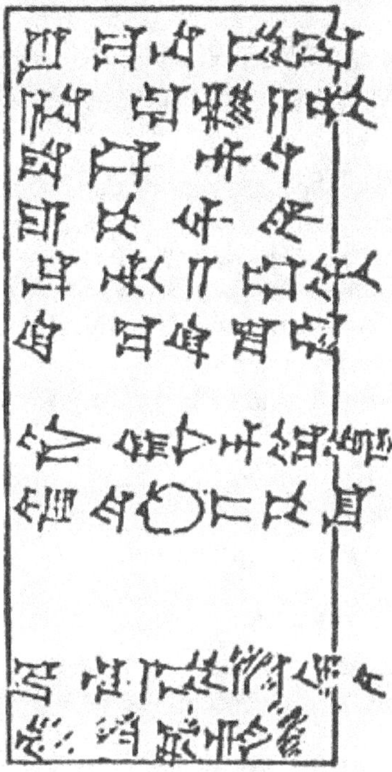

'Two thirds of a mina of wax to make a bronze key for the temple of Shamash received by the metal worker from the temple treasury

In the presence of Silli-nin-karrak and of the storekeeper, his colleagues

On the nineteenth day of the month of Arabsammu in the year of building the temple wall'

Hieroglyphs of Indus Script Cipher are sitnified on the Shahi Tump leopard weight which has been produced using the lost-wax casting method. The hieroglyphs are: 1. leopard; 2. ibex or antelope; 3. bees (flies). The rebus-metonymy readings in Meluhha are:

karaḍa 'panther'; karaḍa tiger (Pkt); खरडा *[kharaḍā] A leopard.* खरड्या *[kharaḍyā] m or* खरड्यावाघ *m A leopard (Marathi). Kol. keḍiak tiger. Nk. khaṛeyak panther. Go. (A.) khaṛyal tiger; (Haig) kariyāl panther Kui kṛāḍi, krānḍi tiger, leopard, hyena. Kuwi (F.) kṛani tiger; (S.) klā'ni tiger, leopard; (Su. P. Isr.) kṛa'ni (pl. -ŋa) tiger. / Cf. Pkt. (DNM) karaḍa- id. (DEDR 1132).Rebus:* करडा *[karaḍā]* Hard from alloy--iron, silver &c. (Marathi) *kharādī* ' turner, a person who fashions or shapes objects on a lathe' (Gujarati)

Hieroglyph: *miṇḍāl* 'markhor' (Tōrwālī) *meḍho* a ram, a sheep (Gujarati)(CDIAL 10120)
Rebus: *mẽṛhẽt, meḍ* 'iron' (Munda.Ho.) *mṛeka, melh* 'goat' (Telugu. Brahui)
Rebus: *melukkha* '*milakkha*, copper'. If the animal carried on the right hand of the Gudimallam hunter is an antelope, the possible readings are: ranku 'antelope' Rebus: ranku 'tin'.

Ka. mēke she-goat; mē the bleating of sheep or goats. Te. mē̃ka, mēka goat.

Kol. me·ke id. Nk. mēke id. Pa. mēva, (S.) mēya she-goat. Ga. (Oll.)mēge, (S.) mēge goat. Go. (M) mekā, (Ko.) mēka id. ? Kur. mēxnā (mīxyas) to call, call after loudly, hail. Malt. méqe to bleat. [Te. mṛēka (so correct) is of unknown meaning. Br. mēl̲h̲ is without etymology; see MBE 1980a.] / Cf. Skt. (*lex.*) *meka-* goat. (DEDR 5087). Meluhha, mleccha (Akkadian. Sanskrit). *Milakkha, Milāca 'hillman' (Pali) milakkhu 'dialect' (Pali) mleccha 'copper'* (Prakritam).

The bees are metaphors for wax used in the lost-wax casting method.

Hieroglyph: *माक्षिक [p= 805,2] mfn.* (fr. मक्षिका) coming from or belonging to a bee Rebus: 'pyrites': *माक्षिक [p= 805,2] n.* a kind of honey-like mineral substance or pyrites MBh. उपधातुः An inferior metal, semi-metal. They are seven; सप्तोपधातवःस्वर्णं माक्षिकं तारमाक्षिकम् । तुत्थं कांस्यं च रातिश्च सुन्दूरं च शिलाजतु ॥ उपरसः upara sḥउपरसः 1 A secondary mineral, (red chalk, bitumen, माक्षिक, शिलाजित &c).(Samskritam)

mákṣā f., *mákṣ* -- m. f. ' fly ' RV., *mákṣikā* -- f. ' fly, bee ' RV., *makṣika* -- m. Mn.Pa. *makkhikā* -- f. ' fly ', Pk. *makkhiā* -- f., *macchī* -- , *°chiā* -- f.; Gy. hung. *makh* ' fly ', wel. *makhī* f., gr. *makí* f., pol. *mačin*, germ. *mačlin*, pal. *mắki* ' mosquito ',*mắkī´la* ' sandfly ', *mắkī´li* ' house -- fly '; Ash. *mačī˜´* ' bee '; Paš.dar. *mēček* ' bee ', weg. *mečī´k* ' mosquito ', ar. *mučək*, *mučag* ' fly '; Mai. *mā̆çhī* ' fly '; Sh.gil.*mā̆ṣī´* f., (Lor.) *m**lçī* ' fly ' (→ Ḍ. *m**lçhi* f.), gur. *mā̆çhī´* ' fly ' (' bee ' in gur. *mā̆çhịkraṇ*, koh. *mā̆çhi* -- *gŭn* ' beehive '); K. *mạ̄chi* f. ' fly, bee, dark spot '; S. *makha,makhi* f. ' fly, bee, swarm of bees, sight of gun ', *makho* m. ' a kind of large fly '; L. (Ju.) *makhī* f. ' fly ', khet. *makkī´*; P. *makkh* f. ' horsefly, gnat, any stinging fly ', m. ' flies ', *makkhī* f. ' fly '; WPah.rudh. *makkhī* ' bee ', jaun. *mākwā* ' fly '; Ku. *mākho* ' fly ', gng. *mā̃kh*, N. *mākho*, A. *mākhi*, B. Or. *māchi*, Bi. *māchī*, Mth. *māchī,mā̃chī*, *makhī* (← H.?), Bhoj. *māchī*; OAw. *mākhī*, lakh. *māchī* ' fly ', *ma* -- *mākhī* ' bee ' (mádhu --); H. *māchī*, *mākhī*, *makkhī* f. ' fly ', *makkhā* m. ' large fly, gadfly '; G. *mākh*, *mākhī* f. ' fly ', *mākhɔ* m. ' large fly '; M. *mās* f. ' swarm of flies ', n. ' flies in general ', *māśī* f. ' fly ', Ko. *māsu*, *māśi*; Si. *balu* -- *mäkka*, st. -- *mäki* -- ' flea ', *mässa*, st. *mäsi* -- ' fly '; Md. *mehi* ' fly '.*makṣātara -- , *mākṣa -- , mākṣiká -- ; *makṣākiraṇa -- , *makṣācamara -- , *makṣācālana -- , *makṣikākula -- ; *madhumakṣikā -- . Addenda: mákṣā -- : S.kcch. *makh* f. ' fly '; WPah.kṭg. *mákkhɔ*, *máṅkhɔ* m. ' fly, large fly ', *mákkhi* (kc. *makhe*) f. ' fly, bee ', *máṅkhi* f., J. *mākhī* f.pl., Garh. *mākhi.* (CDIAL 9696) mākṣiká ' pertaining to a bee ' MārkP., n. ' honey ' Suśr. 2. *mākṣa -- . [mákṣā --] 1. WPah.bhad. *mā̇chī* ' bee ', khaś. *mākhī*; -- Pk. *makkhia* -- , *macchia* -- n. ' honey '; Ash. *mači*, *mačik* ' sweet, good ', *mačianá* ' honey '; Wg. *máçi*, *mäç* ' honey ', Kt. *maçī˜*, Pr. *maṭék*, Shum. *maçhī*, Gaw. *māçhī*, Kal.rumb. Kho. *maçhí*, Bshk. *mẽçh*, Phal. *mn/açhī*, *mḗçhī*, Sh. *mā̆çhī´* f., S. L. *mākhī* f., WPah.bhiḍ. *mā̇chī* n., H.*mākhī* f.2. K. *mā̃ch*, dat. *°chas* m. ' honey ', WPah.bhal. *māch* n. -- For form and meaning of Paš. *māš*, *mōṣ* ' honey ' see NTS ii 265, IIFL iii 3, 126.*mākṣakulika -- , *mākṣikakara -- , *mākṣikamadhu -- .Addenda: mākṣika -- : Kho. *maçhi* ' honey ' BKhoT 70.(CDIAL 9989)*mākṣikakara or **mākṣakara* -- ' bee '. [Cf. *madhu- kara* -- m. ŚārṅgP., *°kāra* -- m. BhP., *°kārī* -- f. R. <-> mākṣiká -- , kará -- 1] Ash. *mačarīk*, *°čerī´k* ' bee ', Wg. *maçarī´k*, Kt. *mačerík* NTS ii 265, *maçe°* Rep1 59, Pr. *mučerík*, *məskerík*, *muṭkurī´k*, Shum. *mā̃çhā´rik*, Kal.rumb. *maçhḗrik*, Bshk.*mā´çēr*, Phal. *mā̆çhurī´* f.; Sh.koh. *mā̆çhāri* f. ' bee ', gil. (Lor.) *m**lçhari* ' bee, wasp, hornet ' (in latter meaning poss. < *makṣātara --); P. *makhīr* m. ' bee ', kgr. ' honey '; -- Gaw. *mā̃ç(h)oṛík* with unexpl. -- *ṛ* -- . (CDIAL 9990) *mākṣikamadhu ' honey '. [mākṣiká -- , mádhu --] P. *mākhyō̃* f., *mākho* m. ' honey, honeycomb '.(CDIAL 9991) مچئي *maça´ī*, s.f. (6th) A bee in general. Sing. and Pl. سره مچئي *sara´h-maça´ī*, s.f. (6th). Sing. and Pl.; or دنداره *ḍanḍḍāra´h*, s.f. (3rd) A hornet, a wasp. Pl. ي *ey*. See دنبره (Pashto) *माक्षिक [p= 805,2] mfn.* (fr. मक्षिका) coming from or belonging to a bee MarkP. मक्षिकः makṣikḥ मक्षि makṣi (क्षी kṣī) का kāमक्षिकः मक्षि (क्षी) का A fly, bee; भो उपस्थितं नयनमधु संनिहिता मक्षिका च M.2.-Comp.-मलम् wax. madhu

मधु *a.* -मक्षः, -क्षा, -मक्षिका a bee. (Samskritam) *माक्षिक [p= 805,2] n.* a kind of honey-like mineral substance or pyrites MBh. उपधातुः An inferior metal, semi-metal. They are seven; सप्तोपधातवः स्वर्णं माक्षिकं तारमाक्षिकम् । तुत्थं कांस्यं च रातिश्च सुन्दूरं च शिलाजतु ॥ उपरसः uparasḥउपरसः 1 A secondary mineral, (red chalk, bitumen, माक्षिक, शिलाजित &c).(Samskritam) மாக்கிகம்

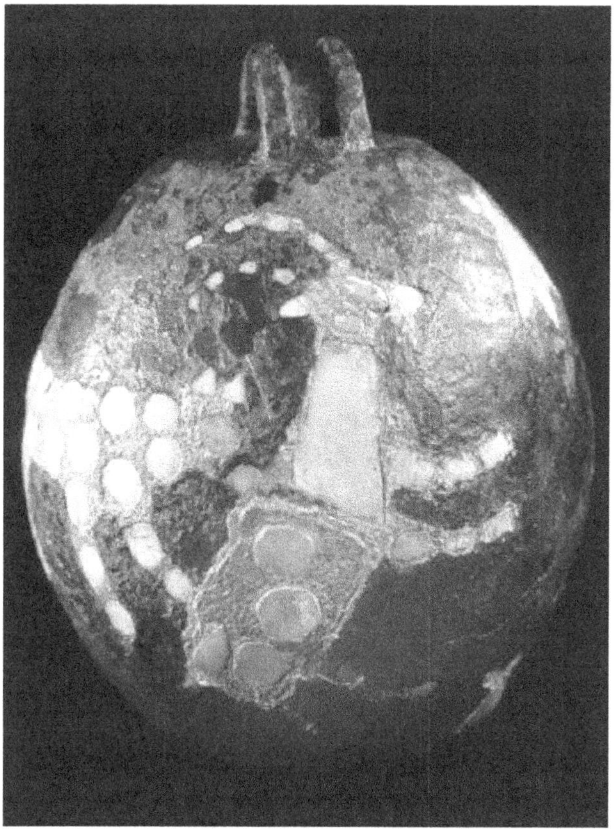

mākkikam, *n*. < *mākṣika*. 1. Bismuth pyrites; நிமிளை. (நாமதீப. 382.) 2. Honey; தேன். (நாமதீப. 410.) *செம்புத்தீக்கல் cempu-t-tīkkal, n. < செம்பு +. Copper pyrites, sulphide of copper and iron; இரும்புஞ்செம்புங்கலந்த உலோகக்கட்டி. Loc.*

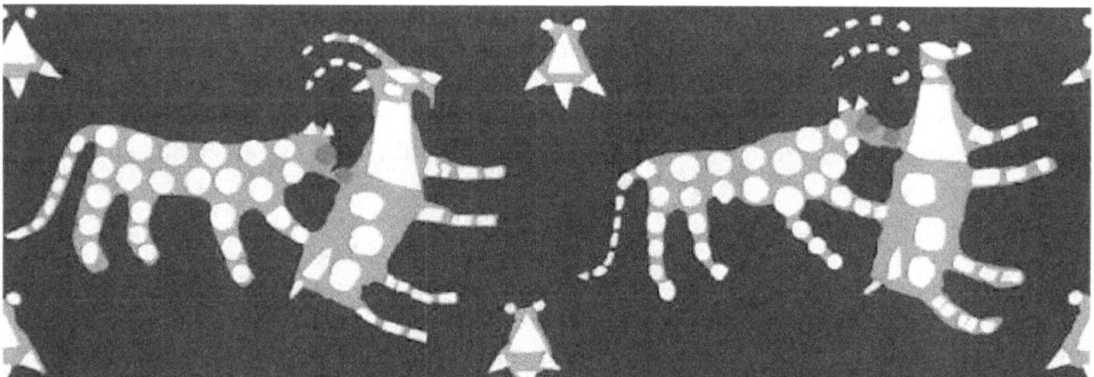

Leopard weight. Shahi Tump. H.16.7cm; dia.13.5cm; base dia 6cm; handle on top. Seashells inlays on frieze. The pair of leopard and ibex is shown twice, separated by stylized flies.

"The artefact was discovered in a grave, in the Kech valley, in eastern Balochistan. It belongs to the Shahi Tump - Makran civilisation (end of 4th millennium -- beginning of 3rd millennium BCe). Ht. 200 mm. weight: 13.5 kg. The shell has been manufactured by lost-wax foundry of a copper alloy (12.6%b, 2.6%As), then it has been filled up through lead (99.5%) foundry. The

shell is engraved with figures of leopards hunting wild goats, made of polished fragments of shellfishes. No identification of the artefact's use has been given. (Scientific team: B. Mille, D. Bourgarit, R. Besenval, Musee Guimet, Paris)."
Source: https://www.academia.edu/8164498/Early_lost-wax_casting_in_Baluchistan_Pakistan_the_Leopards_Weight_from_Shahi Tump Leopard weight of Shahi Tump (Balochistan), National Museum, Karachi. The artefact was discovered in a grave, in the Kech valley, in Balochistan. ca. 4th millennium BCE. 200 mm. h. 13.5kg wt. The shell has been manufactured by lost-wax foundry of a copper alloy (12.6% Pb, 2.6% As), then it has been filled up through lead (99.5%) foundry. The shell is engraved with figures of leopards hunting wild goats, made of polished fragments of shellfishes. No identification of the artefact's use has been given. (Scientific team: B. Mille, D. Bourgarit, R. Besenval, Musee Guimet, Paris.

Meluhha hieroglyphs:
karaḍa 'panther' Rebus: *karaḍa* 'hard alloy'. mlekh 'goat' Rebus: milakkhu 'copper' (Pali)

The pinnacle of achievement in Bronze Age Revolution relates to the invention of *cire perdue* technique of metal castings to produce metal alloy sculptures of breath-taking beauty. This achievement is exemplified by Nihal Mishmar artifacts dated to ca. 5th millennium BCE.

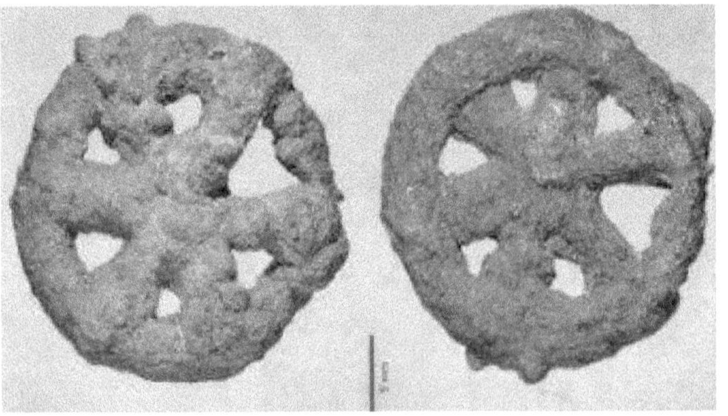

Mehergarh. 2.2 cm dia. 5 mm reference scale. Perhaps coppper alloyed with lead. [quote]Bourgarit and Mille (Bourgarit D., Mille B. 2007. Les premiers objets métalliques ont-ils été fabriqués par des métallurgistes? *L'actualité Chimique* . Octobre-Novembre 2007 - n° 312-313:54-60) have reported the finding (probably in the later still unreported excavation period) of small Chalcolithic "amulets" which they claim to have been produced by the process of Lost Wax. According to them, "The levels of the fifth millennium Chalcolithic at Mehrgarh have delivered a few amulets in shape of a minute wheel, while the technological study showed that they were made by a process of lost wax casting. The ring and the spokes were modelled in wax which was then coated by a refractory mould that was heated to remove the wax. Finally, the molten metal was cast in place of the wax. Metallographic examination confirmed that it was indeed an object obtained by casting (dendrite microstructure). This discovery is quite unique because it is the earliest attestation of this technique in the world." They then, further on, state that "The development of this new technique of lost wax led to another invention, the development of alloys...Davey (Davey C. 2009.The Early History of Lost-Wax Casting, in J. Mei and Th. Rehren (eds), *Metallurgy and Civilisation: Eurasia and Beyond Archetype*, pp.

147-154. London: Archetype Publications Ltd.) relies only upon these Mehrgarh findings , as well as on the Nahal Mishmar hoard, to claim that Lost Wax casting began in the Chalcolithic period before 4000 BCE." [unquote] Shlom Guil

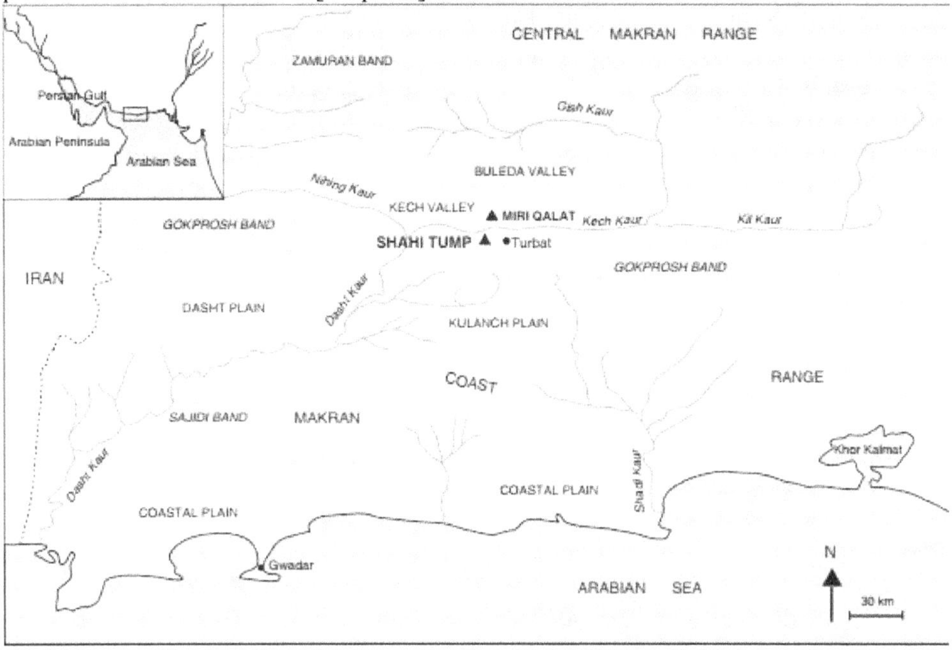

Shahi Tump. Kech valley, Makran division, Baluchistan, Pakistan (After Fig. 1 in Thomas et al) Benoit Mille calls the bronze stamps of Shahi-Tump 'amulets' (made from copper alloyed with lead). Mehrgarh is well recognised as a centre for early pyrotechnologies.The wax models of the stamps would have been solid and may have had a simple core inserted.This is perhaps the first stage in the technology:"Small copper-base wheel-shaped "amulets" have been unearthed from the Early Chalcolithic levels at Mehrgarh in Balochistan (Pakistan), dating from the late fifth millennium B.C. Visual and metallographic examinations prove their production by a lost-wax process—the earliest evidence so far for this metalworking technique. Although a gap of more than 500 years exists between these ornaments from Mehrgarh and the later lost-wax casts known in the Indo-Iranian world, the technological and compositional links between these artefacts indicate a similar tradition. We already know that the lost-wax process was commonly used during the second half of the fourth millenium B.C, as exemplified by figurative pinheads and compartmented seals, the latter of which were produced and distributed across the region until the early second millennium B.C. Most, if not all, of these artefacts were made using the lost-wax technique. This intensive practice of lost-wax casting certainly stimulated the technical development of the process, allowing the elaboration of more complex and heavier objects. The "Leopards Weight" (Balochistan, late fourth or early third millennium B.C.) is one of the best examples of these developments: the lost-wax copper jacket, with its opened hollow shape, constitutes an extraordinary technical achievement.(Mille, B., Bourgarit, D., and Besenval, R. 2005. 'Metallurgical study of the 'Leopards weight' from Shahi-Tump (Pakistan)', in C. Jarrige and V. Lefevre, eds., *South Asian Archaeology 2001*, Editions Recherches sur les Civilisations, Paris: 237-44) True hollow casting does not appear until the third millennium B.C., as illustrated by the manufacture of statuettes, including the Nausharo bull figurine (Balochistan, 2300–2100 B.C.), or those from BMAC sites in Central Asia (based upon analyses of items in the Louvre collections). The birth of the lost-wax casting process can also be paralleled with the first emergence of alloying in South Asia, as many of these early lost-wax cast artefacts were made of a copper-lead alloy (c. 10–40 wt% Pb and up to 4 wt% As). Significantly, it seems that the copper-lead alloy was solely dedicated to artefacts made using the lost-wax technique, a choice no doubt driven by the advantageous casting properties of such an alloy." (Mille, Benoit, On the origin of lost-wax casting and alloying in the Indo-Iranian world, in: Lloyd Weeks, 2007, *The 2007 Early Iranian metallurgy workshop at the University of Nottingham*)

https://www.academia.edu/3858109/The_2007_workshop_on_early_Iranian_metallurgy_at_the_University_of_Nottingham
(Source: B. Mille, R. Besenval, D. Bourgarit, 2004, Early lost-wax casting in Balochistan (Pakistan); the 'Leopards weight' from Shahi-Tump. in: *Persiens antike Pracht, Bergbau-Handwerk-Archaologie*, T. Stollner, R Slotta, A Vatandoust, A. eds., pp. 274-280. Bochum: Deutsches Bergbau Museum, 2004.https://www.academia.edu/5689136/Reflections_Upon_Accepted_Dating_of_the_Prestige_Items_of_Nahal_Mishmar

Section 7. Conclusion & Executive Summary

Two unique discoveries resulted in a breakthrough to confirm the decipherment of Harappa Scrip Cipher as metalwork catalogues recorded in Meluhha language.

The discoveries are:

1. Mohenjodaro three-sided tablet with Harappa Script inscription showing a boat loaded with ox-hide ingots
2. Three pure tin ingots found with Harappa Script inscriptions from a shipwreck in Haifa.

m1429 Mohenjodaro prism tablet. A hieroglyph to signify *bagalo* 'shipping vessel' is *bagala* 'Pleiades'. Such a hieroglyph showing 6 or 7 women as Pleiades is signified on three inscriptions of Harappa Script Corpora. *bagalo* = an Arabian merchant vessel (Gujarati) *bagala* = an Arab boat of a particular description (Ka.); *bagalā* (M.); *bagarige, bagarage* = a kind of vessel (Kannada) Rebus: *bangala = kumpaṭi = angāra śakaṭī* = a chafing dish a portable stove a goldsmith's portable furnace (Telugu) cf. *bangaru bangaramu* = gold (Telugu) *karaṇḍa* 'duck' (Sanskrit) *karaṛa* 'a very large aquatic bird' (Sindhi) Rebus: करडा *[karaḍā]* Hard from alloy--iron, silver &c. (Marathi) A pair of palm trees flanking a pair of oxhide ingots. Hieroglyph: **tāḍa3* ' fan -- palm ', *tāḍī* -- 2 f. in *tāḍī -- puṭa* -- ' palm -- leaf ' Kād., *tāla* -- 2 m. ' *Borassus flabelliformis* ' Rebus: *ḍhālako* = a large metal ingot (Gujarati) ढाळ (p. 204) Cast, mould, form (as of metal vessels, trinkets &c.) करण्ड m. a sort of duck L. కారండవము (p. 274) [*kāraṇḍavamu*] *kāraṇḍavamu.* [Skt.] n. A sort of duck. (Telugu) Rebus: *karaḍā* 'hard alloy'. Thus, the cargo is signified as hard alloy ingots.

kāru a wild crocodile or alligator (Telugu) *ghariyal* id. (Hindi) கராம் *karām*, n. prob. *grāha*. Rebus: *kāru*'artisan' (Marathi) *kāruvu* 'artisan' (Telugu) *khār* 'blacksmith' (Kashmiri) Hieroglyph fish = *aya* (Gujarati); crocodile = *kāru* (Telugu) Rebus: *ayakāra* 'ironsmith' (Pali)
Tin ingots in the Museum of Ancient Art of the Municipality of Haifa, Israel (left #8251, right #8252).

Three pure tin ingots each bear inscribed Harappa Script hieroglyphs; I have argued in a monograph in *Journal of Indo-Judaic Studies*, that the inscriptions were Meluhha hieroglyphs (Harappa Script writing)[19]. A third ingot was found inscribed with an added hieroglyph: moulded head, hieroglyph: *mũhe* 'face' (Santali) Rebus: *mũh* 'ingot'. Thus, the inscriptions on the tingots signify *ranku dhatu mũh,* 'tin mineral ingot'.

These two discoveries PLUS the discovery of Susa pot containing metal implements confirm the function of Harappa Script to document trade transactions related to metalwork of Meluhha artisans. Significantly, they also point to the contributions made by Bharata seafaring merchants for transactions in tin trade to provide the important resource for the Tin-Bronze Revolution. The world's major source of tin is the Mekong River delta of Ancient Far East. This has led to a hypothesis of a Maritime Tin Route from Hanoi to Haifa for further testing and researches.

Databases of Harappa Script[8] inscriptions (which may also be called: *Dharma saṁjñā* – Corporate badges of responsibility or *Bharata hieroglyphs)* narrate the cultural, socio-economic history of a civilization. It is called Harappa Script since the discovery of the first inscription on a seal (surface find) from Harappa in 1872 reported by Cunningham[9]. The number of Harappa script inscriptions total over 8000 (as of 2016). Select inscriptions with illustrations and details of decipherment are presented in a 799-page book: *Harappa Script & Language*[10]. This book is intended to be a basic resource for further historical researches on the Script and Meluhha language and to complement the Mythic Society's multi-volume *History and Culture of Bharata*. Examples presented in this article are taken from this book.

Harappa Script Corpora: an overview

Seafaring Meluhha merchants used Harappa Writing in trade transactions; artisans created metal artifacts, lapidary artifacts of terracotta, ivory for trade. Glosses of the Proto-Indic or Harappa language are used to read rebus the Harappa script inscriptions.

The glyphs of the Harappa script or Harappa Writing include both pictorial motifs and signs. Both categories of glyphs are read rebus. As a first step in delineating the Harappa language, an *Indian lexicon*[11] provides a resource, compiled semantically in clusters of over 1240 groups of words/expressions from ancient Bharata languages as a Proto-Indic substrate dictionary.

Bitumen. Young woman spinning and servant holding a fan. Fragment of a relief known as "The spinner". Bitumen mastic, Neo-Elamite period (8th century B.C.E–middle of the 6th century B.C.). Found in Susa. Height: 9 cm (3.5 in). Width: 13 cm (5.1 in).Louvre Museum. Excavated by Jacques de Morgan.

The legs of the platform and the seat are feline; the fish is ligatured with six blobs to signify the hieroglyphic nature of the orthographic, sculptural frieze composition.

What unites the bizarre components of the composition (for e.g. spinner and fish on platform with feline legs) and a person with winnowing fan, is an example of rebus rendering, the *mlecchita vikalpa,* cryptography. Spinner (*kātī*) lady rebus *khātī* 'wheelwright'; ayo 'fish' rebus: aya 'iron' ayas 'alloy metal' PLUS baTa 'six' rebus: bhaTa 'furnace' kola 'tiger' rebus: kol 'working in iron, blacksmith'. *kulā* 'winnowing fan '(Oriya) rebus: *kol* 'working in iron'._

Framework of data mining to decode inscriptions

Mahadevan concordance of 2906 inscriptions excludes inscribed objects which do not contain 'texts'; for example, this concordance excludes about 50 seals inscribed with the 'svastikā' pictorial motif and a pectoral which contains the pictorial motif of a one-horned bull with a device in front and an over-flowing pot. Parpola concordance has been used to present such objects which also contain valuable orthographic data which may assist in decoding the inscriptions. Many broken objects are also contained in Parpola concordance which are useful, in many cases, to count the number of objects with specific 'field symbols', a count which also provides some valuable clues to support the decoding of the messages conveyed by the 'field symbols' which dominate the object space.

In the process of normalizing the orthography of some glyphs to identify the core 'signs' of the script, some information is lost and at times, the process itself impedes the possibility of decoding the writing system. This can be demonstrated by (1) the 'identification' of a 'squirrel' glyph and (2) the failure to identify 'dotted circle' or 'stars' as glyphs.

It is, therefore, necessary to view the inscribed object as a composite message composed of glyphs: pictorial motifs and signs alike. Many scholars have noted the contacts between the Mesopotamian and Sarasvati Sindhu (Indus) Civilizations, in terms of cultural history, chronology, artefacts (beads, jewellery), pottery and seals found from archaeological sites in the two areas.

Section 8. Some select Critical comments on the decipherment by other leading experts.

Gregory Possehl has provided a bibliography of attempts at decipherment of Harappa Script.[12] This bibliography also includes reference to my preliminary attempts positing the possible economic context of the entire corpora of inscriptions. As research progressed and additional evidences could be collated from the civilizational contact areas of the Ancient Far East and Acient Near East, a new hypothesis has emerged, that of the Maritime Tin Route which linked Hanoi and Haifa, predating the Silk Road by two millennia. This is subject to further detailed researches and further excavations of all the sites of the civilization on the Vedic River Sarasvati Basin.

Historical researches by savants have led to the present state of knowledge which is poised for a paradigmatic shift in our understanding of the civilizational Pre-history. I invite a reference, in particular,to the 9th Maulana Azad Memorial Lecture delivered on November 11, 2016 by Prof. Shivaji Singh on the links between archaeology and the Veda to write afresh, the early history of Bharata. The full text of the lecture is at https://www.academia.edu/30071331/With_Veda_in_One_Hand_and_Spade_in_the_Other_Writing_Early_History_of_India_Afresh_--_Lecture_by_Prof._Shivaji_Singh

Prof. Shivaji Singh notes: "So clear and decisive is this Sarasvati evidence (rediscovery of the Lost River Sarasvati with majority of Harappan sites in its valley and other cultural and chronological facts that show that the Harappan Civilization should better be designated as Sarasvati Civilization) that many scholars, who earlier believed in Vedic-Harappan dichotomy and shared the view of Aryan arrival in India from outside, now, accept Vedic-Harappan identity. A noteworthy example of this shift in perception is presented by Prof. BB Lal who is an internationally recognized archaeologist and well-known for his extremely judicious approach. I name him particularly because as a former Director General, Archaeological Survey of India, he has been an eye witness and himself an active participant in harvesting much of that data and evidence we call the 'Sarasvati evidence'…the Rigveda (3.53.14) refers to a people called KIkaTas and their leader Pramaganda. And, all the scholars including Griffith, the famous English translator of the text, agree that KIkaTas were a people of Magadha (South Bihar)…the episode of Videgha MAthava and his priest Gotama RAhugaNa reaching the banks of SadAnIrA (modern Gandak in Bihar) described in the Satapatha Brahmana (1.4.1.14-17)…Gotama RAhugaNa is a famous Rigvedic Rishi, a scion of AngIrA family, who has composed as many as twenty of the Rigveda (1.74-93). The Satapatha Brahmana itself informs that the event behing described belongs to a bygone age. Is there any room left to doubt that this journey from Sarasvati to SadAnIrA took place in the Rigvedic period itself?...Bharatas were the greatest champions of the Yajnya-based Vedic ideology…it is Kalibanga that was the capital of the Bharatas…Indianness or Bharatiyata cannot be defined in geographical and political terms. It can be articulated only culturally as a set of values based on intuitive recognition of transcendental spirituality…Bharatiyata or Indianness is distinguished by a unique spiritual vision of life which the Rigvedic Rishis have bequeathed to humanity."

With such a resonant call to write the integrated history of Bharata using spiritual texts of the Veda and archaeological evidences, the over 8000 inscriptions of Harappa Script provide a textual framework to outline the urban progress in the sphere of metallurgy, as active participants in the Bronze Age Revolution.

I submit that the decipherment announced in this monograph is consistent with the vision of integrated History and Culture of Bharata framed on the Veda and Archaeology. This integration is dramatically validated by a discovery reported in April 2015 by the Students of Institute of Archaeology, New Delhi of a yajnakunda found in Binjor together with a Harappa Script Seal. The decipherment has shown that a Soma SamsthA Yaga was performed at Binjor. See: अष्टाश्रि यूप in yajna kunda Binjor (4MSR) archaeological evidence of Soma Yaga dated to ca. 2500 BCE. Significance of the discovery for Vedic and civilization studies http://tinyurl.com/zkwyh7s See also 744 monographs posted at https://independent.academia.edu/SriniKalyanaraman

I deem it a privilege that Prof. Shivaji Singh sought it fit to comment in a private communication: “I consider your work extremely important because you have broken the myth that Bharatiya tradition has downgraded the status of the artisans and placed them with the Shudras. In fact, it is this myth on which the entire structure of Subaltern history is based.”

I submit in all humility that if this monograph which outlines the framework of decipherment has contributed to a better understanding of the contributions made by Bharatam Janam to Bronze Age Revolution, it will be a precursor to the paradigm shift in early history of Bharata.

Principal reason for failures in past decipherment efforts

Received wisdom about Aryan Invasion as a ‘linguisticdoctrine’[13] is also a principal reason for the polemics of dubious decipherment claims. It is heartening no note that many linguists now recognize the nature of the Bharata *sprachbund.* A language X is also proposed to explain the nature of lexical, semantic structures with common features noticed in many ancient Bharata languages explained by sustained sustained cultural contacts among the people of Bhāratam.

Since Harappa Script is NOT syllabic, long texts of inscriptions are NOT necessary for decipherment. Now that over 8000 inscriptions are available, there is enough evidence to unravel the cipher of Harappa Script. Recognizing the metalwork catalogues of Harappa Script Corpora, the contributions of *Bhāratam Janam* to the Bronze Age Revolution and as intermeḍiaries along the Maritime Tin Route which preceded the Silk Road by 2 millennia, can be re-evaluated and the Bronze Age interaction maps can be re-drawn.

It is a leap of faith to rush to judgement that Harappa Script is NOT based on language because the length of inscriptions is very short (composed of upto 5 symbols). It has been demonstrated from the evolution of Egyptian hieroglyphs from ca. 32nd cent. BCE that short texts as inscriptions are read rebus. If each symbol or hieroglyph is logo-phonetic, words and combinations of words are identifiable on hypertexts created by Harappa Script.

Momentous, defining discovery of Vedic cultural foundations of Bharata

Discovery of Binjor Vedic fire-altar with an octagonal *Yūpa* and a seal inscribed with Harapp Script by students of Institute of Archaeology of India Museum in 2015 is momentous. It conclusively establishes the Vedic culture continuum in Vedic River Sarasvati Basin because the octagonal *Yūpa* is a signature tune of a *Soma Yaga Samstha.* The seal of Binjor is a metalwork catalogue. Octagonal shape is the *rudra bhāga of linga* metaphored as Pillar of light and fire. cf. *skambha sukta* Atharvaveda (AV X.7,8).

A remarkable archaeological evidence validates the Vedic culture and provides an indication of the spoken language of the people who invented and used Harappa Script. The evidence is from Binjor (near Anupgarh) in an archaeological site called 4MSR. At this site, agni kunda with *aṣṭāśri Yūpa* was found evidencing the performance of a *Soma Yaga*. Vajapeya is one of 7 *samstha* (profession) for processing/smelting soma (a mineral, NOT a herbal): सोमः [सू-मन् Uṇ.1.139]-संस्था a form of the Soma-sacrifice; (these are seven:- अग्निष्टोम, अत्यग्निष्टोम, उक्थ, षोढशी, अतिरात्र, आप्तोर्याम and वाजपेय). The Vajapeya performeḍ in Binjor and Kalibangan should have been related to the *Soma-samstha*: सोमः संस्था specified as वाजपेय with the shape of the *Yūpa* with eight- or four-angles. For every *Soma Yaga* such a *Yūpa* is installed. 19 such Yūpas have been found in Rajasthan, Allahabad, Mathura and East Borneo.[14] At the Vājapeya, the *yūpa* is eight-angled (as in Binjor), corresponding to the eight quarters (Sat.Br. V.2.1.5 *aṣṭāśrir yūpo bhavati*) अश्रि [p= 114,2] f. the sharp side of anything , corner , angle (of a room or house) , edge (of a sword) S3Br. Ka1tyS3r. often ifc. e.g. अष्टा*श्रि , त्रिर्-/अश्रि , च्/अतुर्-श्रि , शता*श्रि q.v. (cf. अश्र)

;([cf. Lat. acies , acer ; Lith. assmu3]). “The first *Yūpa* inscription of Mulavarman (in East Borneo) was erected to commemorate a *bahu-suvarnaka* sacrifice,'that on which gold is spent (used?) in profusion'.”[15] I suggest that *bahu-suvarnaka* refers to the many wealth-giving metals worked in a *Soma Yaga.*

Binjor. The fire altar, with a yasti made of an octagonal brick. Photo: Subhash Chandel, ASI

Binjor seal

Binjor (4MSR) Seal Inscription, decipherment

Fish + scales, *aya ā̃s (amśu)* ‘metallic stalks of stone ore’. Vikalpa: *badhoṛ* ‘a species of fish with many bones’ (Santali) Rebus: *baḍhoe* ‘a carpenter, worker in wood’; *badhoria* ‘expert in working in wood’(Santali) *khambhaṛā* 'fish ***fin***' rebus: *kammaTa* ‘mint, coiner, coinage’

gaṇḍa 'four' Rebus: *khaṇḍa* 'metal implements. Together with cognate ancu 'iron' the message is: native metal implements mint

Thus, the hieroglyph multiplex reads: *aya ancu khaṇḍa kammaṭa* ‘metallic iron alloy implements, mint, coiner, coinage’.

koḍi ‘flag’ (Ta.)(DEDR 2049). Rebus 1: *koḍ* ‘workshop’ (Kuwi) Rebus 2: *khŏḍ* m. ‘pit’, *khŏḍü f.* ‘small pit’ (Kashmiri. CDIAL 3947)

The bird hieroglyph: *karaḍa* करण्ड m. a sort of duck L. కారండవము (p. 274) [*kāraṇḍavamu*] Rebus: *karaḍā* 'hard alloy'

Thus, the text of Harappa Script inscription on the Binjor Seal reads: 'metallic iron alloy implements mint, hard alloy workshop' PLUS the hieroglyphs of one-horned young bull PLUS standard device in front read rebus:

kõda 'young bull, bull-calf' rebus: *kõdā* 'to turn in a lathe'; *kõnda* 'engraver, lapidary'; *kundār* 'turner'.

Hieroglyph: *sãghāṛɔ* 'lathe' (Gujarati) Rebus: *sangara* 'proclamation. सं-ग्रह *saṁgraha, samgaha* ‘a guardian , ruler , manager , arranger’ R. BhP.

Together, the message of the Binjor Seal with inscribed text is a proclamation, 'a metalwork catalogue (of) manager, turner of metallic iron alloy implements, hard alloy workshop'

Harappa Script hieroglyphs on priest statue of Mohenjo-daro signify *dhăvaḍ* 'iron-smelter', potṛ, पोतृ 'purifier'

The 'purifier' is also a *dhăvaḍ* 'iron-smelter'. Metallurgical smelting process is a process of purification. Purification of minerals is achieved through smelting in fire.

The hieroglyph signifiers are related to some etyma of Bharata *sprachbund* in this addendum. *vaṭa-* string, rope, tie (Samskrtam) is signified by the string which ties the 'dotted circle' on the forehead and right-shoulder of the Priest. The rebus reading is: *-vaḍ* వటుగ 'clever, skilful' (Telugu).

Hieroglyph: string, wisp: S. *dhāī* f. ' wisp of fibres added from time to time to a rope that is being twisted ', L. *dhāī̃* f. Rebus: *dhāu* ' ore (esp. of copper) '; Or. *ḍhāu* ' red chalk, red ochre ' (whence *ḍhāuā* ' reddish '; dhā́tu n. ' substance ' RV., m. ' element ' MBh., ' metal, mineral, ore (esp. of a red colour) ' Mn., ' ashes of the dead ' lex., ' *strand of rope ' (cf.*tridhā́tu* -- ' threefold ' RV., *ayugdhātu* -- ' having an uneven number of strands ' KātyŚr.). [√dhā]

Thus, the 'dotted circle' *dhāī̃* PLUS *vaṭa* 'string' is read: *dhăvaḍ* 'smelter'.
The *uttarīyam*˜ worn by the Priest is *potta* -- , °*taga* -- , °*tia* -- n. ' cotton cloth ' (Prakrtam) *potti* 'cloth' (Kannada) Rebus: *Potṛ, पोतृ* 'purifier' Priest (Rigveda). போத்தி *pōtti*, *n*. < போற்றி. 1. Grandfather; பாட்டன். *Tinn*. 2. Brahman temple- priest in Malabar; மலையாளத்திலுள்ள கோயிலருச் சகன். पोतदार (p. 303) *pōtadāra* m (P) An officer under the native governments. His business was to assay all money paid into the treasury. He was also the village-silversmith. (Marathi)

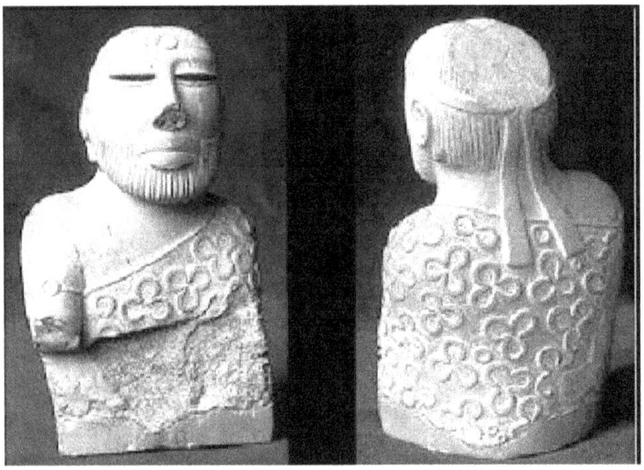

The fillet worn on the forehead and on the right-shoulder signifies one strand; while the trefoil on the shawl signifies three-strands.

 Single strand (one dotted-circle)

 Two strands (pair of dotted-circles)

 Three strands (three dotted-circles as a trefoil)

These orthographic variants provide semantic elucidations for a single: *dhātu, dhāū, dhāv* 'red stone mineral' or two minerals: dul PLUS dhātu, dhāū, dhāv 'cast minerals' or *tri- dhātu, -dhāū, -dhāv* 'three minerals' to create metal alloys'. The artisans producing alloys are *dhăvaḍ* m. 'a caste of iron -- smelters', *dhāvḍī* 'composed of or relating to iron') (CDIAL 6773).
dām 'rope, string' rebus: *dhāu* 'ore' rebus: मेढा [*mēḍhā*] A twist or tangle arising in thread or cord, a curl or snarl (Marathi). Rebus: *meḍ* 'iron, copper' (Munda. Slavic) *mẽṛhẽt*, *meḍ* 'iron' (Munda).

Semantics of single strand of rope and three strands of rope are: 1. Sindhi *dhāī* f. ' wisp of fibres added from time to time to a rope that is being twisted ', Lahnda *dhāī̃* id.; 2. *tridhā'tu* -- ' threefold ' (RigVeda)

Evolution *ḍha-, dha-* in Brahmi script syllables are evocative of 'string' and 'circle, dotted circle' as may be seen from the following orthographic evidence of epigraphs dated from ca. 300 BCE:

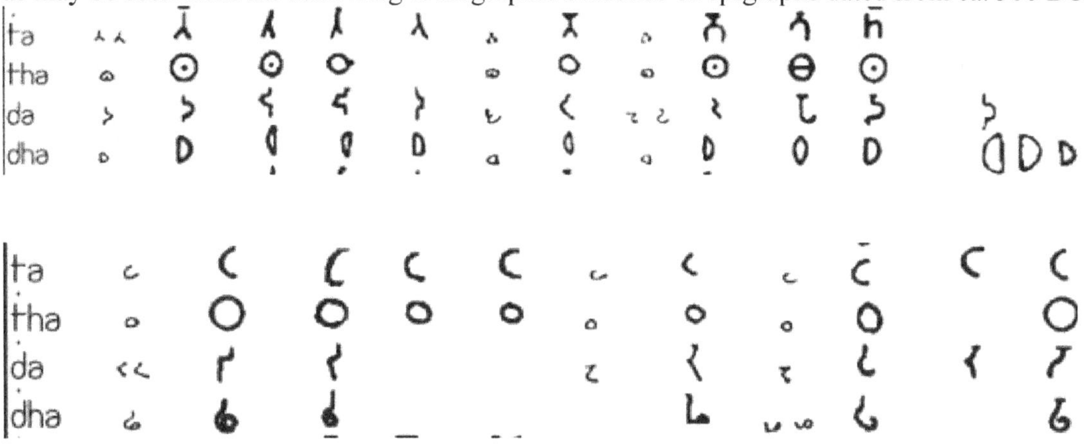

It may be seen from the table of evoution of Brahmi script orthography that

1. a circle signified the Brahmi syllable '*ṭh*a-' and a dotted circle signified the syllable '*tha-*';

2. a string with a twist signified the syllable '*da-*', a string ending in a circled twist signified the syllable '*ḍha-*' and a stepped string signified the syllable '*ḍa-*'.

Dance-step of the dancing girl and wick of lamp of Mohenjo-daro

Lost-wax casting. Bronze statue, Mohenjo-daro. Bronze statue of a woman holding a small bowl, Mohenjo-daro; copper alloy made using *cire perdue* method (DK 12728; Mackay 1938: 274, Pl. LXXIII, 9-11)

Both women in the *cire perdue* (lost-wax) cast bronze sculptures, seem to hold a lamp (bowl) wick in a hand. *karā̃* n. pl. ' wristlets, bangles ' (Gujarati) Rebus: *khār* 'blacksmith'. *bāṭi* 'cup' *bāṭi* 'wick' rebus: *bhaṭṭī* 'furnace, forge' *kola* 'woman' rebus: k*ol* 'working in iron' One is shown with a dance-step: *meḍ* 'dance step' rebus: *meḍ* 'iron'. Thus, iron furnace blacksmith. *loh* 'copper, iron, metal' PLUS *bhaṭṭī* 'furnace, forge' *bhāṭi* ' kiln ' *bhaṭhī, bhaṭṭī* ' *bhāṭhā* ' kiln '; H. *bhaṭṭhā* m. ' kiln', *bhaṭ* f. 'kiln'.

That dance-step is hieroglyph is evident from the inscription on a potsherd, Bhirrana. Hieroglyph: *meṭ* sole of foot, footstep, footprint (Ko.); *meṭṭu* step, stair, treading, slipper (Te.)(DEDR 1557). Rebus: *meḍ* 'iron'(Munda.Ho.); मेढ *meḍh*'merchant's helper'(Pkt.) *meṛed-bica* = iron stone ore, in contrast to bali-bica, iron sand ore (Munda)

मेढो vi. 138 वणिक्सहायः, one who helps a merchant.

Des'īnāmamālā Glossary, p. 71
Desinamamala of Hemacandra ed. R. Pischel (1938)

Thus, the two *cire perdue* statues of women and the hieroglyph on Bhirrana potsherd signify: metal furnace turner, merchant.

These are clear, unambiguous evidences of the spoken language (Mleccha or Proto-Prakritam of Bharata *sprachbund*) which is a continuum of the Vedic culture exemplified by *chandas* of 10,800 Rigveda *rica*-s. Complementing the *rica*-s are the Harappa Script Corpora with over 8000 inscriptions. Both resources are veritable data mines for further historical researches in Vedic culture continuum in Bharata, economic history with particular reference to Bharata's archaeo-metallurgy contributions to Bronze Age Revolution and formation/evolution of Bharata languages. One hypertext comes closest to a metaphor in Vedic cultural tradition. It is in the depiction of three faces on a horned person seated in penance (Seal m0304). Vedic metaphor is related to Vis'vakarma, Tvasta. He may signify *Triśiras*, son of *Tvaṣṭā*, कुबेर. Such a parallel interpretation is consistent with 1) decipherment of hieroglyphs on the seal m0304, 2) the Germanic tradition of Tuisto, Father of Germanic people, 3*)* *tatara* 'smelter' (Japanese), 4) *ṭhaṭherā* 'brassworker' (Sindhi), 5) *kamaḍha* 'penance' Rebus: *kammaṭa* 'mint, coiner, coinage'.

Are the torcs on *Karnonou* (Cernunnos) corporate badges (bracelets) of dharma samjnA? If so, *karā̃* n. pl. ' wristlets, bangles ' (Gujarati) Rebus: *khār* 'blacksmith'

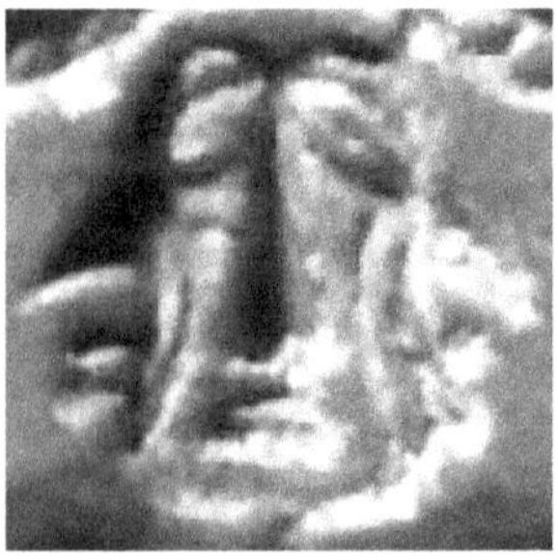

Triśiras 'three heads' of person seated in penance Seal m0304.

Germanic people are divided into three large branches, the Ingaevones, the Herminones and the Istaevones. Their ancestry is derived from three sons of Mannus, son of Tuisto, their common forefather. (Publius Cornelius Tacitus, *Germania*, 98 CE). Tuisto is equated to the Vedic Tvaṣṭṛ.16 It is possible that the three branches of people associated with Tuisto may explain the metaphor of three heads, *triśiras* remembered from Rigvedic tradition*:* त्रि--शिरस् [p= 460,3] *mfn.* three-headed (त्वाष्ट्र, author of RV.x,8) Ta1n2d2yaBr. xvii

Br2ih. KaushUp. MBh. Ka1m. (ज्वर) BhP. x , 63 , 22 कुबेर L (Monier-Williams).

Stone bas-relief from the boatmen's guild pillar from Paris

The *kāraṇī or kāraṇīka*, 'helmsman' signified on Seal m0304 may also relate to *karnonou (Cernunnos)* on the Pillar of Boatmen of 1st cent., CE.

Another possibility is that the three heads signified on seal m0304 may relate to Kubera ~*Triśiras*,son of *Tvaṣṭā* of Rigveda. One meaning of *tvāṣṭra* त्वाष्ट्र is 'copper'. *mũhe* 'face' (Santali) Rebus: *mũh* 'ingot' PLUS *kolmo* 'three' rebus: *kolimi* 'smihy, forge'. Thus, the orthography of 'three faces' may signify pewter, bronze, brass. (three alloys of copper formed by adding zinc and tin in varying proportions). Hence, the rebus reding to signify coppersmith, brass worker, bronze worker: **ṭhaṭṭhakāra ' brass worker '. 2. *ṭhaṭṭhakara -- . [*ṭhaṭṭha -- 1, kāra -- 1]1. Pk. ṭhaṭṭhāra -- m., K. ṭhö̃ṭhur m., S. ṭhā̃ṭhāro m., P. ṭhaṭhiār, °rā m.2. P. ludh. ṭhaṭherā m., Ku. ṭhaṭhero m., N. ṭhaṭero, Bi. ṭhaṭherā, Mth. ṭhaṭheri, H. ṭhaṭherā* m.(CDIAL 5493)

Bharata artisanal competence in metals technologies is exemplified by Wootz (*ukku*) steel sword presented by Purushottama (Porus) to Alexander on the banks of Jhelum river and Delhi (Besnagar) or Kodachadri non-rusting iron pillars.

A painting in Steel Authority of India Institute, Ranchi.

Kodachadri temple, Karnataka and iron pillar. Culturally, smithy/forge was the temple of Ancient Bharata. *kole.l* 'smithy, forge' was also *kole.l* 'temple' a rebus expression repeatedly signified on Harappa Script Corpora points to the roots of *weltanschauung* of *adhyatmika* traditions of Bharata from ca. 8th millennium BCE. A priest of this ancient temple was shown on an exquisite statue of Mohenjo-daro with Harappa Script hieroglyphs (ca. 3rd millennium BCE) to signify that he is smelter and purifier (of smelted minerals). A dance-step and a woman with a wick-lamp signified on bronze statues are exemplars of the Bronze Age Revolution which impacted *Bharatam Janam*, 'metalcaster folk', an expression used by *Viśvamitra* in Rigveda (RV 3.53.12).

Significance of Harappa Script decipherment to explain the wealth of Bharatam ca 1 CE

It is a fact of great historical significance that Bharata accounted for 32.9% of World GDP in 1 CE.[17]

At the turn of the Common Era, Bharata was indeed a land of *bahu-suvarnaka*, riches of gold and metallurgical excellence as evidenced by the decipherment of Harappa Script.

This cultural continuity foundation built over millennia sets the tone and tenor for the *History and Culture of Bharata*.

S. Kalyanaraman
Sarasvati Research Center December 9, 2016

Bibliography

A number of concordances and sign lists have been compiled, by many scholars, for the 'Indus' script:

Dani, A.H., *Indian Palaeography*, 1963, Pls. I-II

Gadd and Smith, *Mohenjodaro and the Indus Civilization*, London,1931,vol. III, Pls. CXIX-CXXIX

Hunter, G.R., *JRAS*, 1932, pp. 491-503

Hunter, G.R., *Scripts of Harappa and Mohenjodaro*, 1934, pp. 203-10

Langdon, in: John Marshall, *Mohenjodaro and the Indus Civilization*, London, 1931, vol. II, pp. 434-55

Koskenniemi, Kimmo and Asko Parpola, *Corpus of texts in the Indus script,* Helsinki, 1979; *A concordance to the texts in the Indus script*, Helsinki, 1982

Mahadevan, I., *The Indus Script: Texts, concordance and tables*, Delhi, 1977, pp. 32-35

Parpola et al., *Materials for the study of the Indus script, I: A concordance to the Indus Inscriptions*, 1973, pp. xxii-xxvi

Vats, *Excavations at Harappa*, Calcutta, 1940, vol. II, Pls. CV-CXVI

Amiet, P., Age of inter-Iranian Trade, Paris: Meeting of National Museums, 1986, p.125-126; Fig. 96, 1-9, (Notes and documents of the Museums of France).

Amiet, P., Susa 6000 years of history, Paris: Meeting of National Museums, 1988, p.64-65; Fig. 26.

Benoit, A., The Civilizations of the former Prochre East Paris: Ecole du Louvre, 2003, p.252-253; Fig. 109 (Manuals Ecole du Louvre, Art and Archaeology).

Corpus of Indus Seals and Inscriptions, 1. Collections in India, Helsinki, 1987 (eds. Jagat Pati Joshi and Asko Parpola)

Corpus of Indus Seals and Inscriptions, 2. Collections in Pakistan, Helsinki, 1991 (eds. Sayid Ghulam Mustafa Shah and Asko Parpola)

Corpus of Indus Seals and Inscriptions 3.1 supplement to Mohenjoo-daro and Harappa, Helsinki, Suomalainen Tiedeakatemia Paropola, Asko, BM Pande and Petteri Koskikallio, 2010.

Kalyanaraman, S., 1988, Harappa Script: A bibliography, Manila.

Kalyanaraman, S, 1995, SarasvatiSindhu civilization: evidence from the veda, archaeology, geology and satellite, 10th Wold Sanskrit Conference, Bangalore.

Kalyanaraman S. 1997, A project to revive the Sarasvati river: Role of GIS, National Seminar on Geographic Information Systems for Development Planning, Chennai, 10-12 January, 1997, Renganathan Centre for Information Studies

Kalyanaraman S, 1999, SarasvatiRiver, Godess and Civilization, in: Memoir 42, Vedic Sarasvati, Geological Survey of India, Bangalore, pp. 25-29.

Kalyanaraman, S, 2000, River Sarasvati: Legend, Myth and Reality, All India Sarasvat Association, Mumbai

Kalyanaraman S., 2001, Sarasvati, Babasaheb Apte Smarak Samiti, Bangalore (1100 pages, 600 illustrations); part of 6 vol. Encyclopaedia on Sarasvati (Other 5 vols. in press).

Kalyanaraman, S., 2003, National River Network, An overview, Bangalore, Rashtrotthana Research and Communication Centre

Kalyanaraman S, 2004, Indian Alchemy: Soma in the Veda, Munshiram Manoharlal, Delhi

Kalyanaraman S., 2004, Sarasvati (an encyclopaedic work in 7 volumes: Civilization, River, Bharati, Technology, Epigraphs, Language), Bangalore, Babasaheb Apte Smarak Samiti, Bangalore

Kalyanaraman S., 2007, Rama Setu, Chennai, Rameswaram Ramsetu Protection Movement

Kalyanaraman S., 2008, Harappa Script encodes mleccha speech (5 vols.: Language, Writing, Epigraphica Sarasvati, Dictionary, Indian Lexicon), Chennai, Jayalakshmi Book Stores, 6 Apparsami Koil St., Mylapore, Chennai 600004

Kalyanaraman, S., 2010, Harappa Script Cipher: Hieroglyphs of Indian Linguistic Area, Amazon.

Kalyanaraman, S., 2011, Rastram – Hindu history in United Indian Ocean States, Amazon.

Kalyanaraman, S., 2012, Indian Hieroglyphs – Invention of Writing, Amazon.

Kalyanaraman, S., 2013 Indus writing in Ancient Near East – Corpora and a Dictionary, Amazon.

Kalyanaraman, S., 2011, Rastram – Hindu history in United Indian Ocean States, Amazon.

Kalyanaraman, S., 2012, Indian Hieroglyphs – Invention of Writing, Amazon.

Kalyanaraman, S., 2013 Indus writing in Ancient Near East – Corpora and a Dictionary, Amazon.

Kalyanaraman, S., 2014, Sagan finds Saravati (Novel), Amazon.

Kalyanaraman, S., 2014, A theory for Wealth of nationa, Amazon.

Kalyanaraman, S., 2014, Indian Ocean Community, Amazon.

Kalyanaraman, S., 2014, Harosheth Hagoyim, Amazon.

Kalyanaraman, S., 2015, Outrage for Dharma, Amazon.

Kalyanaraman, S., 2015, Akkadian Rising sun (Novel) , Amazon

Kalyanaraman, S., 2015, Harappa Script Deciphered, Amazon.
Kalyanaraman, S., 2015, Philosophy of Symbolic Forms in Meluhha, Amazon.

Kalyanaraman, S., 2015, Meluhha, a Visible language, Amazon.

Kalyanaraman, S., 2015, Meluhha, tree of life (Novel), Amazon.

Kalyanaraman, S., 2016, Harappa Scrip & Language -- Data mining of Corpora, *tantra yukti* & knowledge discovery of a civilization

http://groups.yahoo.com/group/hinducivilization

Kalyanaraman, S., 2016, 642 monographs/papers at https://independent.academia.edu/SriniKalyanaraman

Mahadevan, I., 1986, A Computer Study of the Harappa Script by I. Mahadevan, International Association of Tamil Research. Madras. (Residence: Vyjayanthi. 112. Chamit!'s Road. Nandanam. Madras 600035. Indio)in: Current Science, January 20, 1986, Vol.55, No.2, pp. 77-79.

Pande, BM, 'Inscribed copper tablets from Mohenjo-daro: a preliminary analysiś in: D. Agrawal/A. Ghosh eds., Radiocarbon and Indian Arcaheology, Bombay 1973, tablet no. 38.

Pande, B. M. 1979 Inscribed Copper Tablets from Mohenjo daro: A Preliminary Analysis. In Ancient Cities of the Indus, edited by G. L. Possehl, pp. 268-288. New Delhi, Vikas Publishing House PVT LTD.

Pande, B. M. 1991 Inscribed Copper Tablets from Mohenjo-daro: Some Observations.Puratattva (21): 25-28.

Parpola, A. 1992 Copper Tablets from Mohenjo-daro and the study of the Harappa Script. In: Proceedings of the Second International Conference on Moenjodaro, edited by I. M. Nadiem, pp. Karachi, Department of Archaeology.

Asko Parpola, 2008, Copper tablets from Mohenjo-daro and the study of the Harappa Script, pp. 132-139 in: Eri Olijdam & Richard H Spoor (eds.), Intercultural relations between south and southwest Asia: Studies in commemoration of ECL During Caspers (1234-1996). BAR Interntional Series 1826. Oxford: Archaeopress.

Shinde, V. & Willis, R.J., (2014). A New Type of Inscribed Copper Plate from Indus Valley (Harappa) Civilisation. Ancient Asia. 5, p.Art. 1. http://www.ancient-asia-journal.com/articles/10.5334/aa.12317/ The paper analyzes a group of nine Indus Valley copper plates (c. 2600–2000 BC), discovered from private collections in Pakistan

Yule, Paul, 1988, A new copper tablet from Mohenjo-daro (DK 11307), in: M. Jansen and G. Urban (eds.), *Reports on field work carried out at Mohenjo-daro, Interim Reports, Vol.2,Pakistan 1983-84*, German Research Project Mohenjo-daro RWTH Aachen, Istituto Italiano per ilmedio ed estremo oriente, Roma
https://www.academia.edu/737300/A_New_Copper_Tablet_from_Mohenjo-daro_DK_11307_ Paul Yule analyses the stratigraphic and archaeological context of this find.

List of Harappa Script 'text signs'

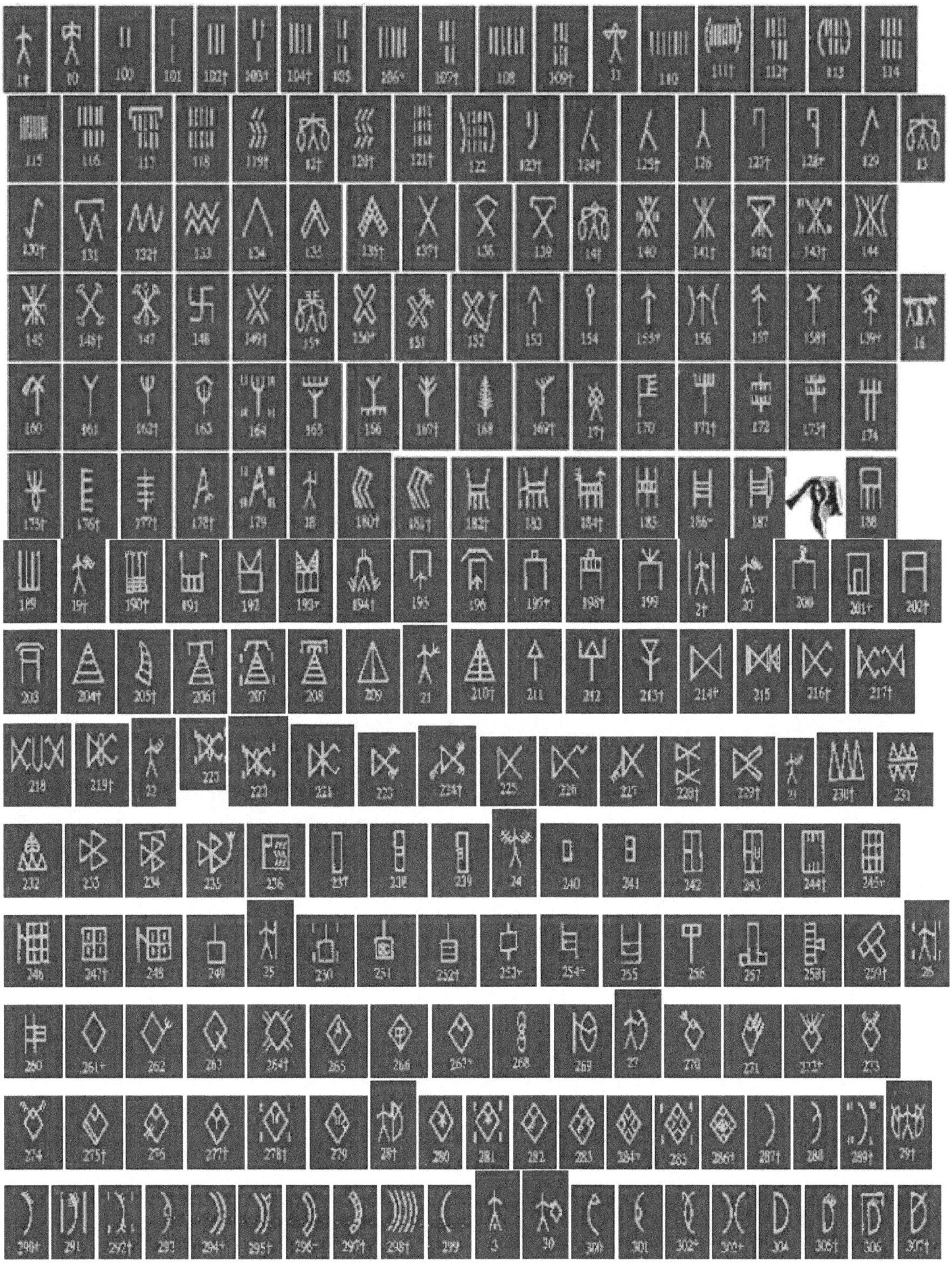

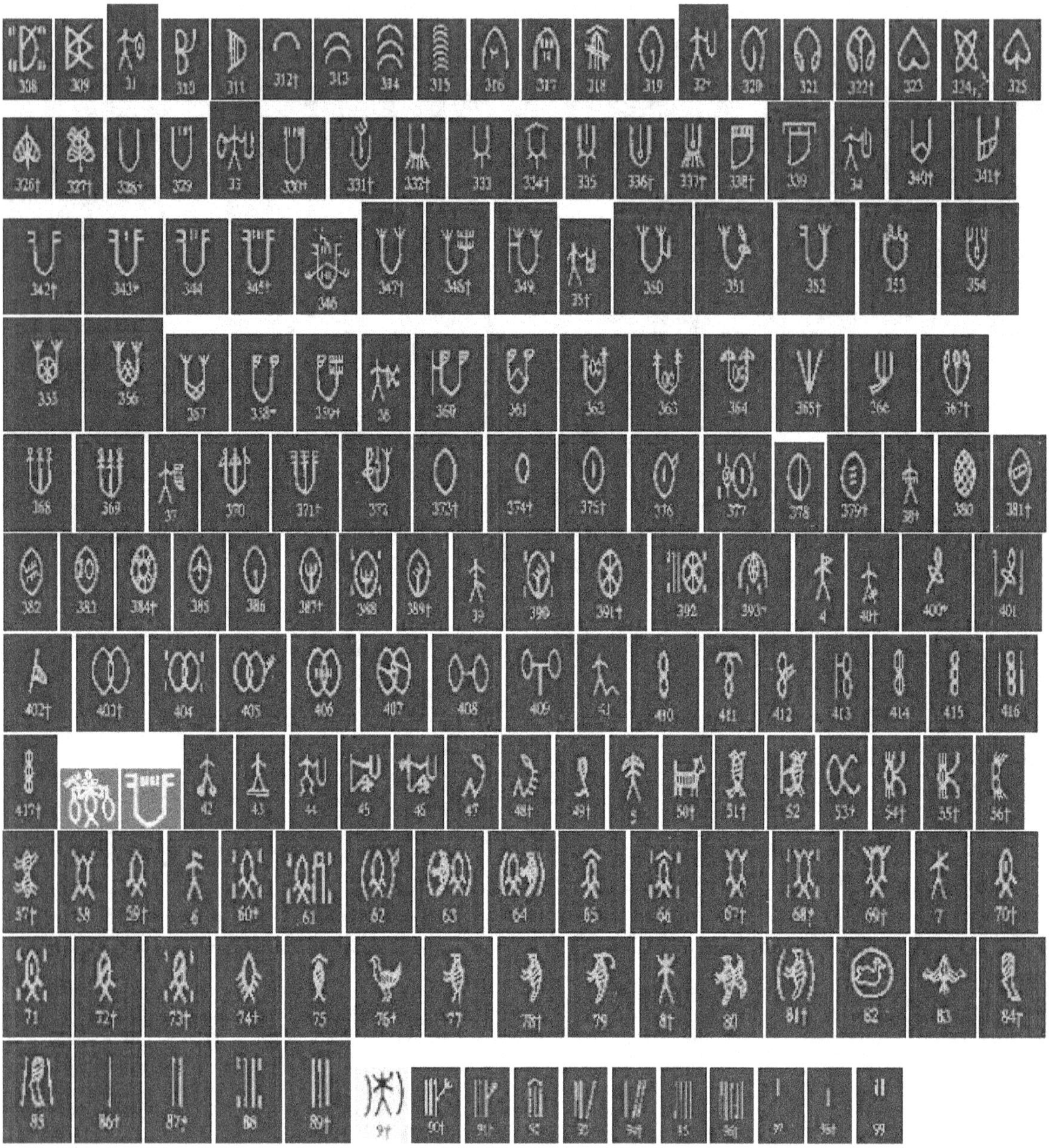

Sign Variants

V001

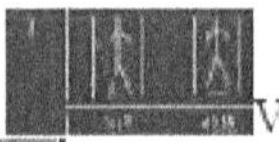
V002

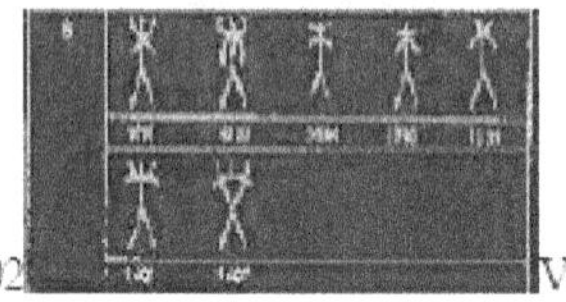
V008

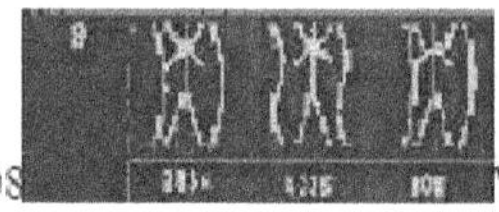
V009

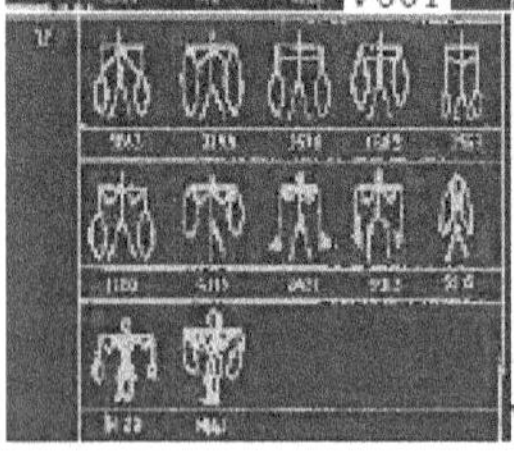
V012

V014

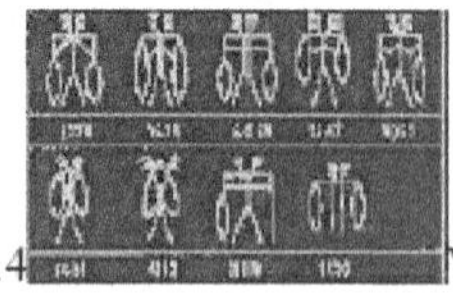
V015

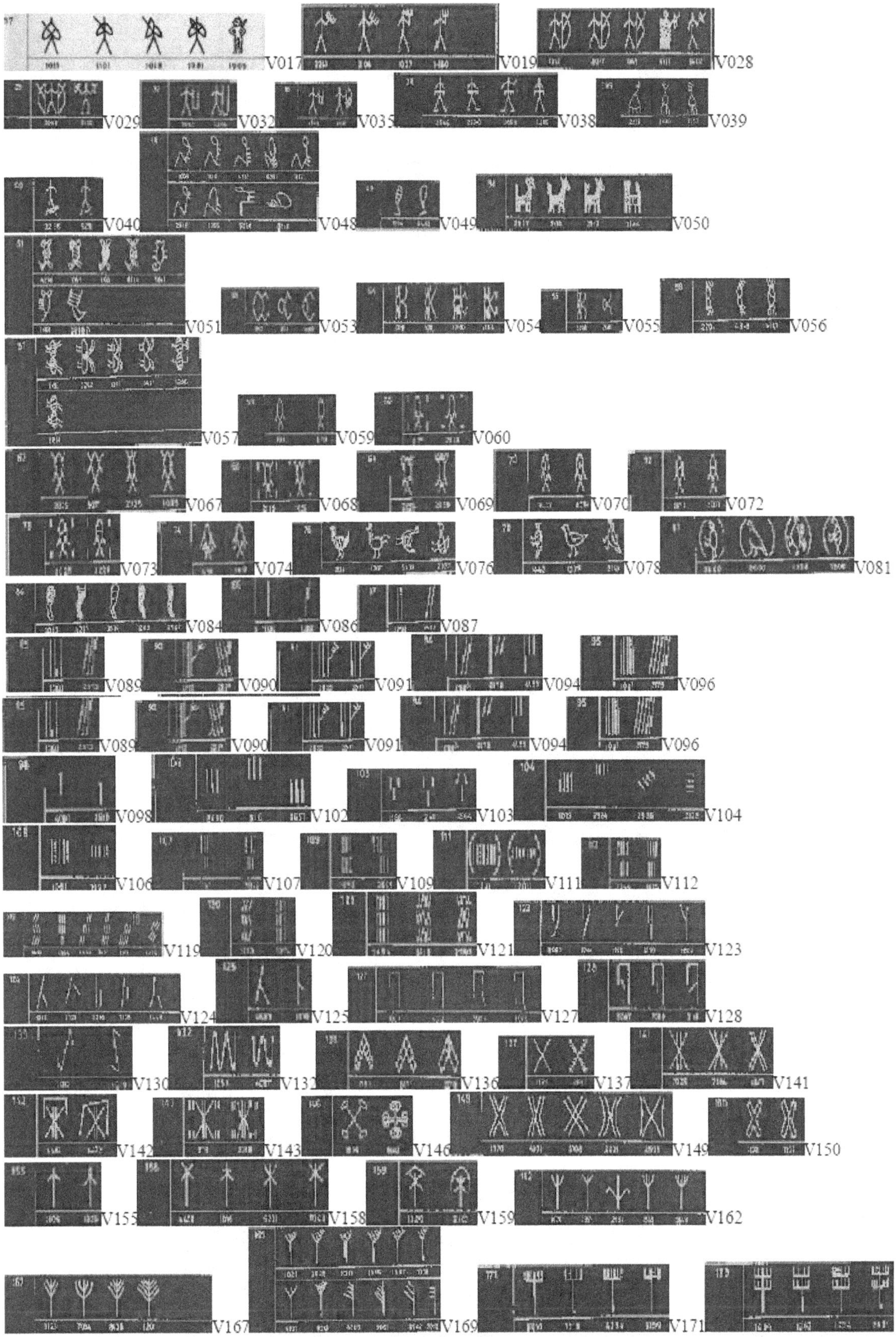
V017 V019 V028
V029 V032 V035 V038 V039
V040 V048 V049 V050
V051 V053 V054 V055 V056
V057 V059 V060
V067 V068 V069 V070 V072
V073 V074 V076 V078 V081
V084 V086 V087
V089 V090 V091 V094 V096
V089 V090 V091 V094 V096
V098 V102 V103 V104
V106 V107 V109 V111 V112
V119 V120 V121 V123
V124 V125 V127 V128
V130 V132 V136 V137 V141
V142 V143 V146 V149 V150
V155 V158 V159 V162
V167 V169 V171

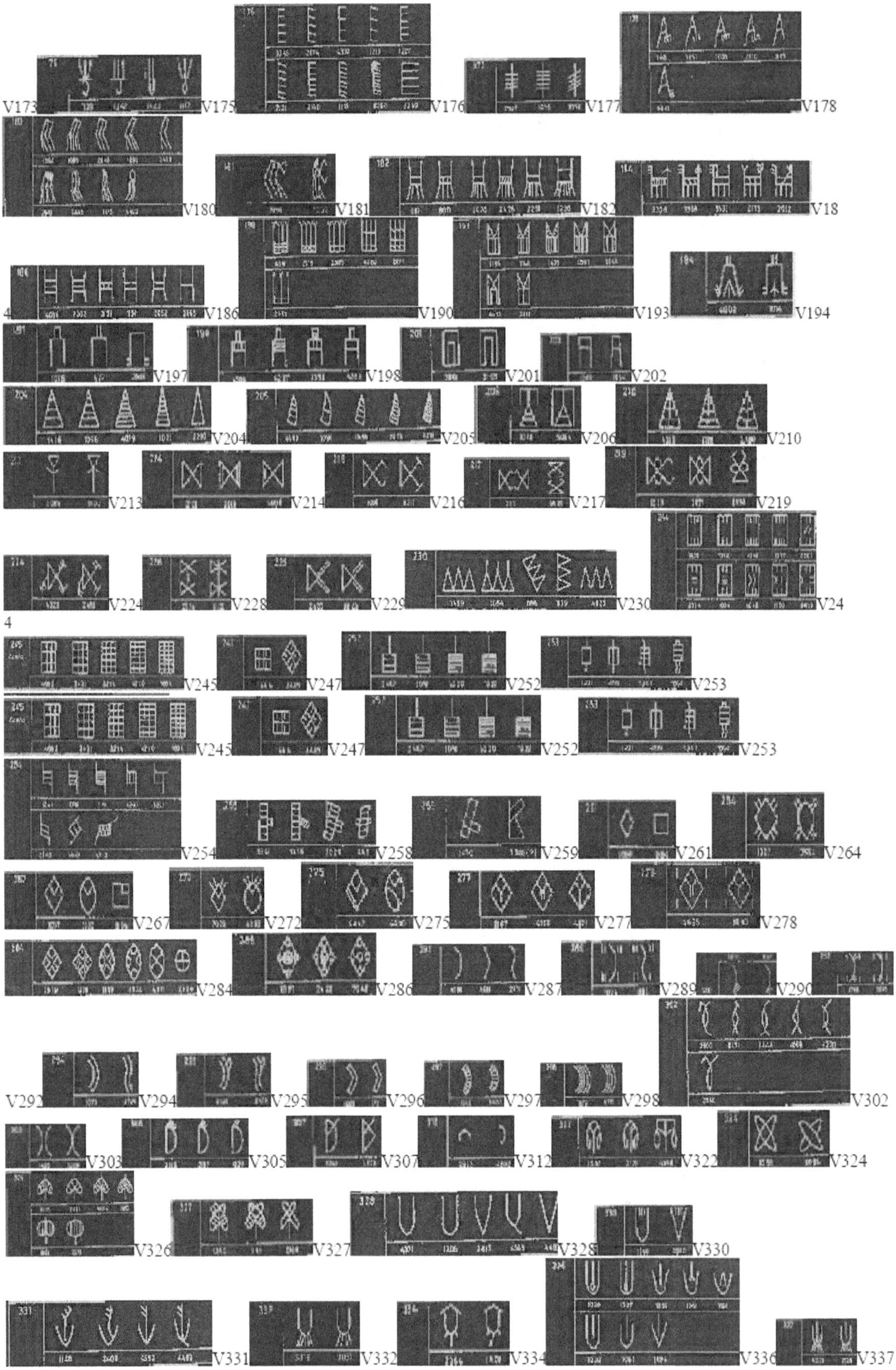

V173 V175 V176 V177 V178
V180 V181 V182 V18
4 V186 V190 V193 V194
V197 V198 V201 V202
V204 V205 V206 V210
V213 V214 V216 V217 V219
V224 V228 V229 V230 V24
4
V245 V247 V252 V253
V245 V247 V252 V253
V254 V258 V259 V261 V264
V267 V272 V275 V277 V278
V284 V286 V287 V289 V290
V292 V294 V295 V296 V297 V298 V302
V303 V305 V307 V312 V322 V324
V326 V327 V328 V330
V331 V332 V334 V336 V337

 V338 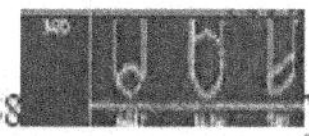 V340 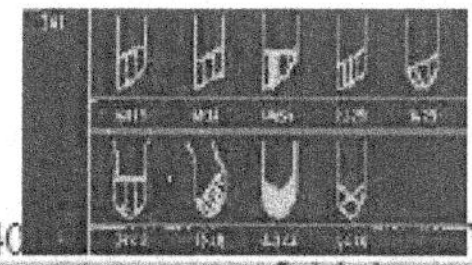 V341 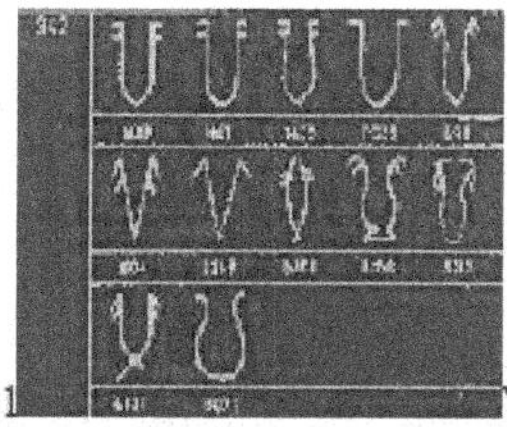 V342  V343

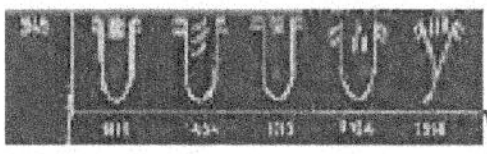 V345 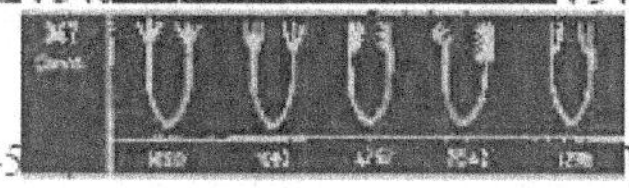 V347  V348 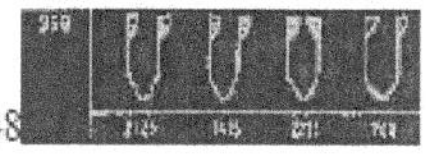

V358

 V359 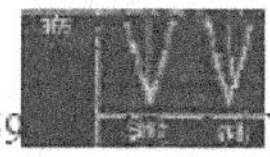 V365 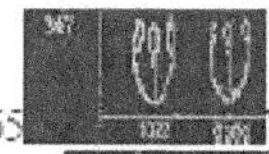 V367 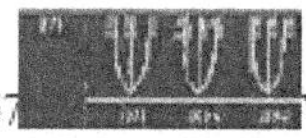 V371 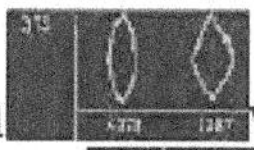 V373

 V374  V375 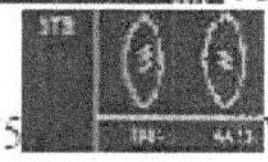 V379  V381 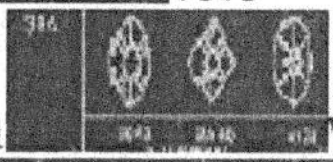 V384

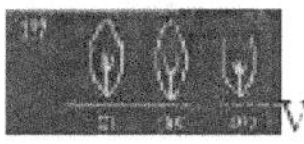 V387 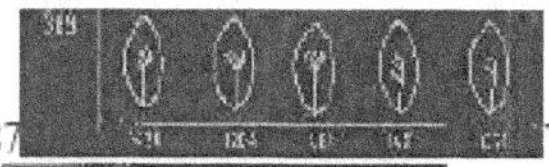 V389  V391  V393 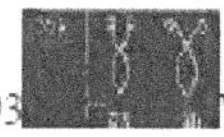 V39

4 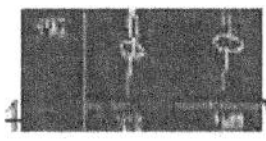 V400 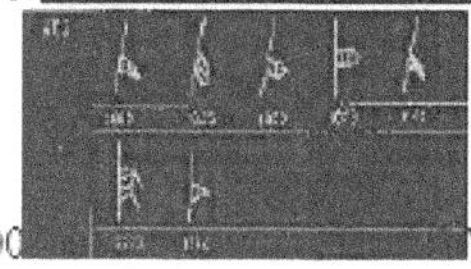 V402  V403 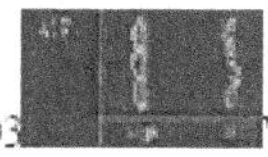 V417

Select inscriptions of Harappa Script Corpora

Background

Seafaring Meluhha merchants used Harappa writing in trade transactions; artisans created metal artifacts, lapidary artificats of terracotta, ivory for trade. Glosses of the Proto-Indic or Harappa language are used to read rebus the Harappa Script inscriptions.

The glyphs of the Harappa script or Harappa writing include both pictorial motifs and signs and both categories of glyphs are read rebus. As a first step in delineating the Harappa language, an *Indian lexicon*[18] provides a resource, compiled semantically cluster over 1240 groups of glosses from ancient Indian languages as a Proto-Indic substrate dictionary.

The evidence is remarkable that many single glyphs or glyptic elements of the Harappa writing can be read rebus using the repertoire of artisans (lapidaries working with precious shell, ivory, stones and terracotta, mine-workers, stone-masons, metal-smiths working with a variety of minerals, furnaces and other tools) who created the inscribed objects and used many of them to authenticate their trade transactions. Many of the inscribed objects are seen to be calling cards of the professional artisans, listing their professional skills and repertoire. Many are veritable mining- and metal-work catalogs.

Continuing legacies of glyptic art noted by Huntington: "There is a continuity of composite creatures demonstrable in Indic culture since Kot Diji ca. 4000 BCE."19

The identification of glosses from the present-day languages of India on Sarasvati river basin is justified by the continuation of culture evidenced by many artifacts evidencing civilization continuum from the Vedic Sarasvati River basin, since language and culture are intertwined, resulting in a unique, logo-semantic writing system. .

Harappa writing in Ancient Near East is a tribute to the Meluhha artisans who have established an expansive contact area in Eurasia and left for posterity the bronze-age *harosheth hagoyim*, 'the smithy of nations.'

Concordance lists for epigraphs

Abbreviations and references to heiroglyphs and text transcripts

m-Mohenjodaro

h-Harappa

ABCDE at the end of a reference number indicate side numbers of an inscribed object. Multiple seal impressions on the same object are numbered 1 to 4.

At the end of the reference number:

'a' sealing; 'bangle' inscription on bangle or bangle fragment; other objects: shell, ivory stick, ivory plaque, ivory cube, faience ornament, steatite ornament; 'ct' copper tablet; 'Pict-'Pictorial motifs (0 to 145) described as illustrations of field-symbols in Appendix III of Mahadevan concordance (pp. 793 to 813); 'it' inscribed tablet; 'si' seal impression; 't' tablet.

Illegible inscribed objects are excluded in the following tabulations. Many potsherds Rahmandheri and Nausharo are excluded since the 'signs' are considered to be potters' marks; only those inscriptions which appear to have parallels of field symbols or 'signs' in the corpora are included.

Based on a number of resources and from the collections of inscribed objects held in many museums of the world, such as the Metropolitan Museum of Art, the Harappa writing Corpora include Sarasvati heiroglyphs, representing many facets of glyptic art of Sarasvati Civilization. The corporas also includes many texts of inscriptions, corresponding to the epigraphs inscribed on objects. The compilation is based mostly on published photographs in archaeological reports right from the days of Alexander Cunningham who discovered a seal at Harappa in 1875, of Langdon at Mohenjodaro (1931) and of Madhu Swarup Vats at Harappa (1940). The corpus includes objects collected in India, Pakistan, other countries and the finds of the excavations at Harappa by Kenoyer and Meadow during the seasons 1994-1995 and 1999-2000.

Framework for decoding epigraphs of Sarasvati Sindhu Civilization

This is also intended to serve as a pictorial and text index to Mahadevan Concordance and to the three volumes published so far of pictorial corpus of Parpola et al.

Many texts are indexed to the text numbers of Mahadevan concordance. The choice of this concordance is based on four factors: (a) the concordance is priced at a reasonable cost; (b) it is a true concordance for every sign of the corpus to facilitate an analysis of the frequency of occurrence of a sign and the context of other sign clusters/ sequences in relation to a sign and for researchers to cross-check on the basic references for the inscribed objects; (c) the exquisite nature of orthography is notable and 'readings' are authentic, even for very difficult to read inscriptions; and (d) signs and variants of signs have been delineated with cross-references to selected text readings.

Mahadevan concordance excludes inscribed objects which do not contain 'texts'; for example, this concordance excludes about 50 seals inscribed with the 'svastikā' pictorial motif and a pectoral which contains the pictorial motif of a one-horned bull with a device in front and an over-flowing pot. Parpola concordance has been used to present such objects which also contain valuable orthographic data which may assist in decoding the inscriptions. Many broken objects are also contained in Parpola concordance which are useful, in many cases, to count the number of objects with specific 'field symbols', a count which also provides some valuable clues to support the decoding of the messages conveyed by the 'field symbols' which dominate the object space.

Cross-references to excavation numbers, publications, photographs and the museum numbers based on which these texts have been compiled are provided in Appendix V: List of Inscribed Objects (pages 818 to 829) in Iravatham Mahadevan, 1977, *The Indus Script: Texts, Concordance and Tables*, Memoirs of the Archaeological Survey of India No. 77, New Delhi, Archaeological Survey of India, Rs. 250. In most cases, these text numbers are matched with the inscribed objects after Asko Parpola concordance [Two volumes: Rs. 21,000: 1. Jagat Pati Joshi and Asko Parpola, eds., 1987, *Corpus of Indus Seals and Inscriptions: 1. Collections in India*, Memoirs of the Archaeological Survey of India No. 86, Helsinki, Suomalainen Tiedeakatemia; 2. Sayid Ghulam Mustafa Shah and Asko Parpola, eds., 1991, *Corpus of Indus Seals and Inscriptions: 2. Collections in Pakistan*, Memoirs of the Department of Archaeology and Museums, Govt. of Pakistan, Vol. 5, Helsinki, Suomalainen Tiedeakatemia]. *Memoir of ASI No. 96 Corpus of Indus Seals and Inscriptions, Vol. II* by Asko Parpola, B.M. Pande and Petterikoskikallio (containing copper tablets) is in press (December 2001).

The debt owed to Iravatham Mahadevan, Asko Parpola, Archaeological Survey of India, Department of Archaeology and Museums, Govt. of Pakistan and Finnish Academy for making this presentation possible is gratefully acknowledged. I am grateful to Iravatham Mahadevan who made available to me his annotated personal copy of a document which helped in collating the texts with the pictures of inscribed objects. [Kimmo Koskenniemi and Asko Parpola, 1980, Cross references to Mahadevan 1977 in: *Documentation and Duplicates of the Texts in the Indus Script, Helsinki*, pp. 26-32].

Four epigraphs from Bhirrana from ASI website http://asi.nic.in and five epigraphs from Bagasra (Gola Dhoro) reported by VH Sonawane in *Puratattva*, Number 41, 2011 have also been included.

Pitfalls of normalising orthography of some glyphs

Parpola (1994) identifies 386 (+12?) signs (or graphemes) and their variant forms. Mahadevan (1977) identifies 419 graphemes; out of these 179 graphemes have variants totalling 641 forms.

Parpola observes: "...the grapheme count might be as low as 350...The total range of signs once present in the Indus script is certain to have been greater than is observable now, for new signs have kept turning up in new inscriptions. The rate of discovery has been fairly low, though, and the new signs have more often been ligatures of two or more signs already known as separate graphemes than entirely new signs." (Parpola, 1994, p. 79)

Many 'signs' are ligatures of two or more 'signs'.

In the process of normalizing the orthography of some glyphs to identify the core 'signs' of the script, some information is lost and at times, the process itself impedes the possibility of decoding the writing system. This can be demonstrated by (1) the 'identification' of a 'squirrel' glyph and (2) the failure to identify 'dotted circle' or 'stars' as glyphs.

It is, therefore, necessary to view the inscribed object as a composite message composed of glyphs: pictorial motifs and signs alike. Many scholars have noted the contacts between the Mesopotamian and Sarasvati Sindhu (Indus) Civilizations, in terms of cultural history, chronology, artefacts (beads, jewellery), pottery and seals found from archaeological sites in the two areas.

An outstanding contribution to the study of the script problem is the publication of the Corpus of Indus Seals and Inscriptions (CISI) Three volumes have been published so far:

> *Corpus of Indus Seals and Inscriptions, 1. Collections in India, Helsinki, 1987 (eds. Jagat Pati Joshi and Asko Parpola)*
>
> *Corpus of Indus Seals and Inscriptions, 2. Collections in Pakistan, Helsinki, 1991 (eds. Sayid Ghulam Mustafa Shah and Asko Parpola)*
>
> *Corpus of Indus Seals and Inscriptions, 3. 1 Supplement to Mohenjo-daro and Harappa, 2010 (eds. Asko Parpola, B.M. Pande and Petteri Koskikallio) in collaboration with Richard H. Meadow and Jonathan Mark Kenoyer. (Annales Academiae Scientiarum Fennicae, B. 239-241.) Helsinki: Suomalainen Tiedeakatemia.*

These volumes in which Asko Parpola is the co-author constitute the photographic corpus. The CISI contains all the seals including those without any inscriptions, for e.g. those with the geometrical motif called the 'svastika'. Parpola's initial corpus (1973) included a total number of 3204 texts. After compiling the pictorial corpus, Parpola notes that there are approximately 3700 legible inscriptions (including 1400 duplicate inscriptions, i.e. with repeated texts). Both the concordances of Parpola and Mahadevan complement each other because of the sort sequence adopted. Parpola's concordance was sorted according to the sign following the indexed sign. Mahadevan's concordance was sorted according to the sign preceding the indexed sign. The latter sort ordering helps in delineating signs which occur in final position. With the publication of CISI Vol. 3, Part 1, the total number of inscriptions from Mohenjo-daro totals 2134 and from Harappa totals 2589; thus, these two sites alone accounting for 4,723 bring the overall total number of inscriptions to over 6,000 from all sites (even after excluding comparable inscriptions on 'Persian Gulf type' circular seals from the total count).

Compendia of the efforts made since the discovery by Gen. Alexander Cunningham, in 1875, of the first known Indus seal (British Museum 1892-12-10, 1), to decipher the script appear in the references listed in the Bibliography.:

Alamgirpur

Late Harappan

pottery, a three-legged chakala_(After YD Sharma)

Alamgirpur Agr-1 a(2) graffiti

9062

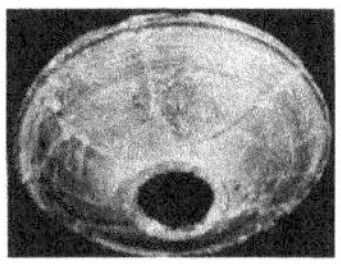

9063

Alamgirpur: Late Harappan pottery (After YD Sharma)

Alamgirpur2

Allahdino (Nel Bazaar)01

Allahdino (Nel Bazaar)02

Allahdino (Nel Bazaar)03

Allahdino (Nel Bazaar)04

Allahdino (Nel Bazaar)05

Allahdino (Nel Bazaar)06

Allahdino (Nel Bazaar)07

Allahdino (Nel Bazaar)08

Allahdino (Nel Bazaar)09

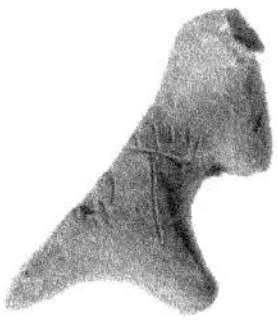

Allahdino (Nel Bazaar)11

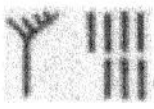9061

Amri

9084

Amri

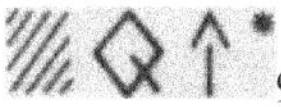9085

Amri06

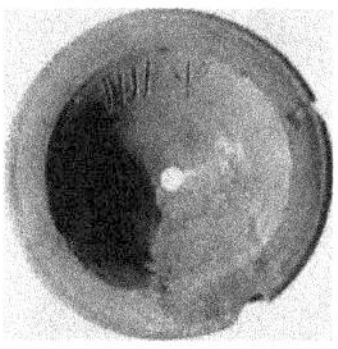

Amri07

Bagasra1 (Gola Dhoro)

Bagasra2 (Gola Dhoro)

Bagasra3 (Gola Dhoro)

Bagasra4 (Gola Dhoro)

Bagasra5 (Gola Dhoro)

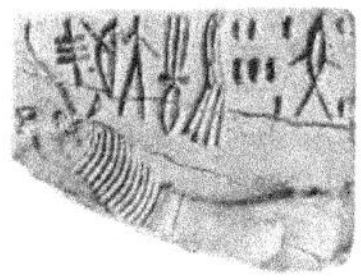

Balakot01

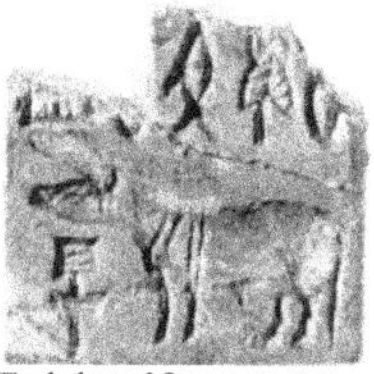

Balakot 02

Balakot 03

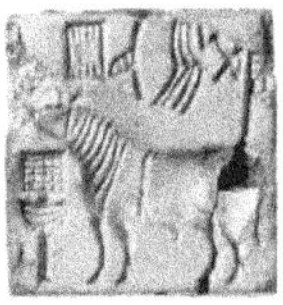

Balakot 04

Balakot 05

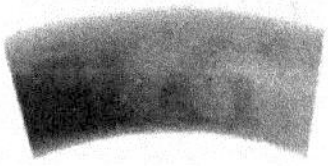

Balakot 06 bangle

Balakot 06bangle

Balakot 06C

Banawali1

Banawali10  9204

Banawali11

Banawali12

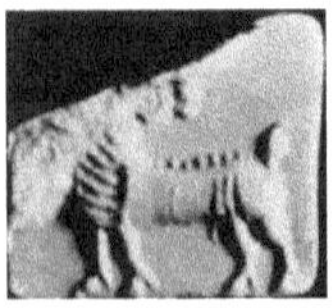

Banawali13a

Banawali14

Banawali15  9203

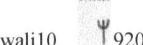

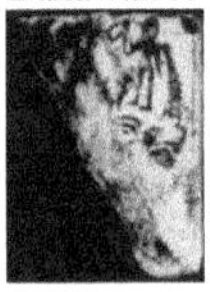

Banawali16

Banawali 17

 9201

Banawali 18a

Banawali19

Banawali2

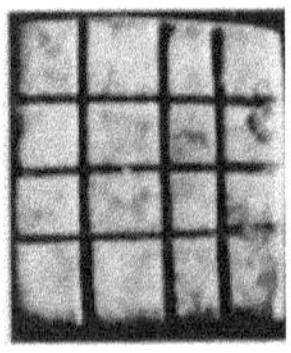

Banawali 20

Banawali 21a

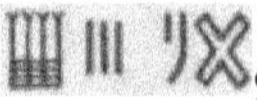 9205

Banawali 23A

Banawali 23B

Banawali 24t  9211

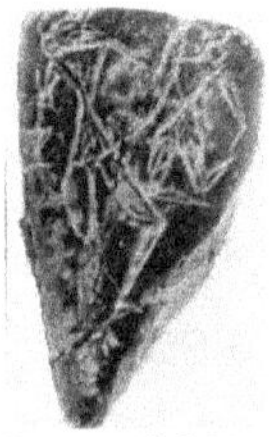

Banawali 26A

Banawali0026a

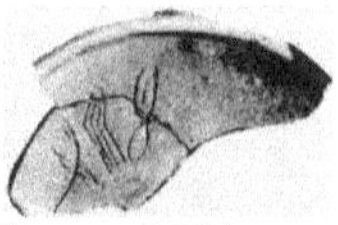

Banawali 28A

 9221

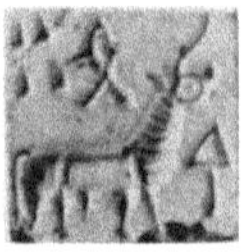

Banawali 3

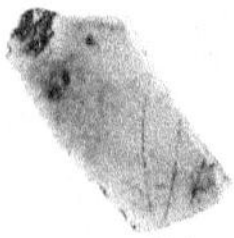

Banawali30

Banawali 4

Banawali 5

 9203

Banawali 6

Banawali 7

Banawali 8

Banawali 9C

Bet Dwaraka 1

S'ankha seal. One-horned bull, short-horned bull looking down and an antelope looking backward.

Bhirrana1

Bhirrana2

Bhirrana3

Bhirrana4

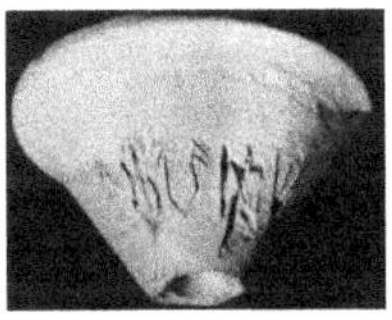

Chandigarh01
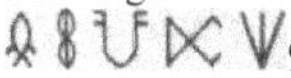 9101

Chandigarh02
 9102

Chandigarh

 9103

Chandigarh
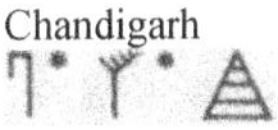 9104

Chanhudaro10

6129

Chanhudaro 11
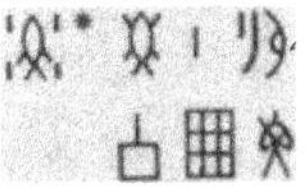 6220

Chanhudaro12a
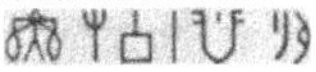

6231

Chanhudaro13
 6221

Chanhudaro14a
 6108

Chanhudaro15a
 6213

Chanhudaro16a
 6222

Chanhudaro17a
6122

Chanhudaro18a
 6216

Chanhudaro1a

6125

Chanhudaro2
 6128

Chanhudaro20
 6210

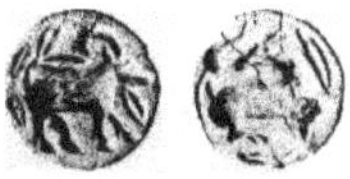

Chanhudaro Seal obverse and reverse. The oval sign of this Jhukar culture seal is comparable to other inscriptions. Fig. 1 and 1a of Plate L. After Mackay, 1943.

Chanhudaro21a
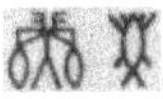 6209

Chanhudaro22a
6115

Chanhudaro23
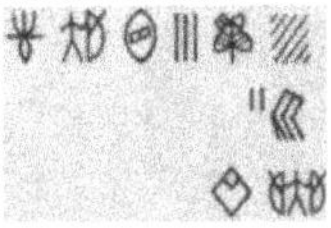

6402 Goat-antelope with a short tail.

The object in front of the goat-antelope is a double-axe.

Chanhudaro24a
 6116

Chanhudaro25

Chanhudaro26 
6405

Chanhudaro27

Chanhudaro28

Chanhudaro29

6403

Chanhudaro3

6230

Chanhudaro30

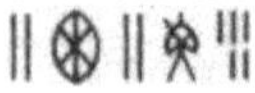
6111

Chanhudaro32a

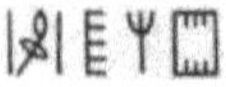
6123

Chanhudaro33a

6104

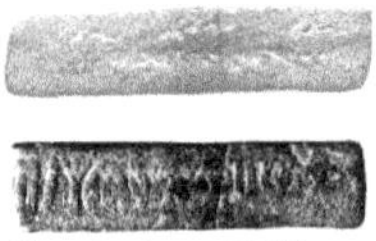

Chanhudaro. Tablet. Obverse and reverse. Alligator and Fish. Fig. 33 and 33a. of Plate LII. After Mackay, 1943.

6233 Pict-67: Gharial, sometimes with a fish held in its jaw and/or surrounded by a school of fish.

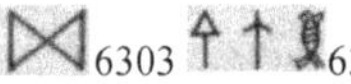
6303 6304

6301

6305 6109

6112

6113 Pict-98

It is seen from an enlargement of the bottom portion of the seal impression that the 'prostrate person' may not be a person but a ligature of the neck of an antelope with rings on its necks or of a post with ring-stones. The head of the 'person' is not shown. So, I would surmise that this is an artist's representation of an act of copulation (by an animal) + a ligatured neck of another bovine or alternatively, a pillar with ring-stones ligatured to the bottom portion of a body. It is not uncommon in the artistic tradition to ligature bodies to the rump of, for example, a bull's posterior ligatured to a horned woman (Pict. 103 Mahadevan) or standing person with horns and bovine features (hoofed legs and/or tail) -- Pict. 86-88 Mahadevan.

Bison (gaur) trampling a prostrate person (?) underneath. Impression of a seal from Chanhujodaro (Mackay 1943: pl. 51: 13). The prostrate 'person' is seen to have a very long neck, possibly with neck-rings, reminiscent of the rings depicted on the neck of the one-horned bull normally depicted in front of a standard device. 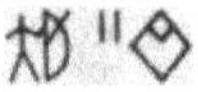

6114 Pict-108

Person kneeling under a tree facing a tiger. [*Chanhudaro Excavations*, Pl. LI, 18]

6118

Chanhudaro Seal obverse and reverse. The 'water-carrier' and X signs of this so-called Jhukar culture seal are comparable to other inscriptions. Fig. 3 and 3a of Plate L. After Mackay, 1943.

6120

Pict-40

Ox-antelope with a long tail; a trough in front.

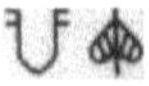
6121

Chanhudaro. Seal impression. Fig. 35 of Plate LII. After Mackay, 1943.

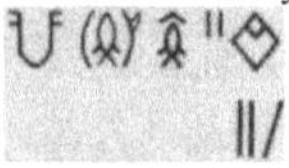

6124

6126 6130 6131 6133

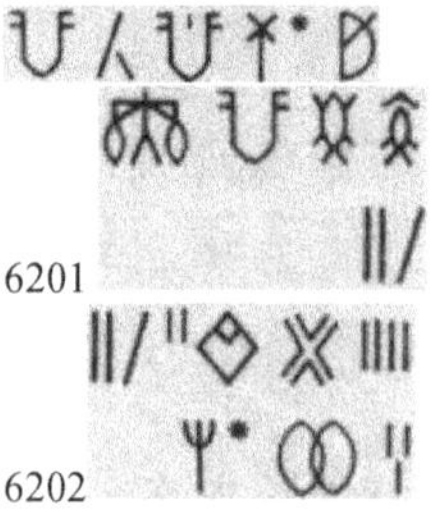
6201 6202

6203

6204 6208

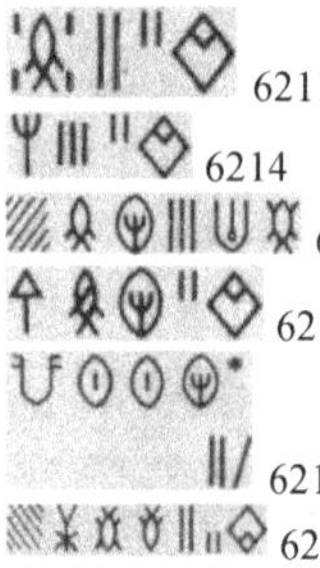
6211 6214 6215 6217 6218 6219

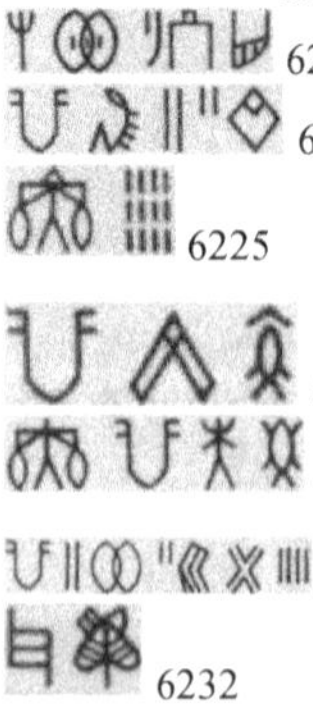
6223 6224 6225 6226 6228 6229 6232

Chanhudaro. Tablet. Fig. 34 of Plate LII. After Mackay, 1943. 

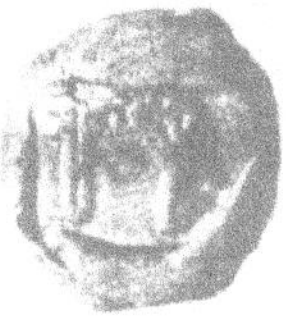

6234
Chanhudaro. Seal impression. Fig. 35 of Plate LII. After Mackay, 1943.  6235

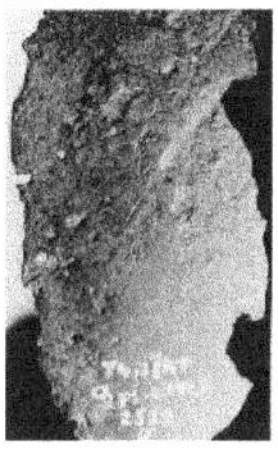

Chanhudaro38A

Chanhudaro

39A1

Chanhudaro

39A2

Chanhudaro4

 6206

Chanhudaro40A

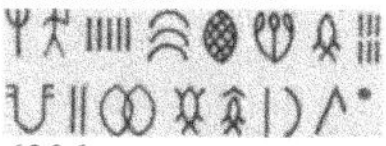

6306

Chanhudaro40B

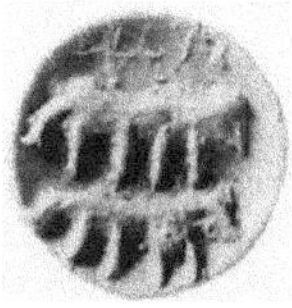

Chanhudaro41a

Chanhudaro42

Chanhudaro43

Chanhudaro46a

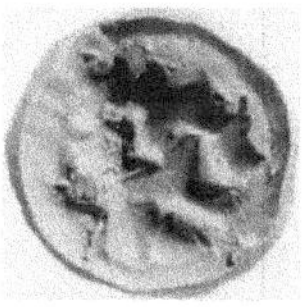

Chanhudaro46b

Chanhudaro47

Chanhudaro 48

Chanhudaro49A

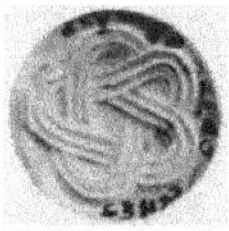

Chanhudaro49B

Chanhudaro 5

 6132

Chanhudaro50A

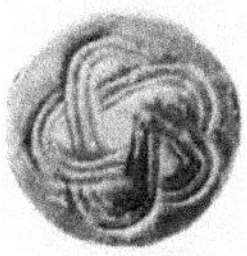

Chanhudaro50B

Chanhudaro 6

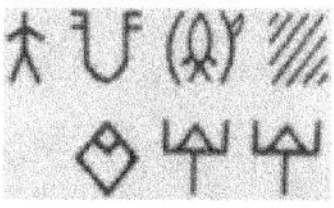

6205

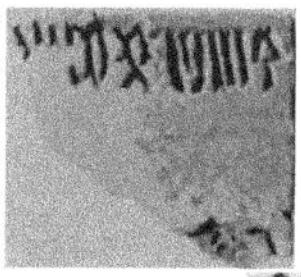

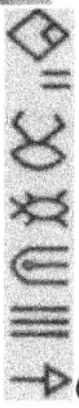

Chanhudaro 7 6207

Chanhudaro 8 6227

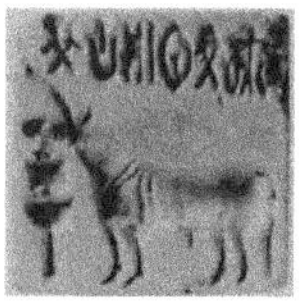

Chanhudaro 9

6127

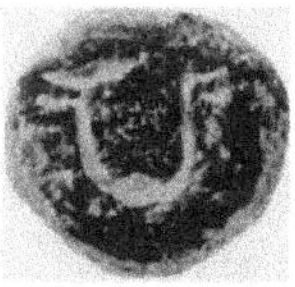

Daimabad1
Sign342

Daimabad
2a

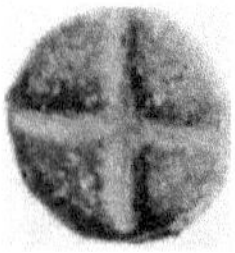 Daimabad
3A

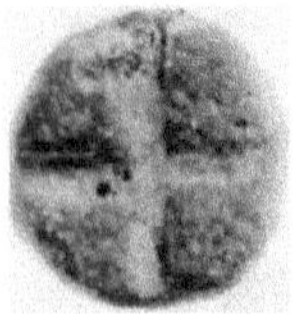
Daimabad 3B

Daimabad 4

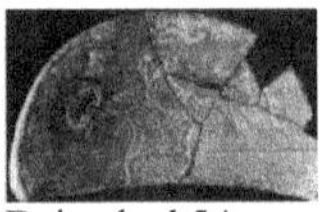
Daimabad 5A

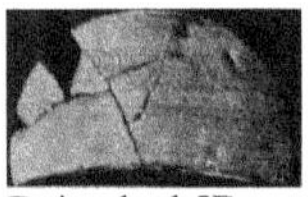
Daimabad 5B

Desalpur1a
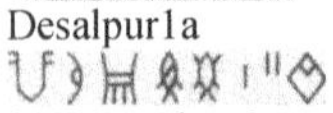
9071Desalpur2

esalpur3
 9073

Dholavira Sign-board mounted on a gateway.

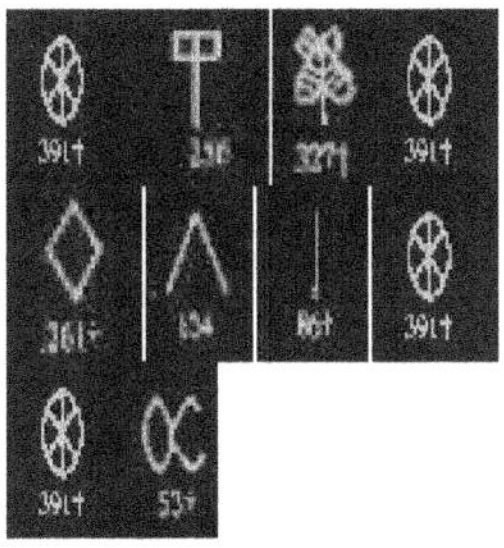

Dholavira (Kotda) on Kadir island, Kutch, Gujarat; 10 signs inscription found near the western chamber of the northern gate of the citadel high mound (Bisht, 1991: 81, Pl. IX).

Dholavira: Seals (Courtesy ASI)

Dholavira1a

9121

Dholavira 2a

Gharo Bhiro (Nuhato) 01

Gumla10a

Gumla8a

h001a

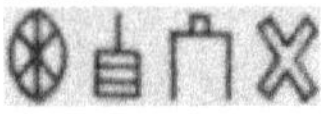

4010

h002

4012

h003

4002

h004

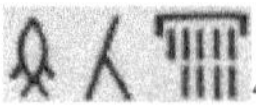

4693

h005

4004

h006a

4006

h007

4008

h008 4001

h009

4009

h010a

4003

h011a

4038

h012

4005

h013

5055 h014

4106

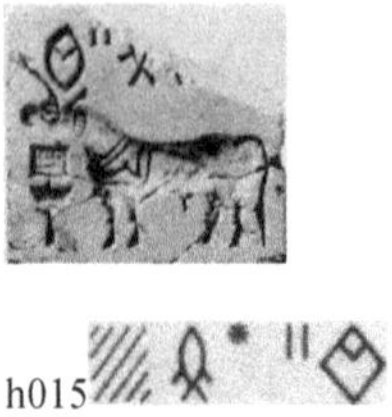

h015

4053

h017

4052

h018

4071

h019

4694

h020

4019

h021

4022

h022

4023

h023

4047

h024

4013

h025

4081

h026

4016

h027

4017

h028

4040

h029

4042

h030

4049

h031

4103

h032

4018

h033

5059

h035

5083

h036

4113

h037

4031

h038

4029

h039

h040

4072

h041

4178

h042

**4057

h043

4077

h044

4028

h045

4043

h046

4076

h047

4030

h048

4091

h049

4133

h050
4131

h051
4090

h052
4109

h053
5089

h054
4085

h055
4107

h056
4110

h057
4086

h058
4105

h059
5120

h060
5119

h061
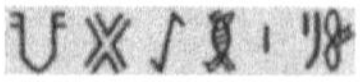
4118

h062
4128

h063
4142

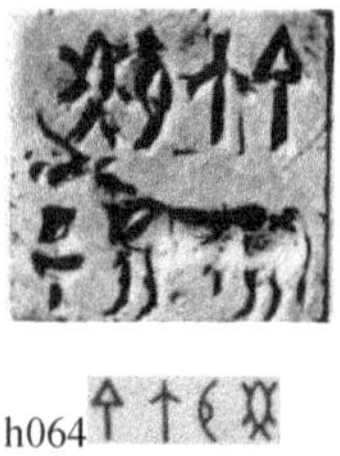
h064

h065
4094

h066
4130

h067
4115

h068
4141

h069
4146

h070
4122

h071
5054

h072
4120

h073
4617 [An orthographic representation of a water-carrier].

h074
4135

h075 

4161

h076

4241

h077

h078 

4244

h079

 5060

h080

4245

h081

 5063

h082a Text
4238

h083
4236

h084

h085 

4232

h086

4233

h087

 4240

h088
4253

h089

090
227

h091 

4230

h092

4229

h093a

 4231

h094 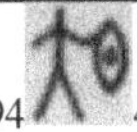 4246

h095

h096 
4249

h097 Pict-95: Seven robed figures (with pigtails, twigs)

 4251

h098

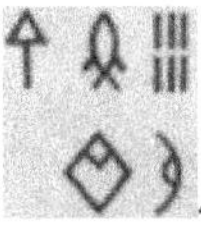 4256 Pict-122 Standard device which is normally in front of a one-horned bull.

h099

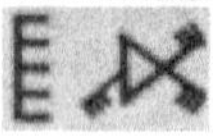

4223

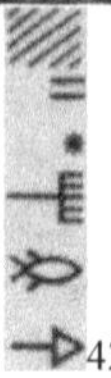

h100 4258 One-horned bull.

h1002

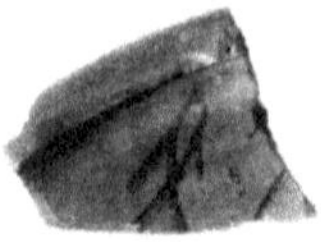

h1007

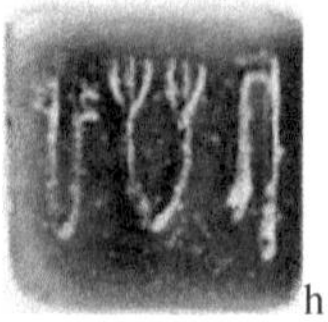

h101

5069

h1010bangle

h1011cone 5103

h1012cone

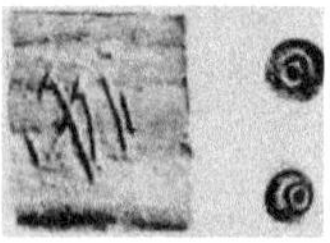

h1017ivorystick

4561

h1018copperobject Head of one-horned bull ligatured with a four-pointed star-fish (Gangetic octopus?)

h102A*

h102B

h102D

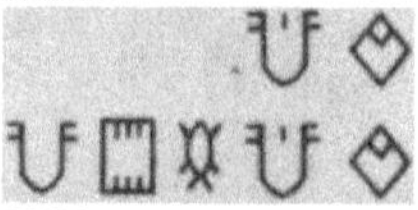

5056

h103

4254

h104

h105

h106

h107

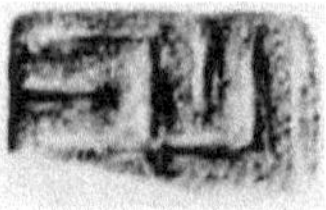

h108

h109

h110

h111

h112

h113

h114

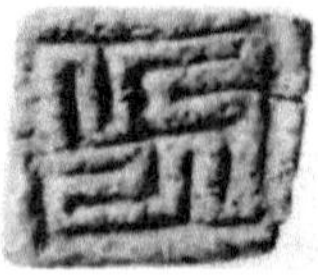

h115

h116

h117

h118
h119

h120

h121

h122

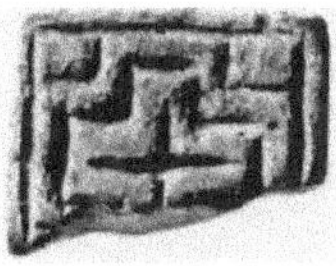

h123

h124

h125

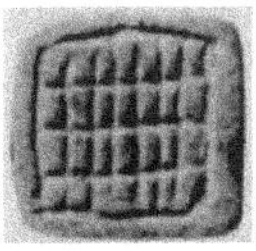

h126

h127

h130

5096

h131 4271

h128

h129A
h129E
4269

h132
5052

h133
4261

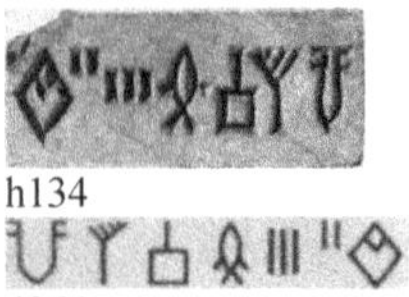
h134
4264

h135
4270

h136
4288

h137a
5058

h138a
5072

h139

4267

h140
4268

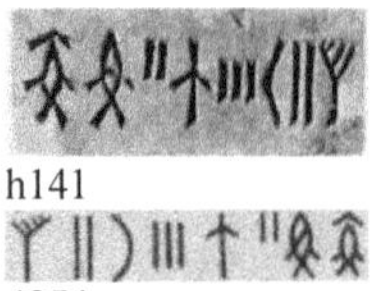
h141
4274

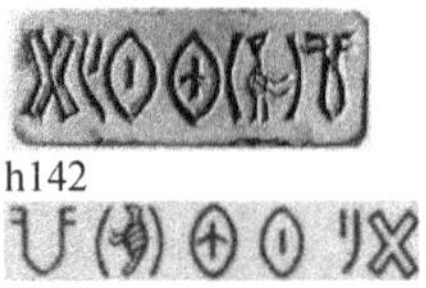
h142
4272

h143a
5101

h144
4280

h145
5067

h146
4628

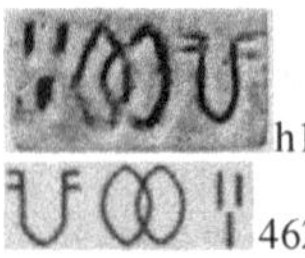
h147
4629

h148
4285

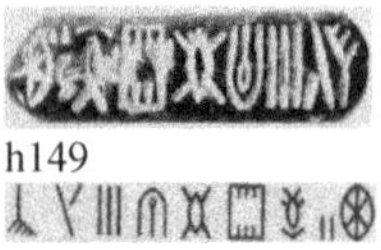
h149
4275

h150
4283

h151
5057

h152
5016

h153
4627

h154
4282

h155

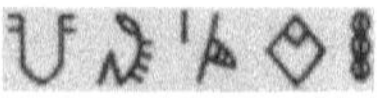
4630

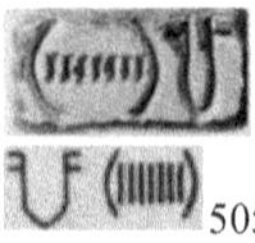
h156
5051

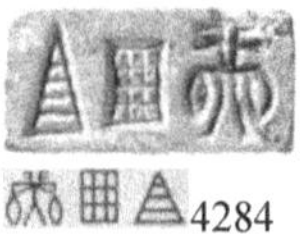
h157
4284

h158
4297

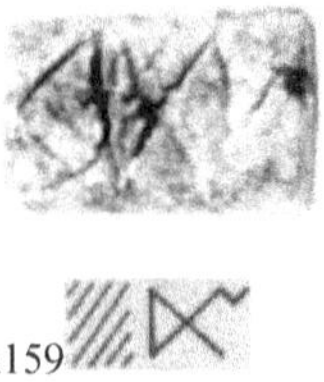
h159
4633

h160A

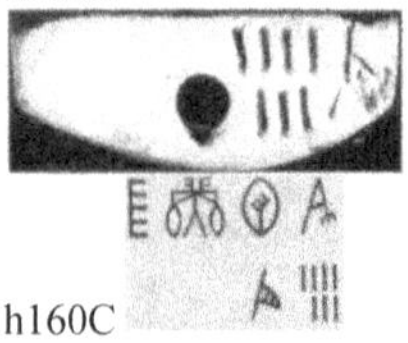
h160C
4276

h161
4262

h162
4294

h163

h164

5046

h165

h166A

h166B

h167A

h167A2

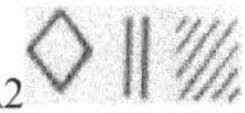
5225

h168

h169A

h169B

5298

h170A

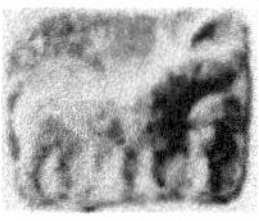
h170B

4701

h171A

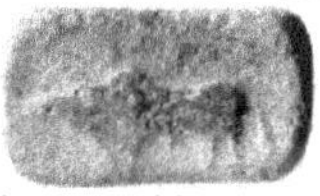
h171B tablet

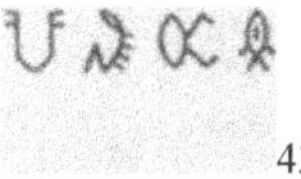
4312
Buffalo.

h172A

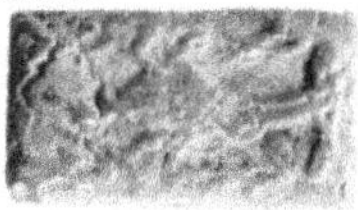

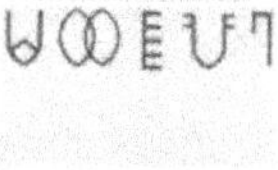
h172B
5305 Pict-66: Gharial, sometimes with a fish held in its jaw and/or surrounded by a school of fish.

h173A

h173B

4333

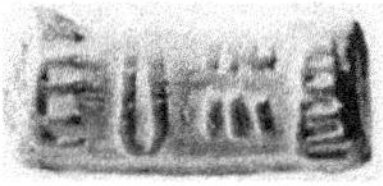
h174A

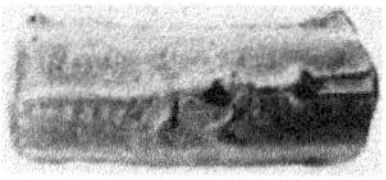
h174B

4338

h175A

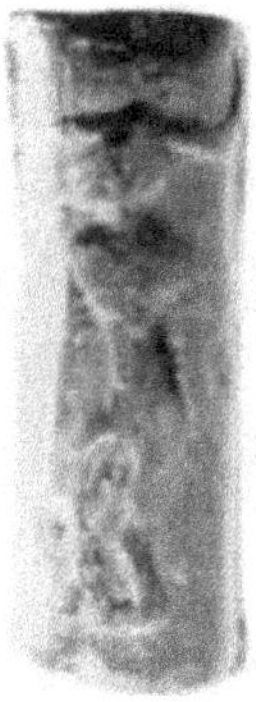
h175B

Pict-87

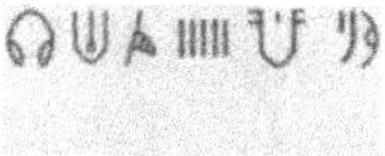
4319 Standing person with horns and bovine features (hoofed legs and/or tail).

h176A

h176B

h176bb

4303 Tablet in bas-relief h176a
Person standing at the centerbetween a two-tiered structure at R., and a short-horned bull (bison) standing near a trident-headed post at L. h176b
From R.—a tiger (?); a seated, pig-tailed person on a platform; flanked on either side by a person seated on a tree with a tiger, below, looking back. A hare (or goat?) is seen near the platform.

h177A

h177B

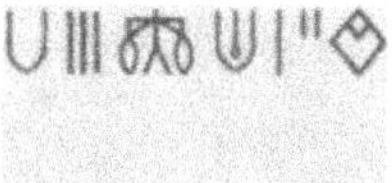
4316 Pict-115: From R.—a person standing under an ornamental arch; a kneeling adorant; a ram with long curving horns.

h178A

h178B 4318
Pict-84: Person wearing a diadem or tall head-dress (with twig?) standing within an arch or two pillars?

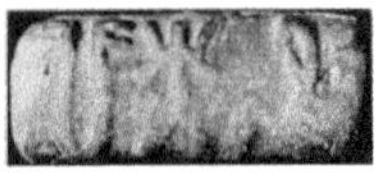

h179A

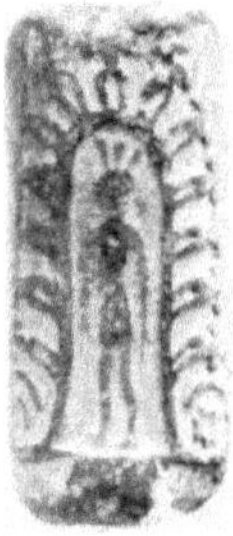

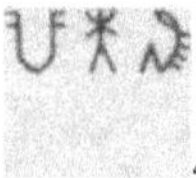

h179B 4307
Pict-83: Person wearing a diadem or tall head-dress standing within an ornamented arch; there are two stars on either side, at the bottom of the arch.

h180A

h180B

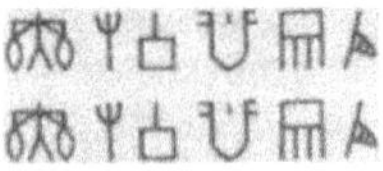

4304 Tablet in bas-relief
h180a Pict-106: Nude female figure upside down with thighs drawn apart and crab (?) issuing from her womb; two tigers standing face to face rearing on their hindlegs at L.
h180b
Pict-92: Man armed with a sickle-shaped weapon on his right hand and a cakra (?) on his left hand, facing a seated woman with disheveled hair and upraised arms.

h181A

h181B

h182A

h182B
4306Tablet in bas-relief
h182a Pict-107: Drummer and a tiger.
h182b Five svastika signs alternating right- and left-handed.

h183A h183B

4327

h184A

h184B

h185A

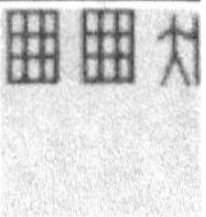

h185B
5279

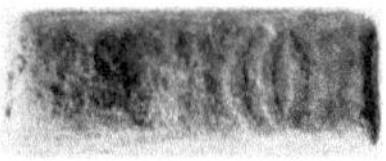

h186A

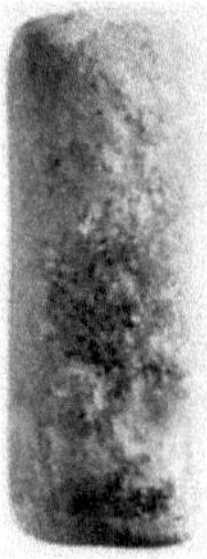

h186B

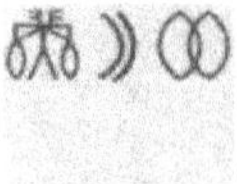

4329

h187A

h187B

5282
Pict-75: Tree, generally within a railing or on a platform.

h188A

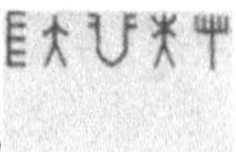

h188B
4325

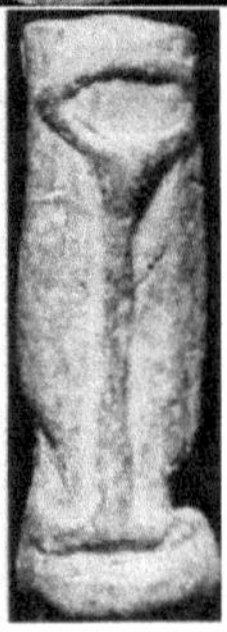

h189A

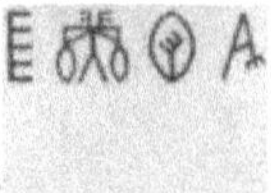

h189B
4341 Pict-126: Anchor?

h190A

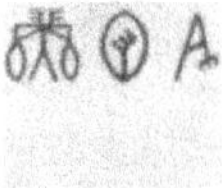

h190B
4323

h191A

h191B 4332

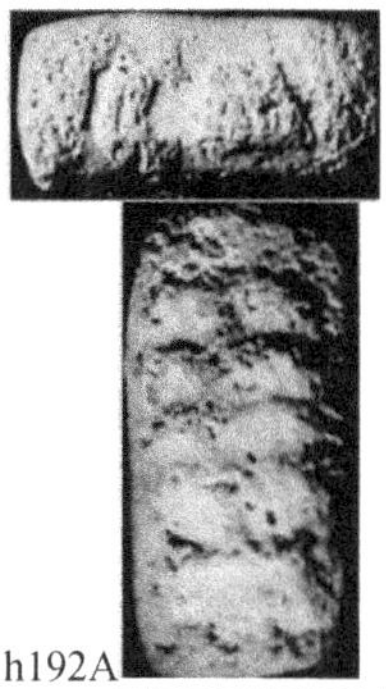

h192A

h192B
5340

h193A

h193B
5332

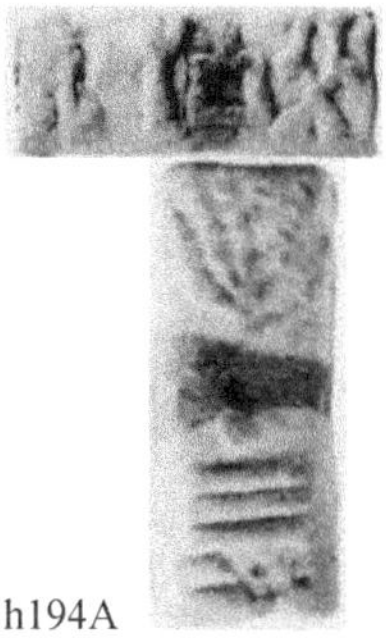

h194A
h194B

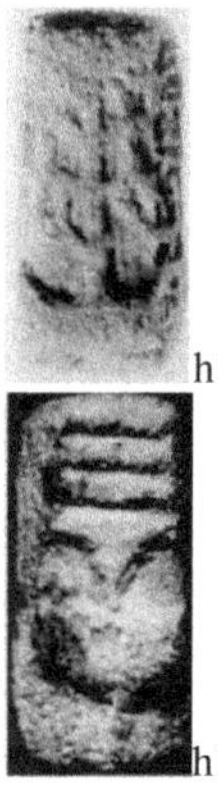

h195A
h195B

h196A

h196B

4309 Tablet in bas-relief h196b

Pict-91: Person carrying the standard. h196a The standard.

h197A

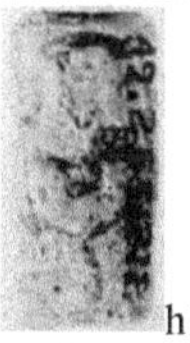

h197B

5333

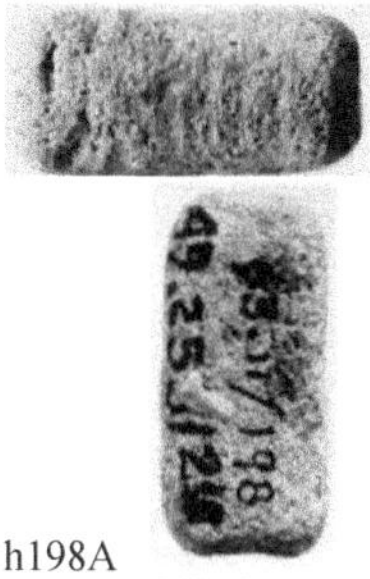

h198A

h198B 5331

h199A

h199B
5252

h200A

h200B
4321

h201A

h201B
5289

h202A

h202B

5334

h203A

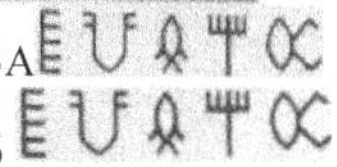

5226
5236

h204A

h204B

5211

h205A
h205B

5254

h206A

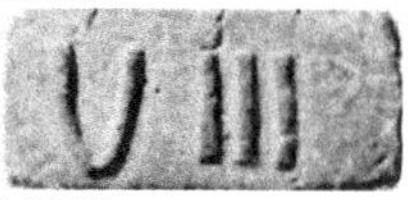

h206B

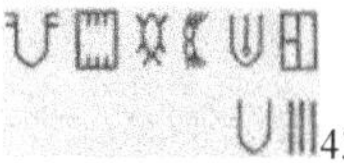

4345

h207A

5297

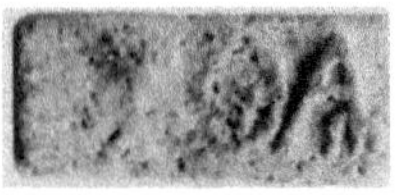

h208A

h208B
5296

h209A

h209B

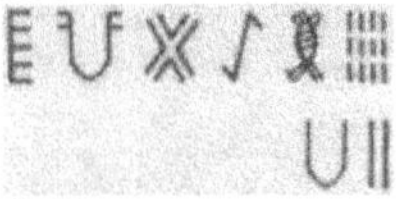

4348

h210A

h210B
4355

h211A

h211B
5274

h212A

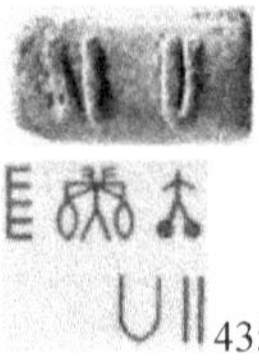

h212B

4357

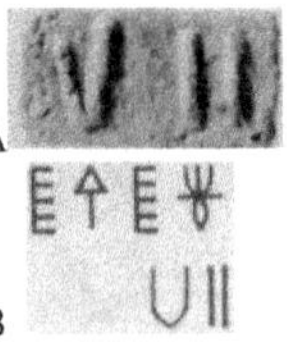

h213A

h213B
5270

h214A

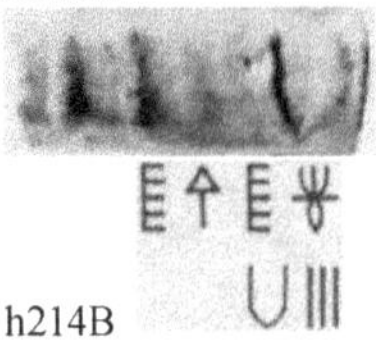

h214B
4684

h215A

h215B

5271

h216A

h216B
5335

h217A

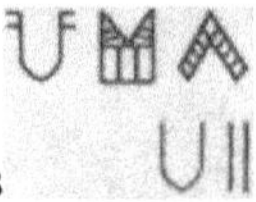

h217B

5336

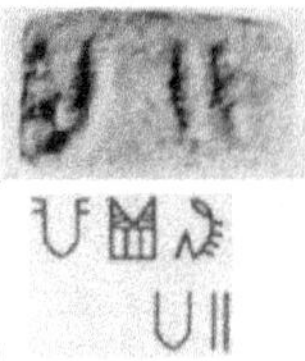

h218A

h218B
5293

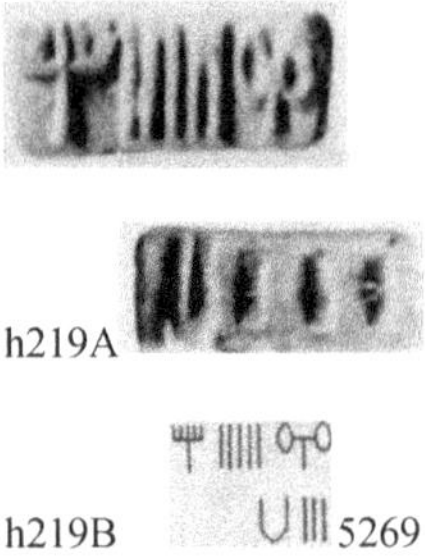

h219A

h219B

5269

h220A
5267

h221A

h221B
5265

h222A

h222B

5339

h223A

h223B
5221

h225A

h226A

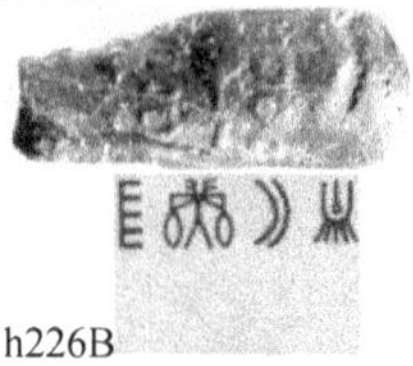

h226B
5243 Standard.

h227A

h227B
4322 Standard. Pict-123

Standard device which is normally in front of a one-horned bull. The device is flanked by columns of dotted circles.

h228A

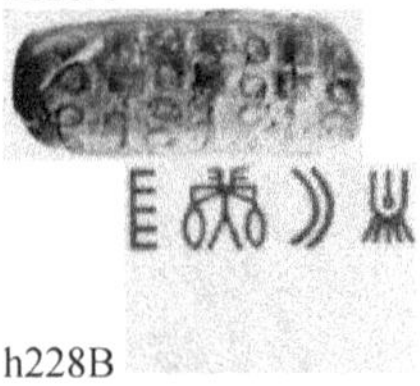

h228B
5244 Standard.

h229A

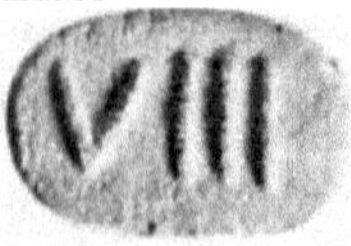

h229B

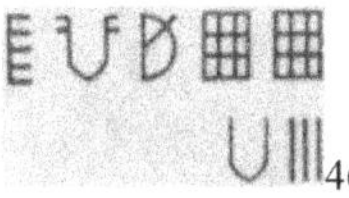

4674

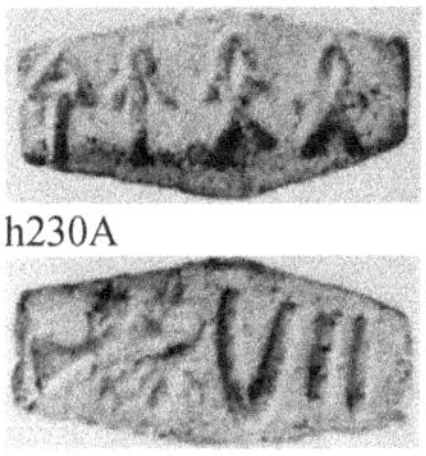

h230A
h230B

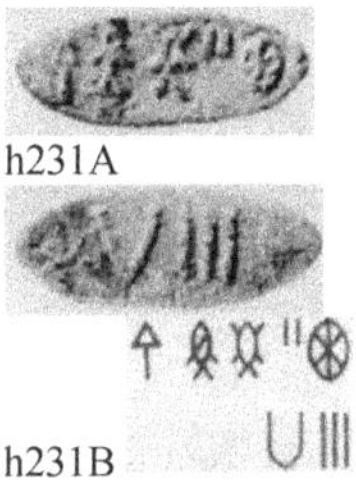

h231A
h231B
4673

h232A
h232B tablet in bas relief

4368
Inscribed object in the shape of a double-axe.

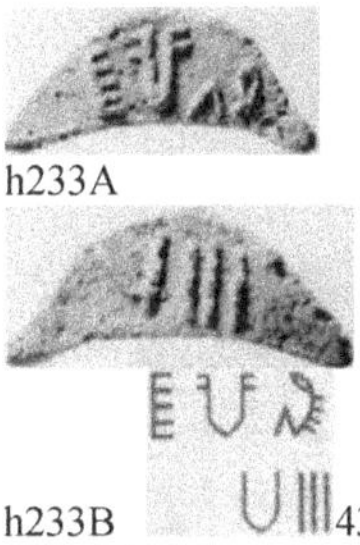

h233A
h233B 4387
Tablet in bas-relief. Sickle-shaped. Pict-131: Inscribed object in the shape of a crescent?

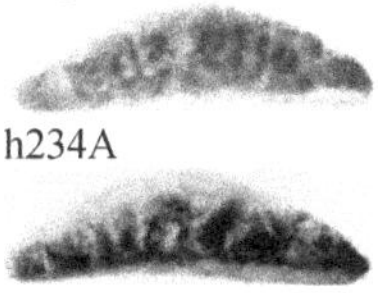

h234A
h234B

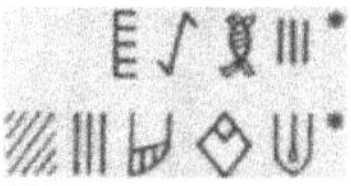

4717

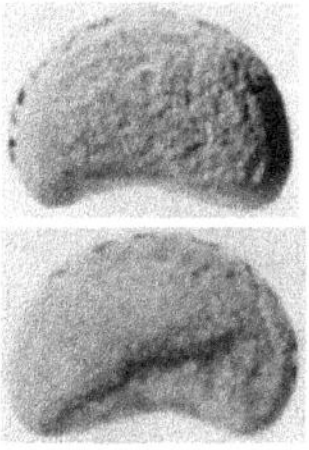

h235A
h235B

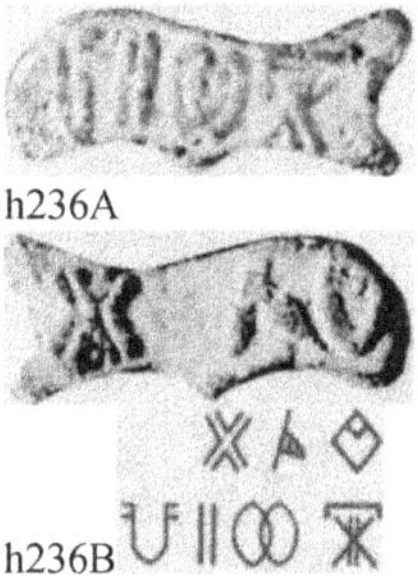

h236A
h236B
4658 Incised miniature tablet.
Object shaped like fish or
sickle? h825A h825B

h237A
h237B
5337

h238A

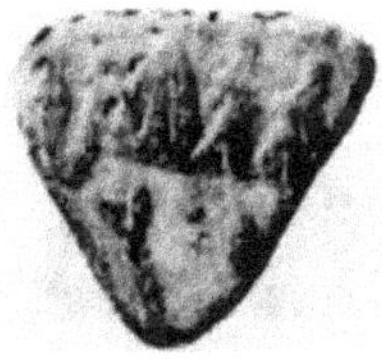

h239A

h239B Tablet in bas

relief
4386

h240
4657

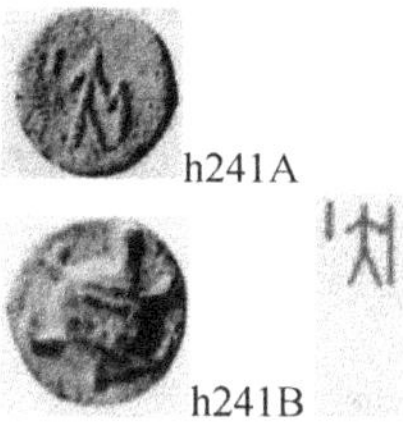

h241A
h241B
4663

Pict-69: Tortoise.

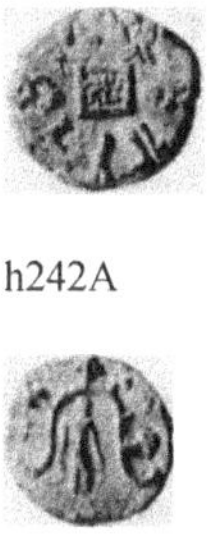

h242A
h242B

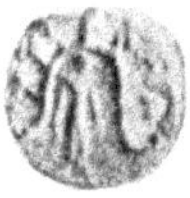

Pict-84

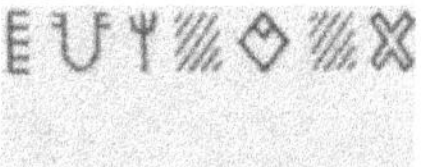

4317

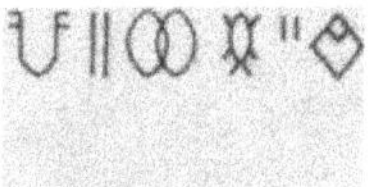

2863

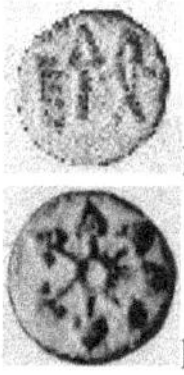

h243A
h243B Tablet in bas-relief
Pict-78: Rosette of seven pipal (?)

leaves. 4664

For See inscription: 4466

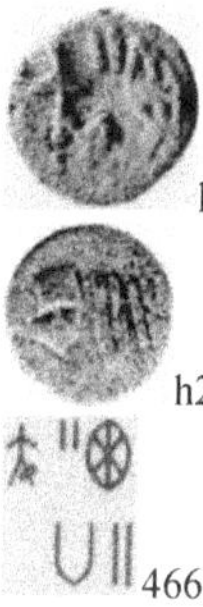

h244A
h244B
4665

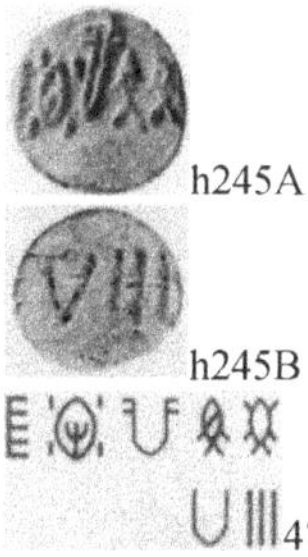

h245A
h245B
4702

h246A

h246B

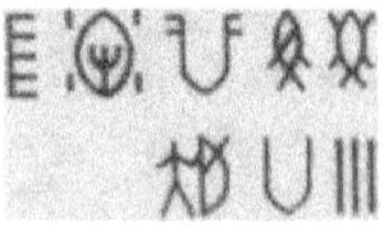

5283

h247A

h247B Tablet in bas-relief

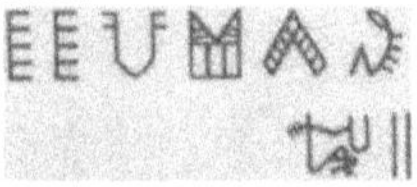

4372

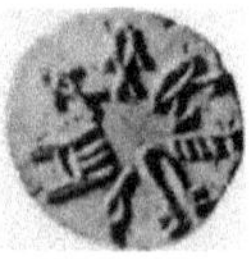

h248A

h248B

Tablet in bas-relief

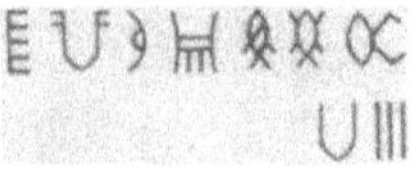

4371

See 3354.

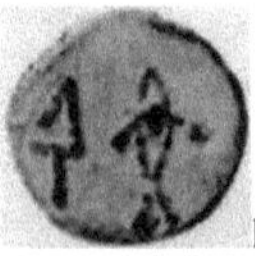

h249A

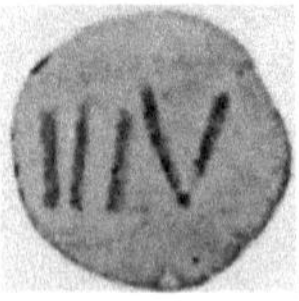

h249B

Tablet in bas-

relief 4374

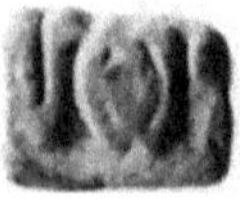

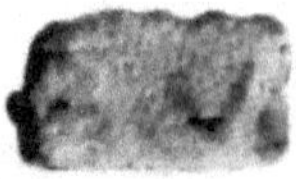

h250A

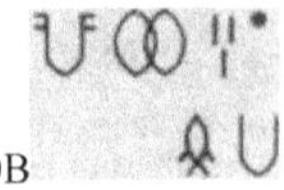

h250B

5250

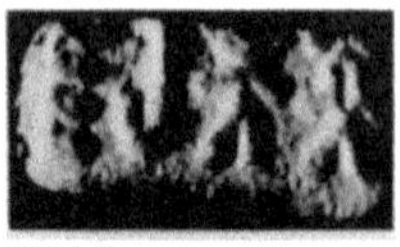

h251A

h251B

h251C

4342 Tablet in bas-relief.

Prism.

Bison (short-horned bull).

h252A

h252B

5215

h253A

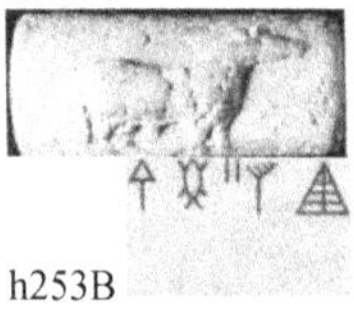

h253B

5219

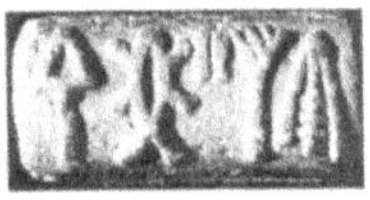

h254A

h254B

5214

h255A

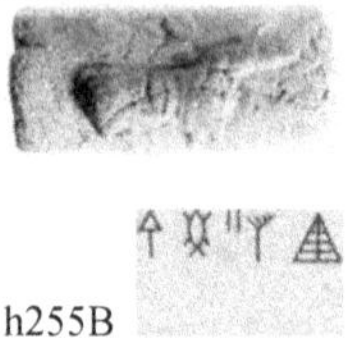

h255B

5208

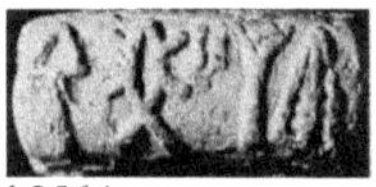

h256A

h256B

5213

h257A

h257B

5216

h258A

h258B 5217

h259A

h259B

5218

h260A

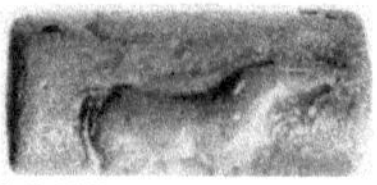

h260B

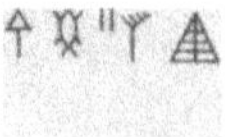

h261

5212

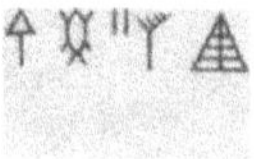

h262

5220

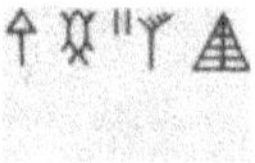

h263

5262

h264

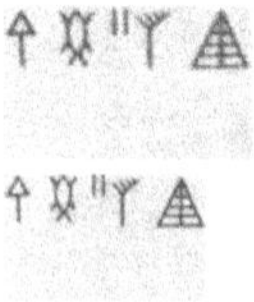

4315

5207, 5208, 5209, 5210, 5212, 5213, 5214,5215, 5216, 5217, 5218,5219, 5220, 5262 Tablets in bas relief. The first sign looks like an arch around a pillar with ring-stones.

One-horned bull.

h252, h253, h255, h256, h257, h258, h259, h260,h261, h262, h263, h264, h265, h276, h277, h859, h860, h861, h862,h863, h864, h865, h866,h867, h868, 869, 870

h266
4011

h267
4007

h268
4020

h269

h270

4014

h271
4069

h272
4619

h273
4176

h274

h275

h276A

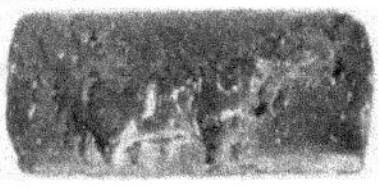
h276B

h277A

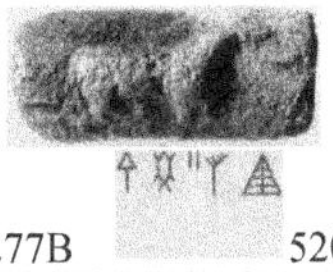
h277B 5207

h278A

h278B

h278C
5205

h279A

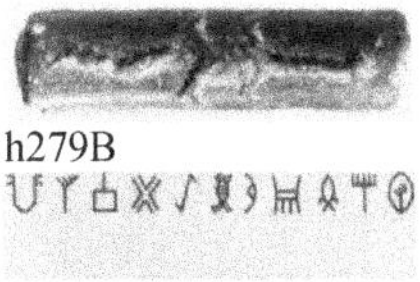
h279B
5256

h280A

h280B
4335

h281A

h281B
4336

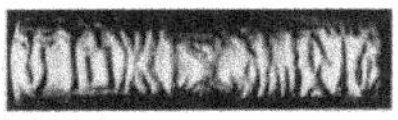
h282A

h282B

h283A

h283B
5253

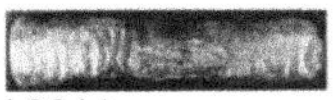
h284A

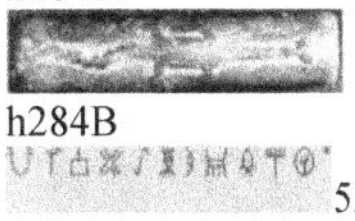
h284B
5229

h285A

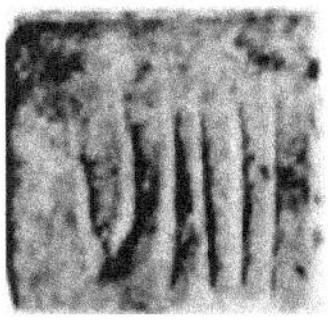
h285B

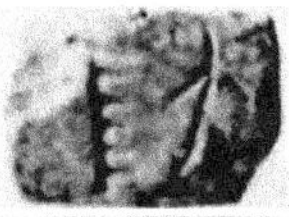
h286A

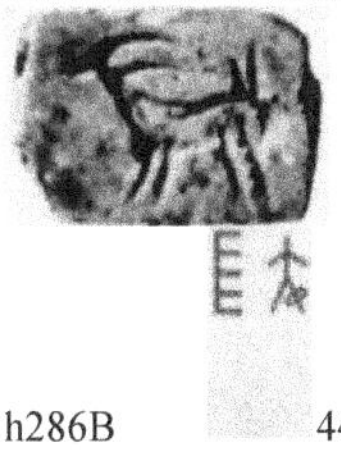
h286B 4429
Incised miniature tablet
Goat-antelope with a short tail

h287A

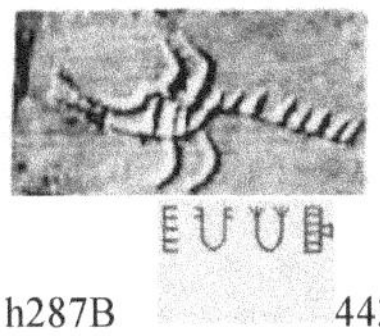
h287B 4430

h288A

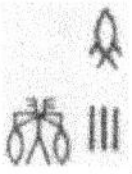
h288B 5463

h289A

h289B

5467

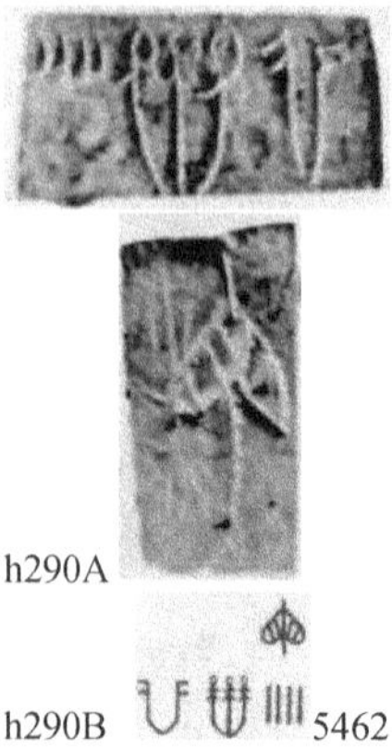

h290A

h290B 5462

h291A

h291B 4440
Standard.

h292A

h292B

4443 Standard.

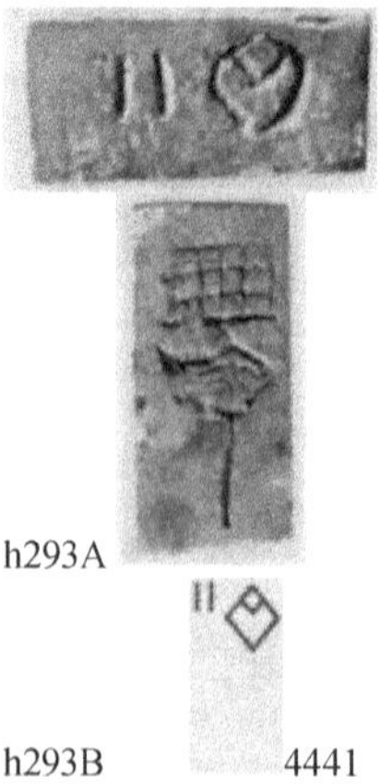

h293A

h293B 4441
Standard.

h294A

h294B 4442

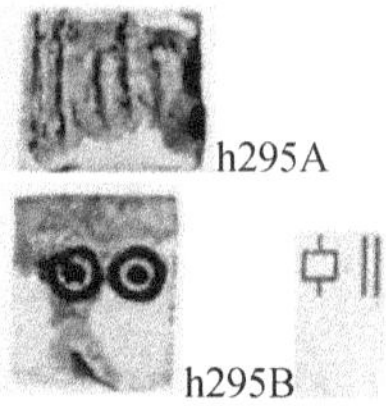

h295A

h295B
4505

h296A

h296B
4457

h297A

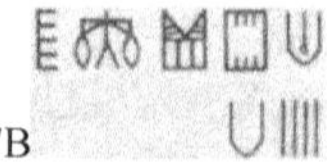

h297B
5497

h298A

h298B
5473

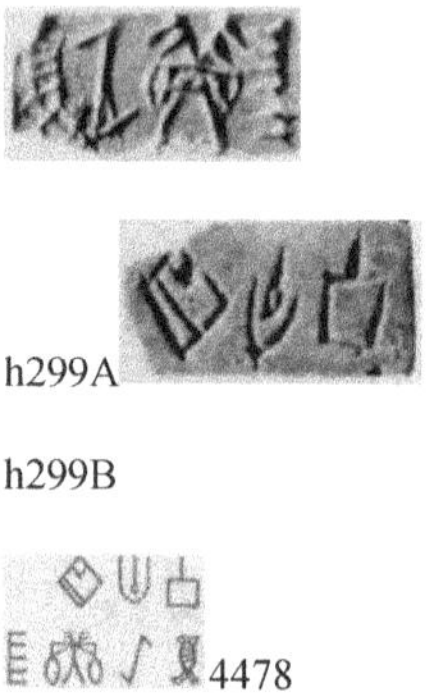

h299A

h299B

4478

h300A

h300B
4454

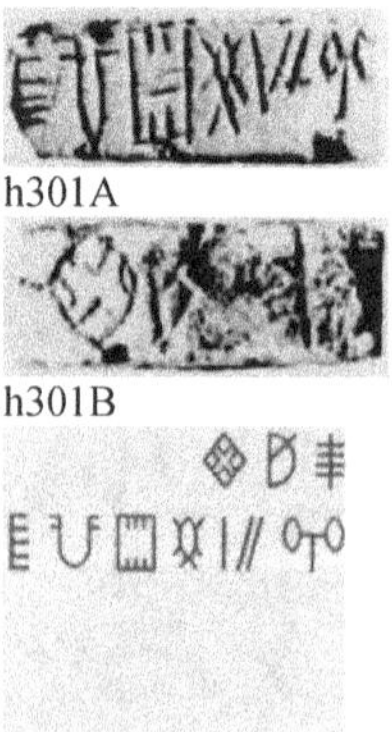

h301A

h301B

4450

h302A

h302B

5460

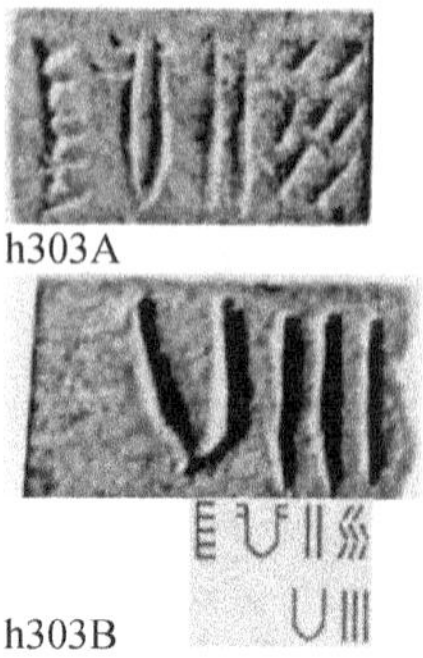

h303A

h303B
4444

h304A

h304B 5401

h305A

h305B Text 5460

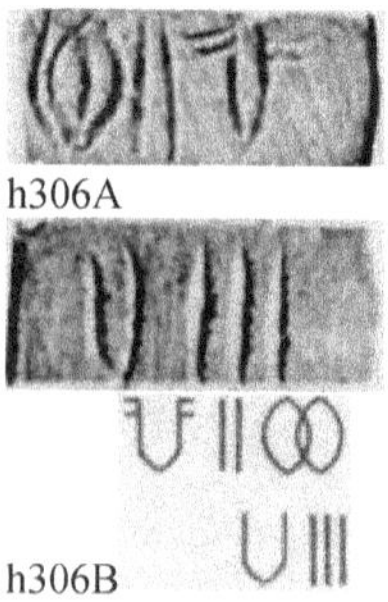

h306A

h306B
5474

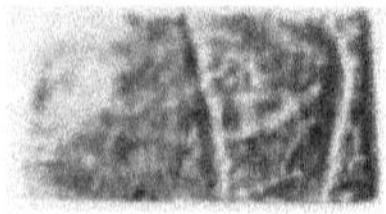

h307A

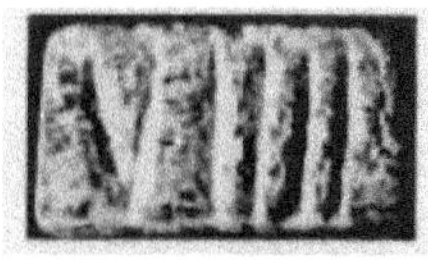

h307B

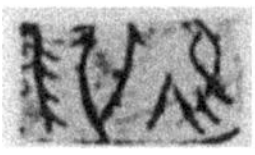

h308A

h308B
5427

h309A
h309B

4403
4405, 4509, 4543, 5419, 5421, 5422, 5423, 5425, 5442, 5449 Incised miniature tablets h309, h311, h317, h932, h959, h935, h960

h310A

h310B
5475

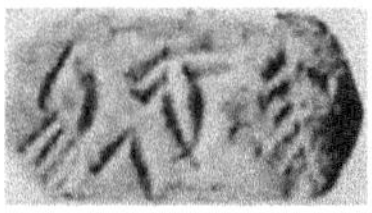

h311A

h311B 5421

h312B

h312Ac

5426

h313A

h313B

5432

5433

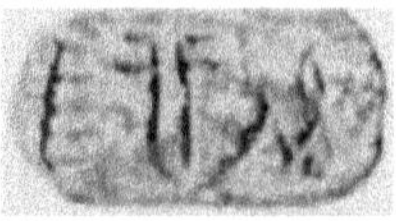

h314A

h314B 5447

h315A

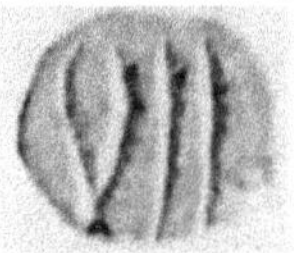

h315B

5464

h316A

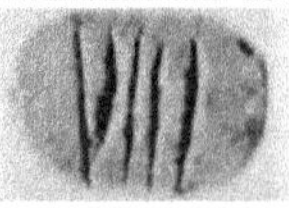

h316B

h317A

h317B 5442

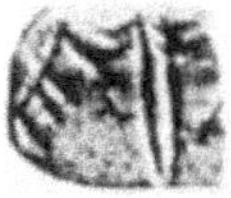

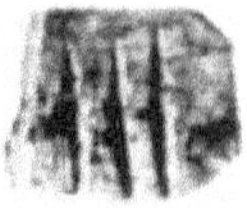

h318A

h318B 5451

h319A

h319B 4544

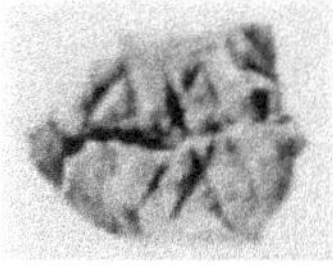

h320A

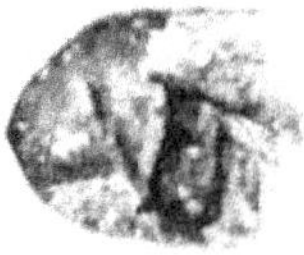

h320B 5450

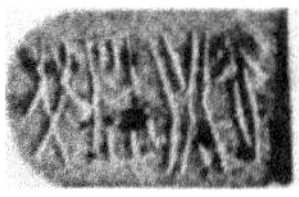

h321A

h321B 5402

h322A

h322B

5498

h323A

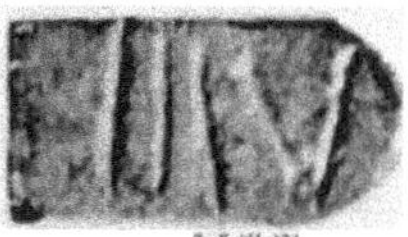

h323B 4497

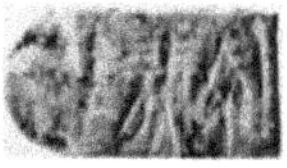

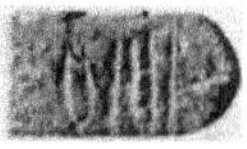

h324A

h324B 4484

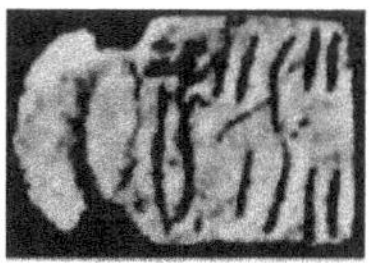

h325A

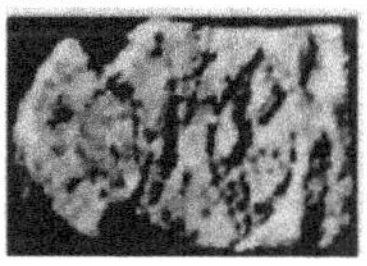

h325B

4416 Pict-130: Inscribed object in the shape of a writing tablet (?)

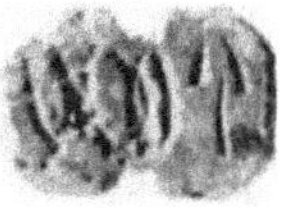

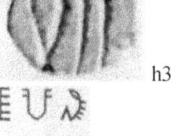

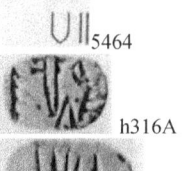

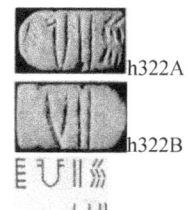

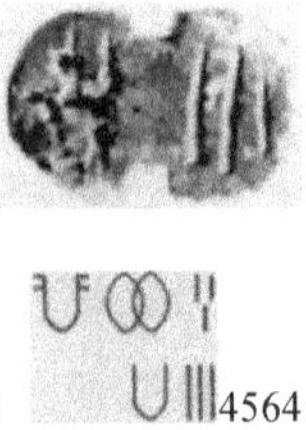
h326A
h326B 4564
Double-axe?

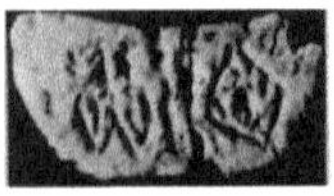

h327A
h327B 5472

5483 Shape of object: Blade of a weapon?

h328a

h328B
4415 Shape of object: Bladeof weapon?

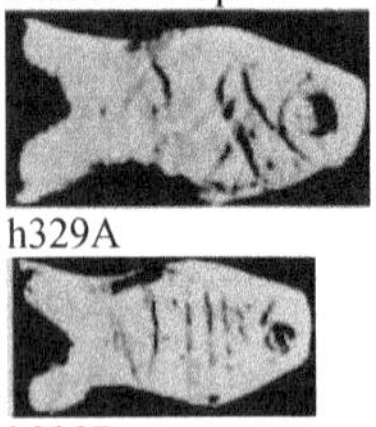
h329A
h329B

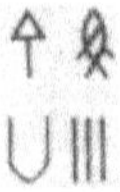
5496 Pict-68: Inscribed object in the shape of a fish.

h330A

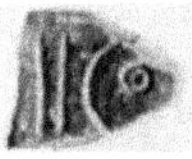

h330B 4560

h331A Incised miniature tablet.

4421, 4422, 4423

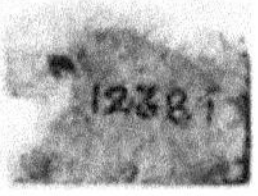

h332C

4885

h333A

h333B

4421

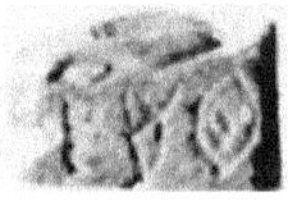
h334A

h334B 4423

h335a

h335B 4425

h336A

h336B

4424

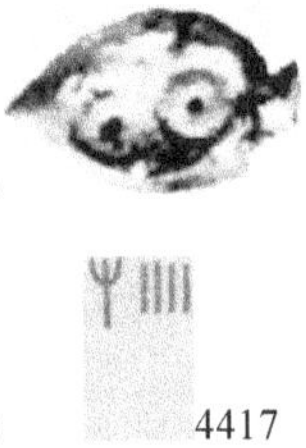
h337A
h337B 4417
Pict-79: shape of a leaf. Dotted circle on obverse.

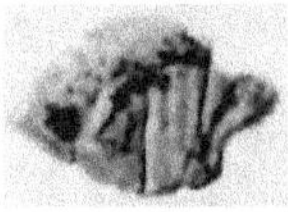

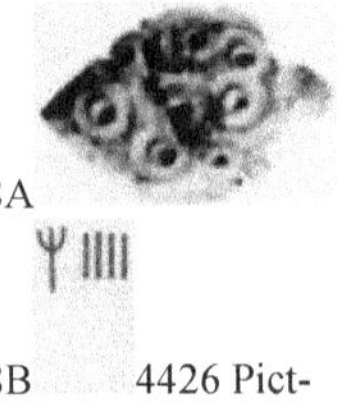
h338A
h338B 4426 Pict-39: Inscribed object in the shape of a tortoise (?) or leaf (?). Dotted circles on obverse.

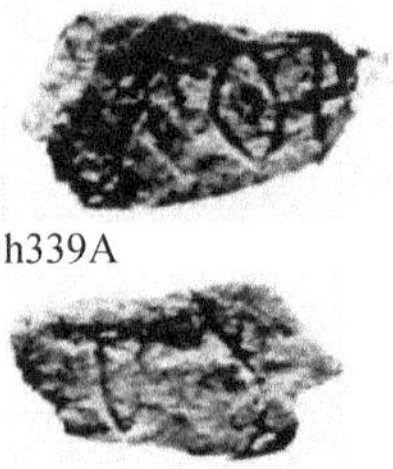
h339A

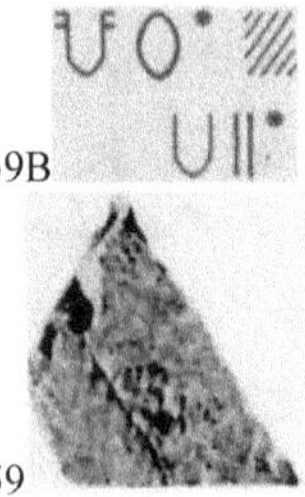
h339B
4559

h340A

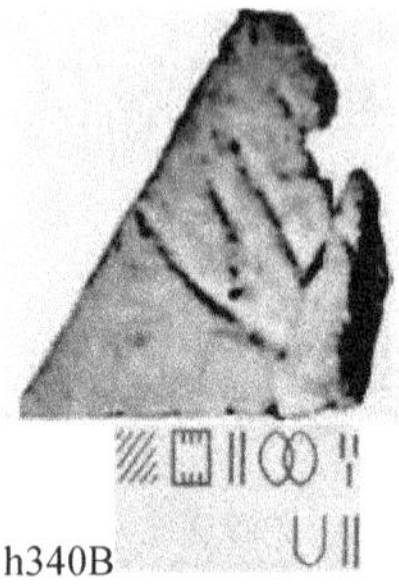
h340B
4420

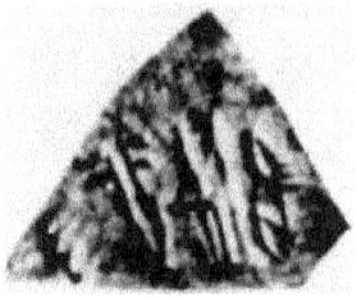
h341A

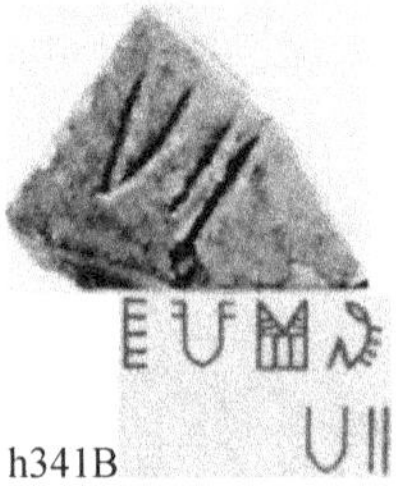
h341B
4419

h342A
h342B

4413

h343A

h343B 4549

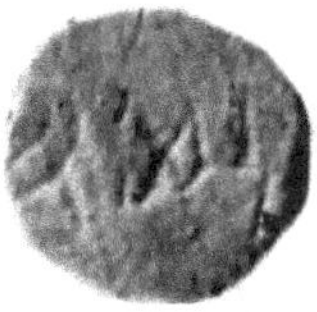

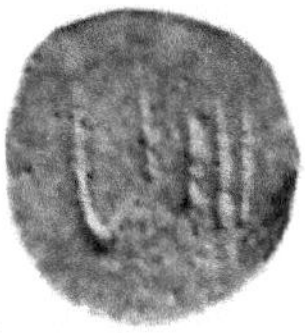

h344A

h344B 4410

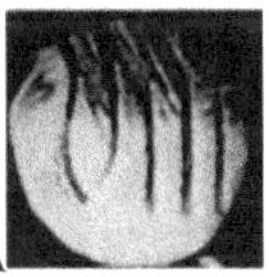

h345A

h345B 4550

h346A

h346B Incised miniature tablet.

4412

h347A  4414

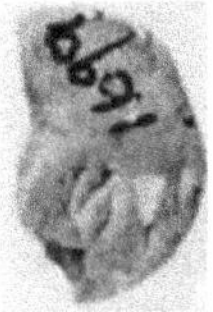

h348A

h348B 4552

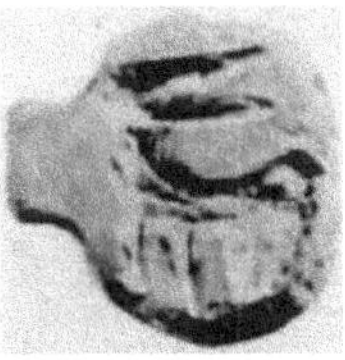

h349A

h349B

h350A

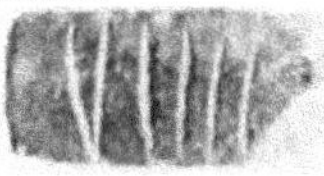

h350B

h350C
4576

h351A

h351B

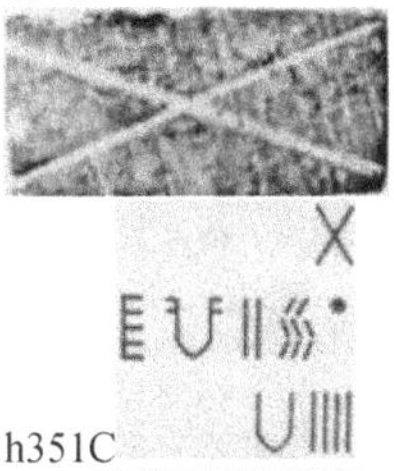

h351C

4581

h352A

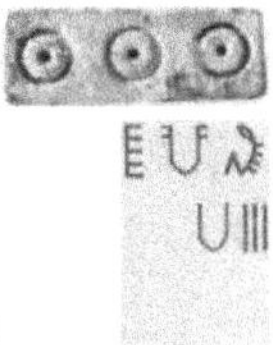

h352B

h352C
4575 Pict-120: One or more dotted circles.

h353A

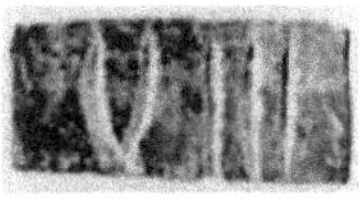

h353B

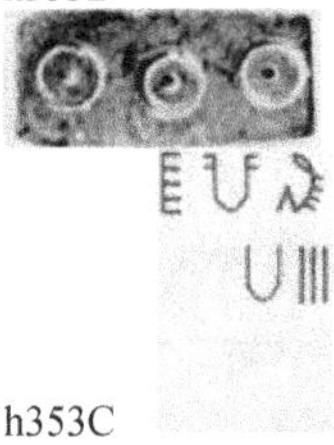

h353C 5416

h354A

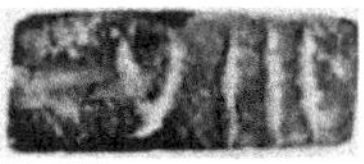

h354B

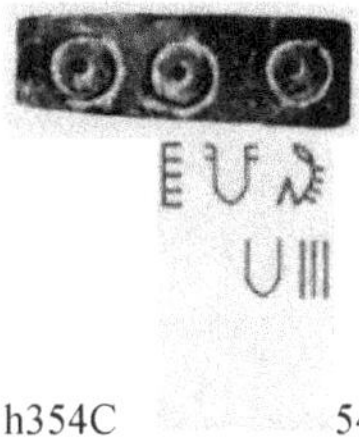

h354C 5499

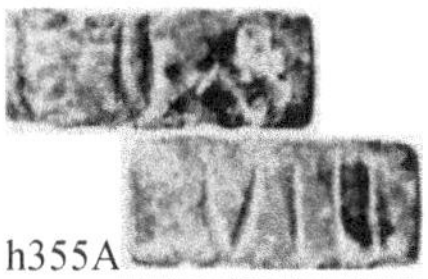

h355A

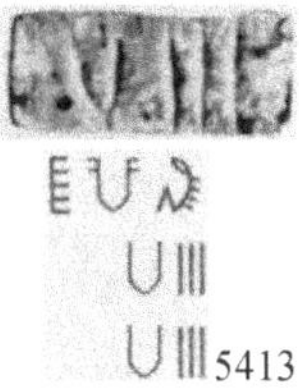

h355B

h355C 5413

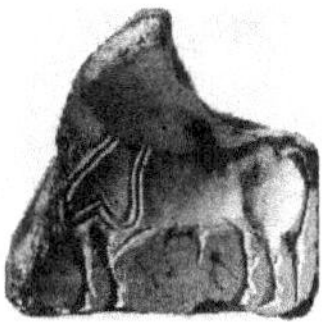

h356

h357

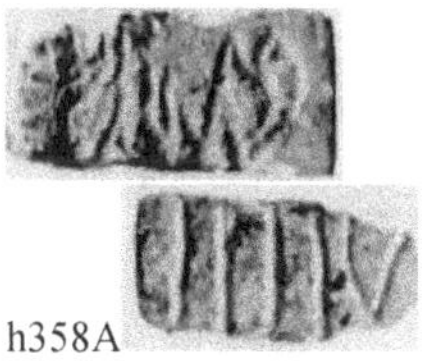

h358A

h358B

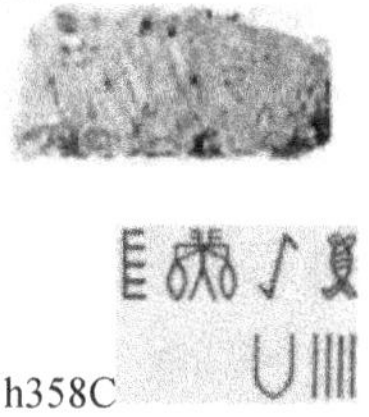

h358C
4579

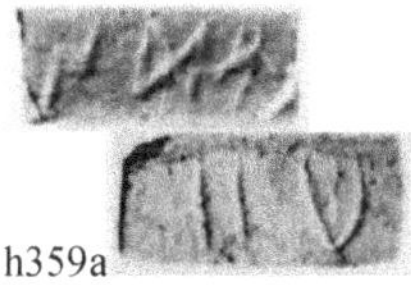

h359a

h359B

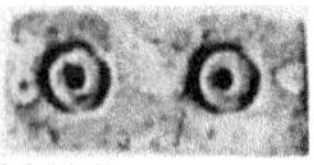

h359C

h360A

h360B

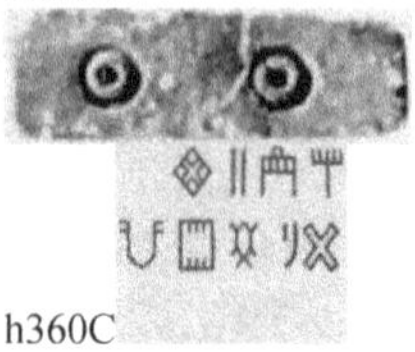

h360C
4584

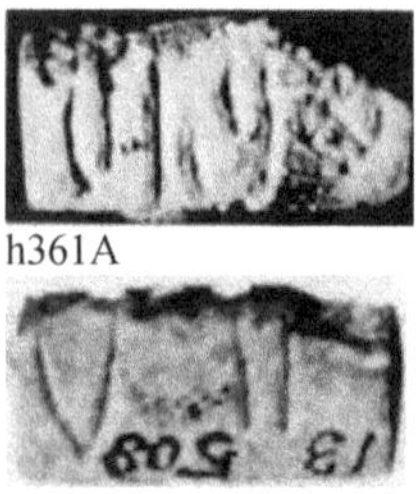

h361A

h361B

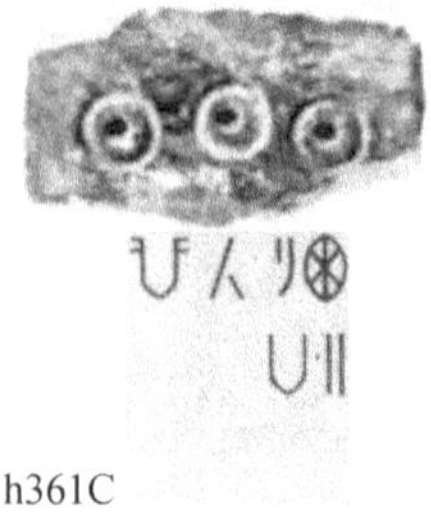

h361C
5476

h362A

h362B

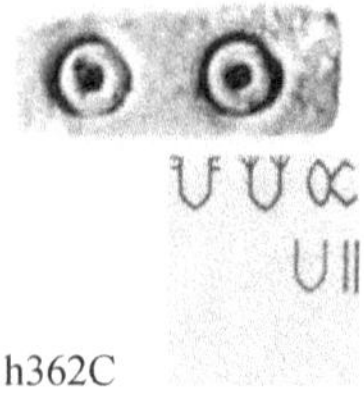

h362C
5466

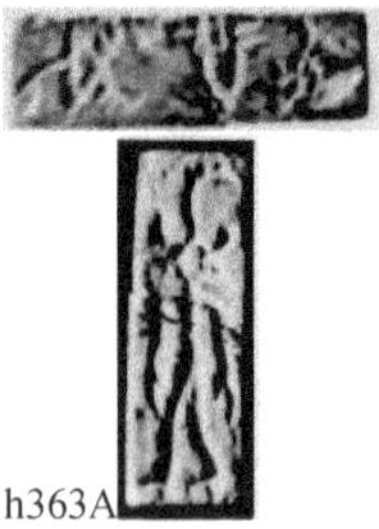

h363A

h363B

h363C

h363E Pict-86

Pict-85 Standing person with horns and bovine features (hoofed legs and/or tail).

5471

h364A

h364B

h364C

h364E 4635

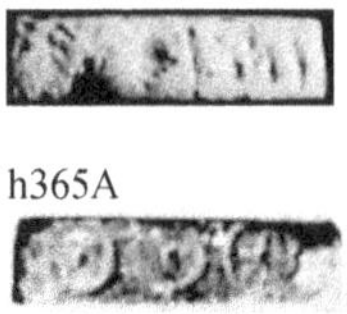

h365A

h365B

h365C

h365E

h366C

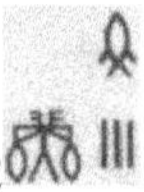

h366E 4590

h367A

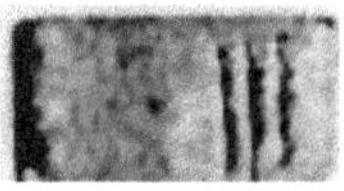

h367B

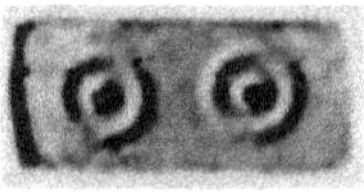

h367C

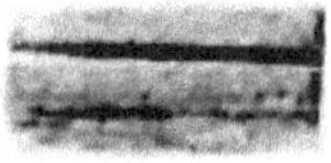

h367E 4401

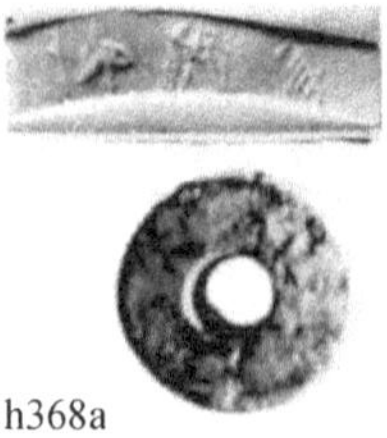

h368a

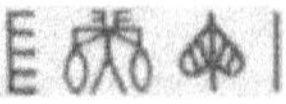

h368E
4409

h369a

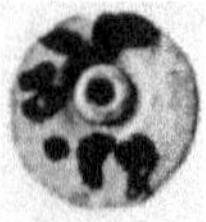

h369C

h369E
4718

h370A

h370A2

h371A

h371A2

h372A

h372A2

h374  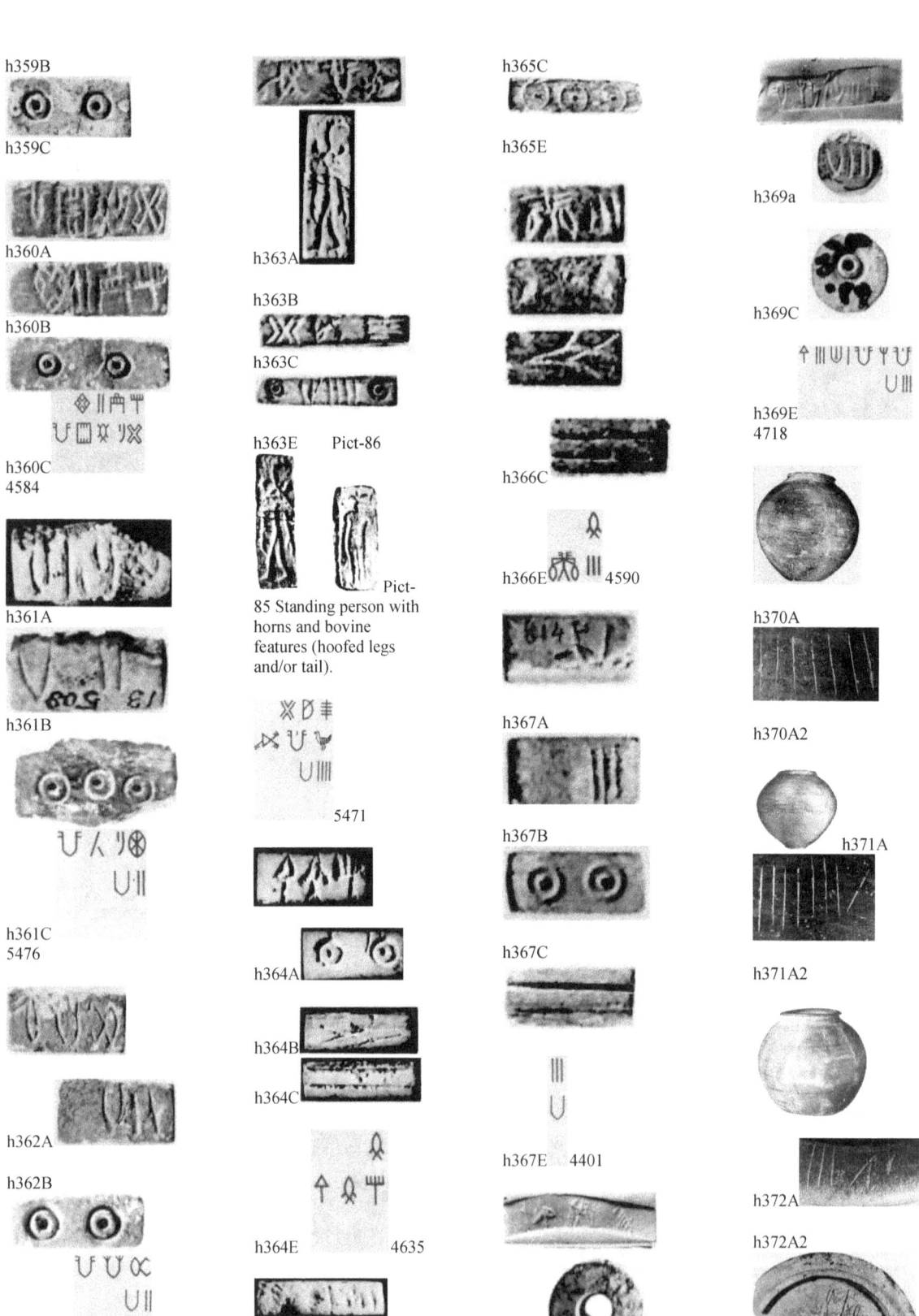 4815

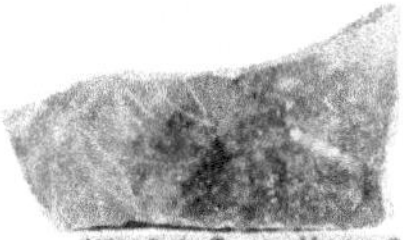
h375 
4812

h377

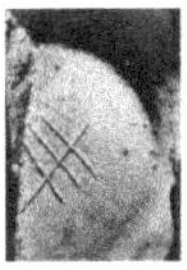

h378

h380 
4902 Bronze dagger

h381  4901
Bronze dagger

h382 
4818

h383 (Not shown).

4021

h384

h385
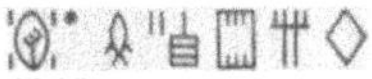
4045

h386

4025

h387

h388

5062

h389
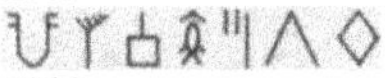
5090

h390 
4024 [The second sign from right appears like a weaver's loom with three looped strings].

h391

5064

h392a  4207

h393

h394a
 5003

h395a

h396

4027

h397

h398

h399

h400

h401
4168

h402

h403

h404

h405

5091

h406  5034

h407
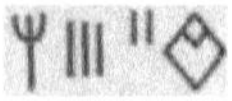
4126

h408

4079

h409

h410

4080

h411 
4078

h412
 4036

h413
 4032

h414

h415
4204

h416  4059

h417

4051

h418

h419 
5092

[The first sign may be a squirrel as in Nindowaridamb 01 Seal].

h420

4614

h421

4026

h422
 4185

h423

4056

h424

h425

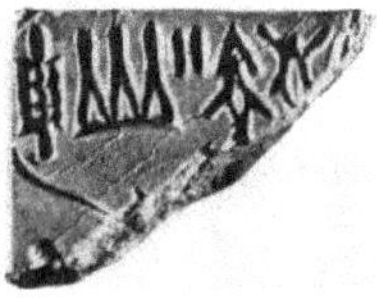
h426

4153

h427
4217

h428

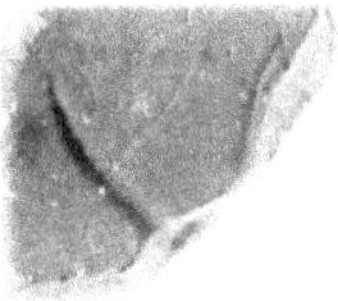

h429

h430

h431

5068

h432

4616

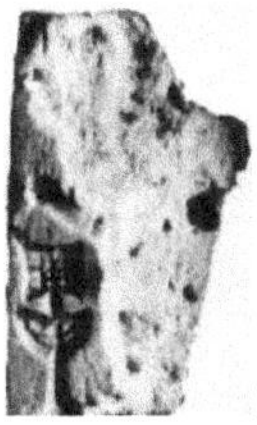
h433

h434

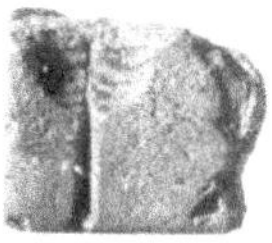
h435
h436

h437

h438

h439

h440
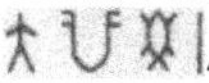4615

h441 
4074

h4424095

h443 4121

h444

h4455110

h446

4034

h447
4089

h448
4054

h449
4082

h450
4084

h451
4137

h452a
4124

h453
4061

h454
4132

h455
4055

h456
4083

h457
5080

h458
4050

h459
4092

h460

h461
4037

h462
4620

h463

h464a
4100

h465
4181

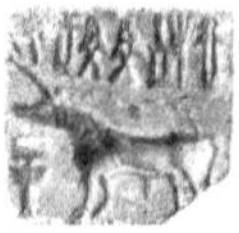
h466
4111

h467
4624

h468
4087

h469
4138

h470
4186

h471
4145

h472
4152

h473
4096

h474
4188

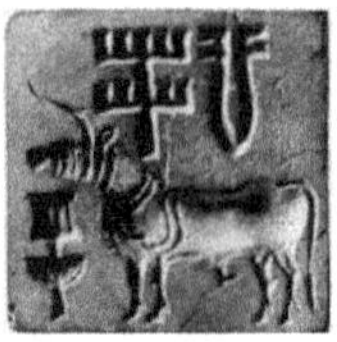
h475
4093

h476
4102

h477

h478
4088

h479
4099

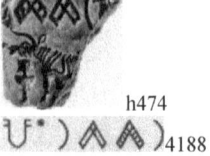

h480 4180

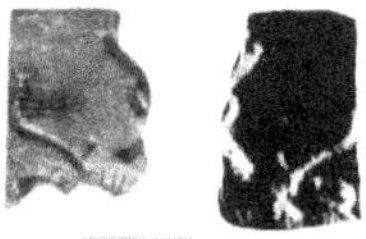

h482 4208

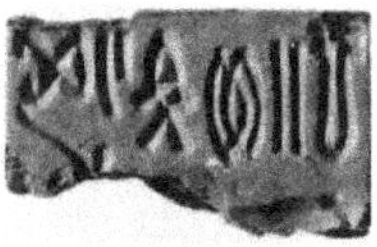

h483

h484

4154

h485

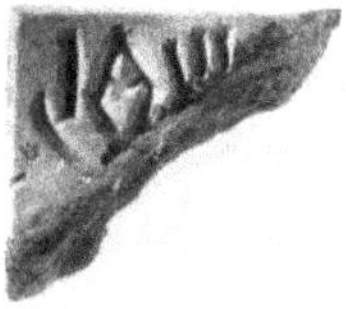

h486

h488
4198

h489

4189

h490 h492

h493

h494

h495

h497

h498

h499

5093

h500

h501
4112

h502
4143

h503

4129

h504 4183

h505

5094

h506
4097

h507 4159

h508

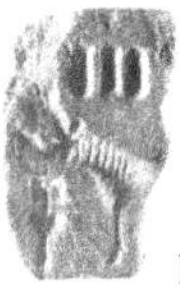

h509

4206

h510
4139

h511

4165

h512a

4618

h513

4163

h514

4116

h515

4162

Text 4166 h516a

h517

h518 4160

h519

4147

h520 4127

h521 4155

h522

h523
5071

h524
4150

h525 4149

h526

h527

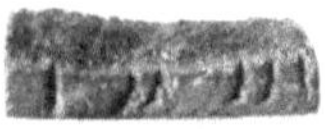

h528

h529

h530 4148
[May have to be arranged from right to left?]

h531
4172

h532

h533
4625

h534

h535

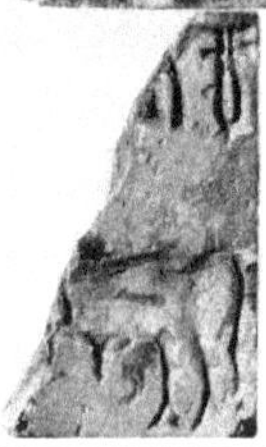

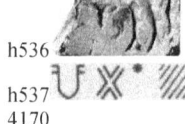

h536
h537
4170

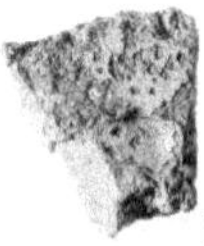

h538

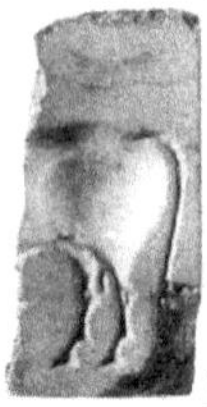

h539

h541 h542

h543
4177

h544
4144

h545 4622

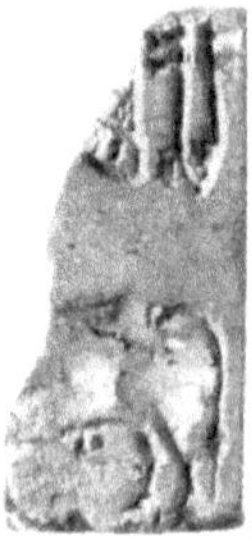

h546
4697

h547

h548

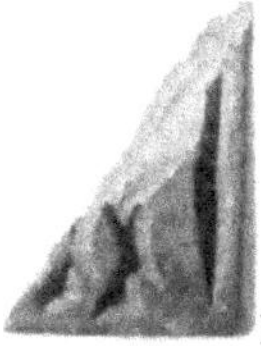

h549

h550 4211

h551 4197

h552

h553

h554

h555

h556

h557

h558 4220

h559
4290

h561

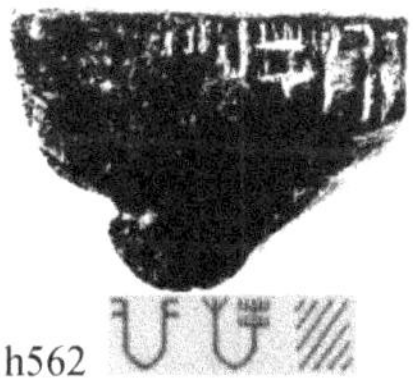

h562
5066

h563 5065

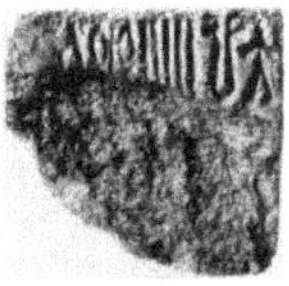

h565

4621

h566 4277

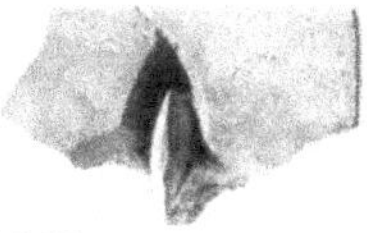

h567

h568

h569

4263

h570

4212

h571

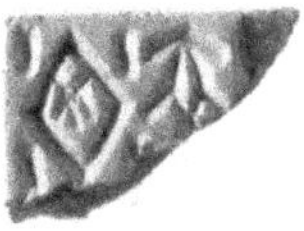

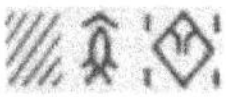

h572
4695

h574
4696

h575

h576

h577
4243

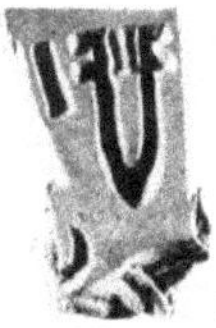

h578

h579

5109

h580

h581
h582

h583

h584  4235 Bison.

h585

h586 4237

h587

h588

h589 
4239

h590

h591 
4228

h592

5081

h593

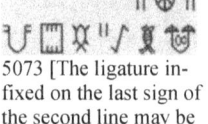

4250 [Composite animal].

h594 [Composite animal].

h595

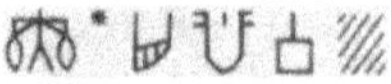
4623

 h596a
4382 [One-horned bull].

h597A

h597D

4075

h598A

h598D

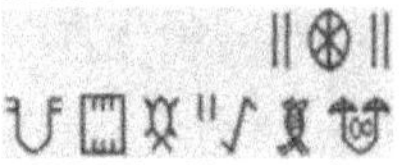
5073 [The ligature infixed on the last sign of the second line may be

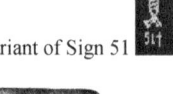

Sign 54 ]

h599A

h599D

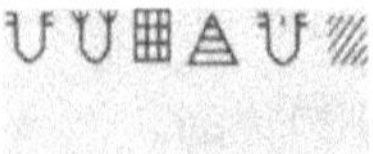
5076

h600  4156
[The last sign may be a

variant of Sign 51 ]

h601

4044

h602a

 4169

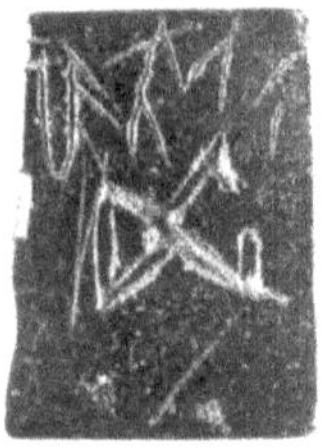

h603 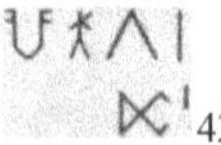 4224

h604

h605

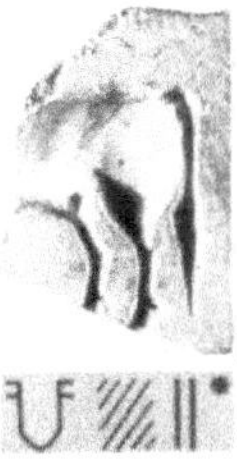
h606
4167

h608

4225

h609

4060

h610

4098
h611

4260 One-horned bull.

h612A
h612B

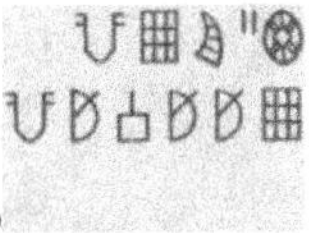
h612D
4123

h613A

h613C

4259 Endless-knot motif?

h614

h616

h617

h618

h619

h620 h621

h622
h623

h624
h625
h626

h627
h628

h629
h630

h631 h632

h633

h634

h635
h636
h637
h638

h639

5061
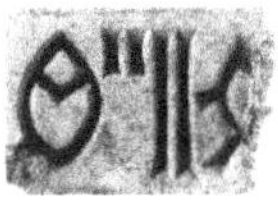
h640

h641A
h641C

4698

h642

4266

h643

4273

h644 4299

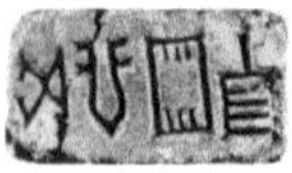

h645

4265

h646

5108

h647

4291

h648

h649

4281

h650A

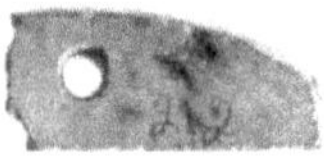

h650C

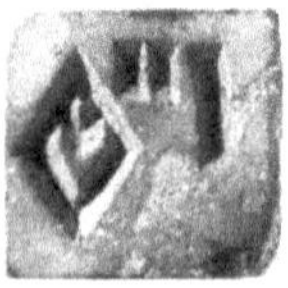

h651 4295

h652

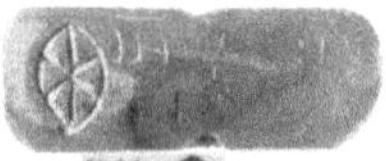

h653 4301

h654

5035

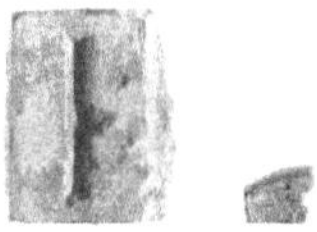

h655AC 4300

h656 4286

h657

4287

h658

4293

h659

5074

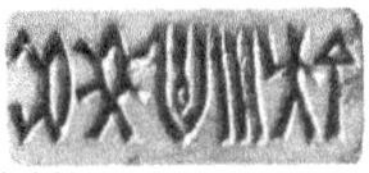

h660

5114

h661 4279

h662a

h663A

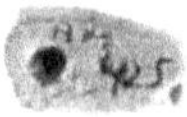

h663C

5006

h664A

h664E

5010

h665

5100

h666

4631

h667A

h667C

4634

h668 5266

h669

4289

h670

h671 4302

h679 4298

h680

5099

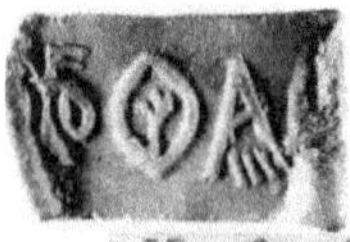

h681a

5105

h682 5078

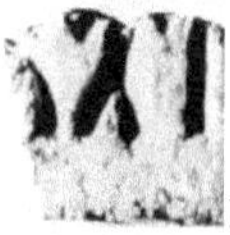

h683

h684

4632

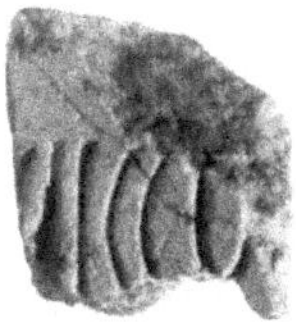
h685

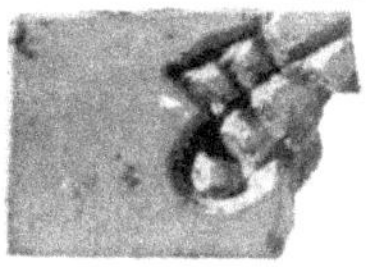
h686

h688A
h688F

h689A

h689B 4222

h690si 5304

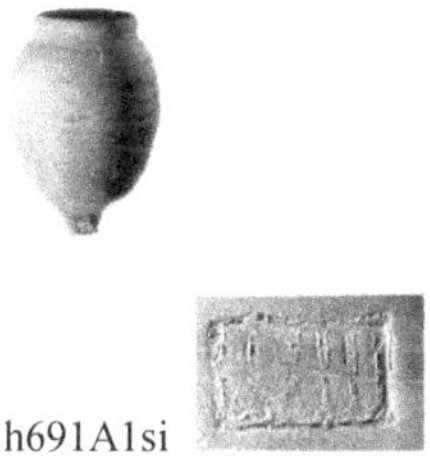
h691A1si
h691A2si

h692A1si
h692A2si

h693t
4707

h694t

h695t

h696At

h696Bt 4677

h697At

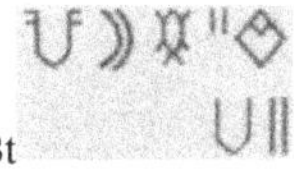
h697Bt
4314

h698At

h698Bt 4659

h699At

h699Bt
5288

h700At
h700Bt

h701At
h701Bt

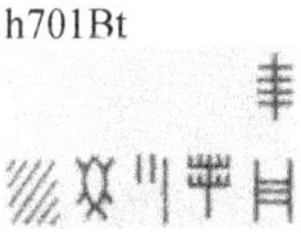
5329

h702At

h702Bt 4601

h703At

h703Bt 4595

h704At
h704Bt

h705At

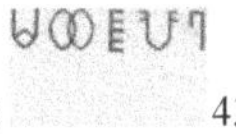
h705Bt 4337

h706At

h706Bt 4340

h707At

h707
4339

h708At
h708Bt 5280

h709 Text 5260

h710 Text 5249

h711 Text
4715

h713At
h713Bt

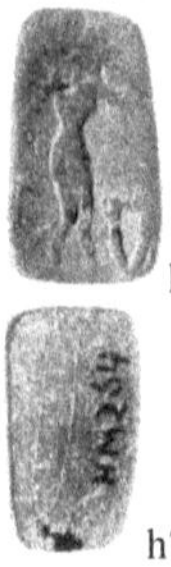
h714At
h714Bt

Standing person with horns and bovine features (hoofed legs and/or a tail) Icon of a person has bull's legs and a raised club.

h715At
h715Bt
5299

h716At

h716Bt h717At

h717Bt

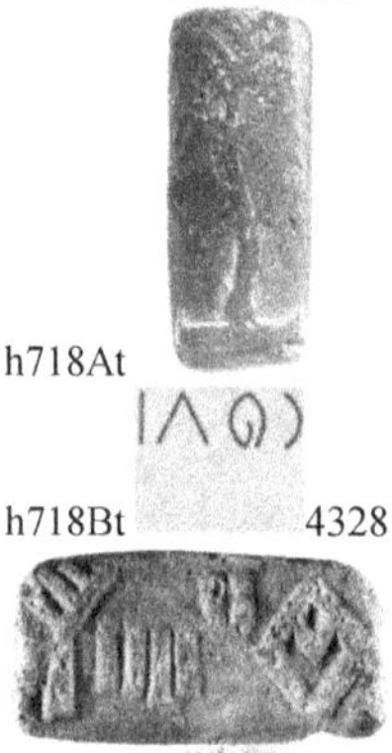
h718At
h718Bt 4328

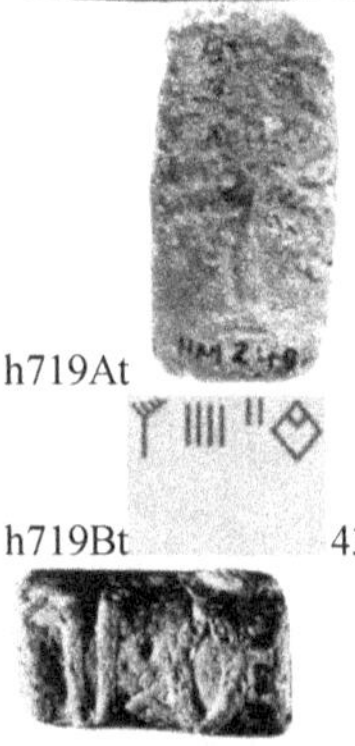
h719At
h719Bt 4326

h720At
h720Bt

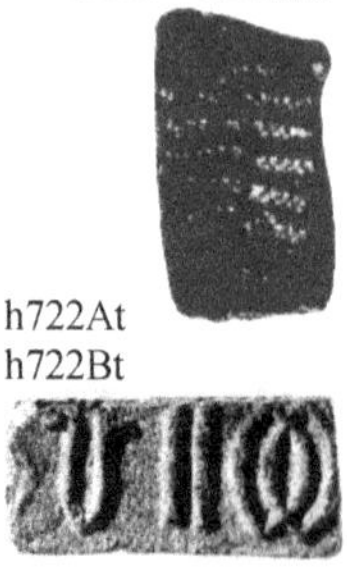
h722At
h722Bt

h723At
h723Bt

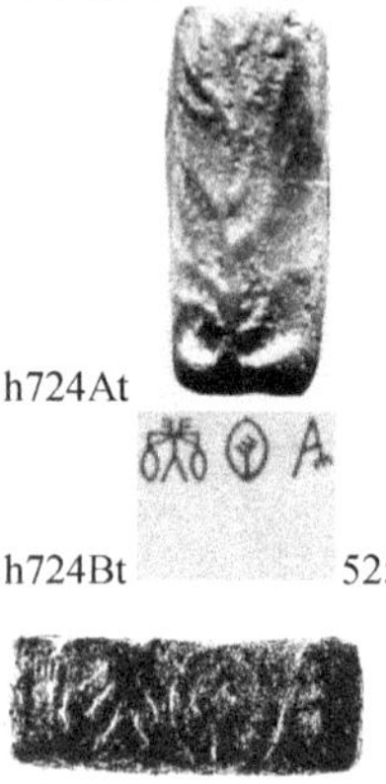
h724At
h724Bt 5255

h725At
h725Bt

h726At
h726Bt

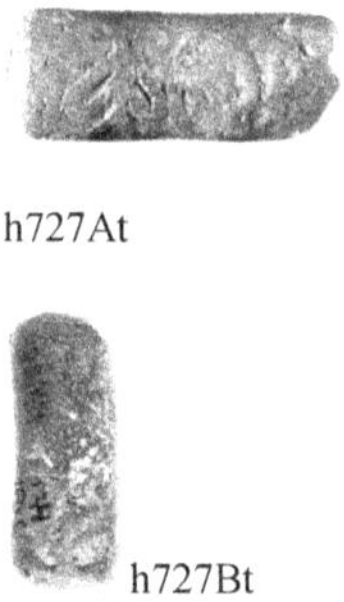
h727At
h727Bt

h728At

h728Bt

h729At
h729Bt
4331

h730At
h730Bt

h731At
h731Bt

h732At
h732Bt

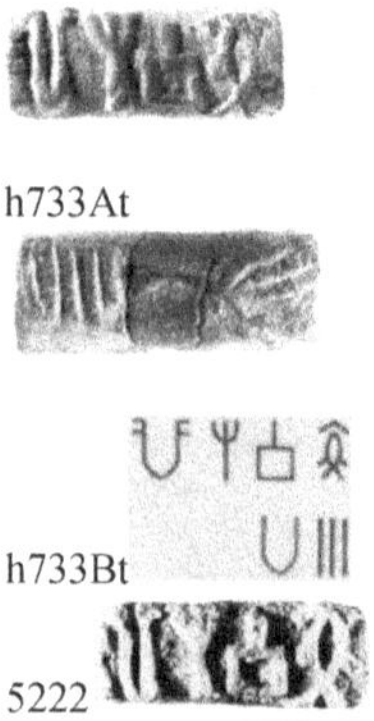
h733At
h733Bt
5222

h734At
h734Bt 5286

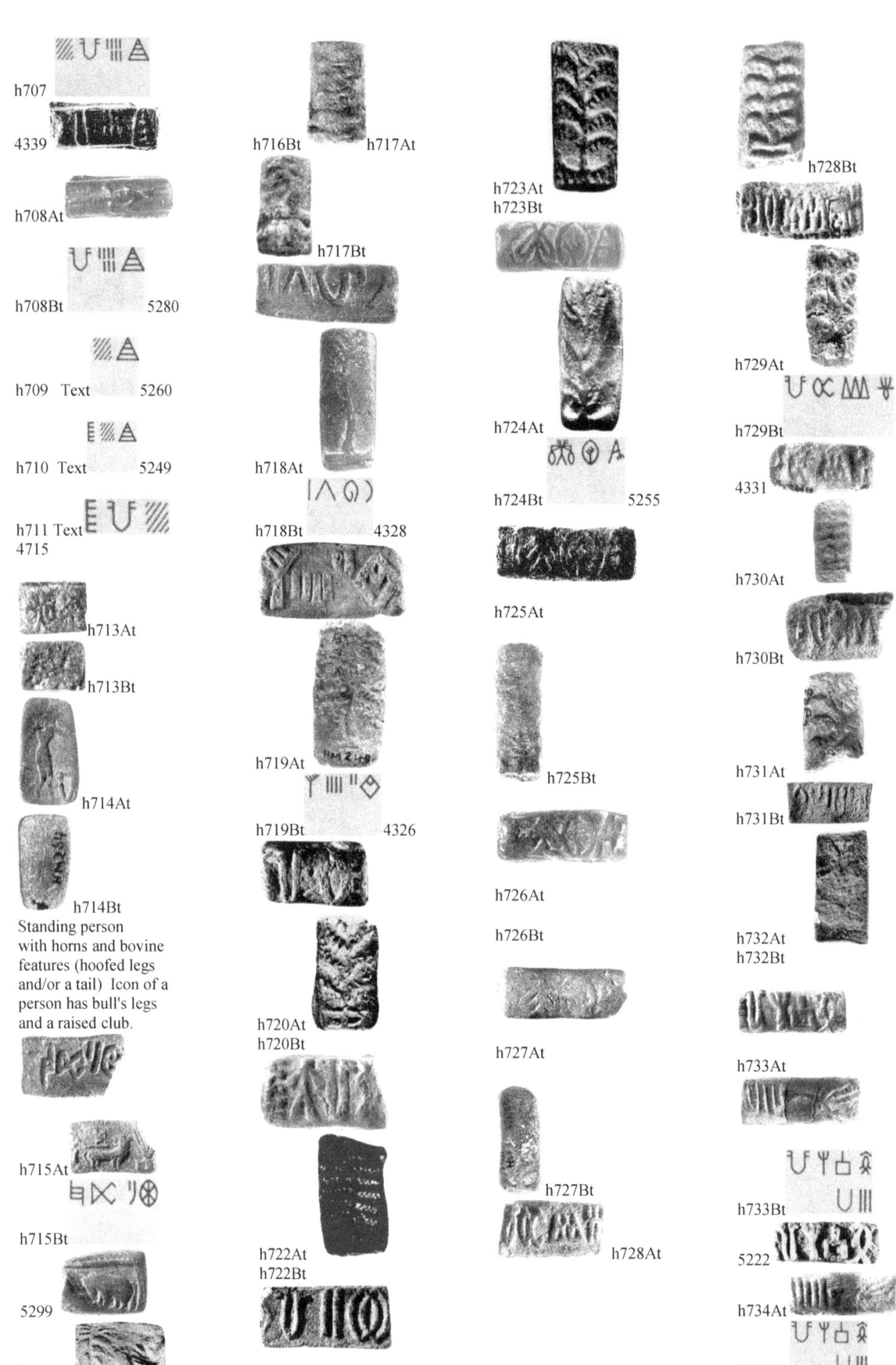

h735At
h735Bt

5310 h736At
h736Bt

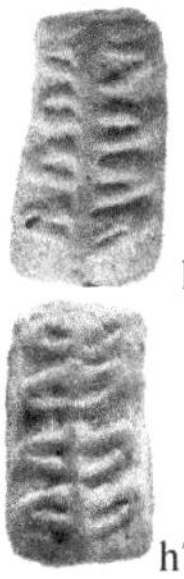
h737At
h737Bt

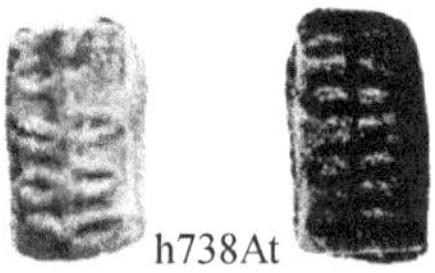
h738At
h738Bt

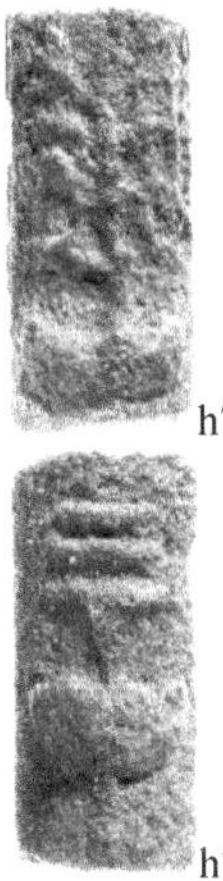
h739At
h739Bt

h740At
h740Bt

h741At
h741Bt
5263

h742At
h742Bt
4320

h743At
h743Bt

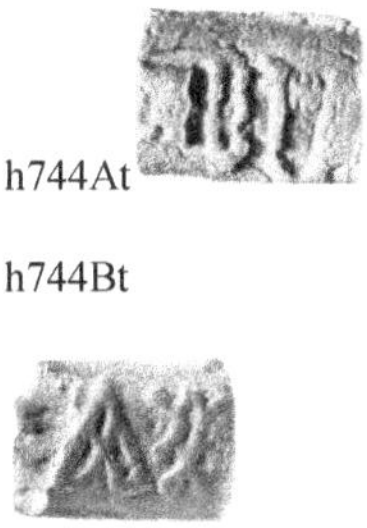
h744At
h744Bt

h745At
h745Bt 5257

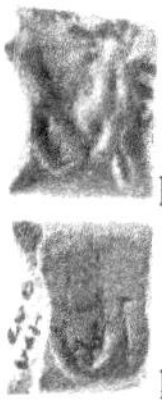
h746At
h746Bt

h747At
h747Bt
4656

h748At
4654

h749At

h750At

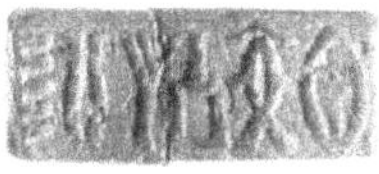
h751At

h752At
5275

h753At
5231

h754At
4716

h755At
5287

h756At
4669

h757At 4655

h758At
h759At

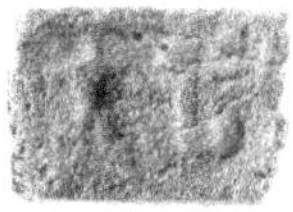

h760At

h760Bt

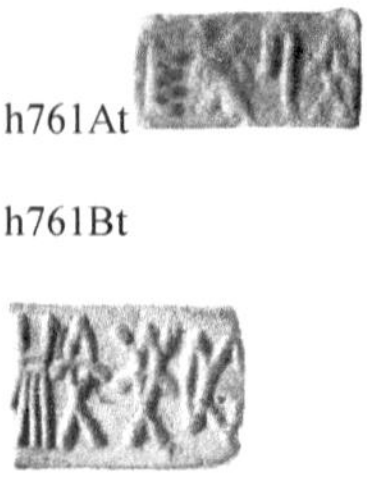
h761At

h761Bt

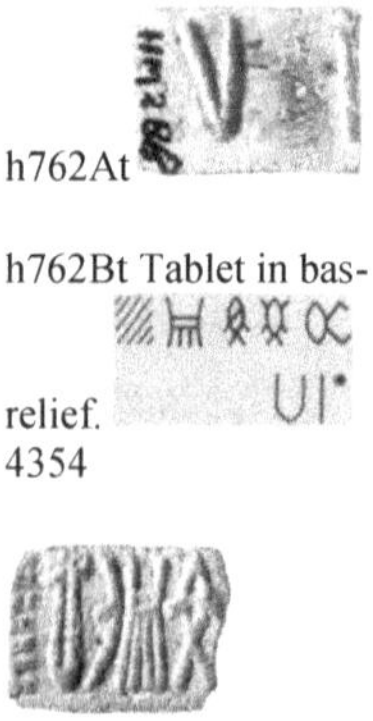
h762At

h762Bt Tablet in bas-relief.
4354

h763At

h763Bt
4661

h764At

h764Bt

h765At

h765Bt 4653

h766At

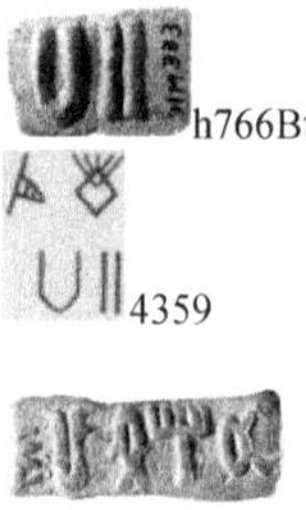
h766Bt
4359

h767At

h767Bt
4352

h768At

h768Bt
4358

h769At

h769Bt
4667

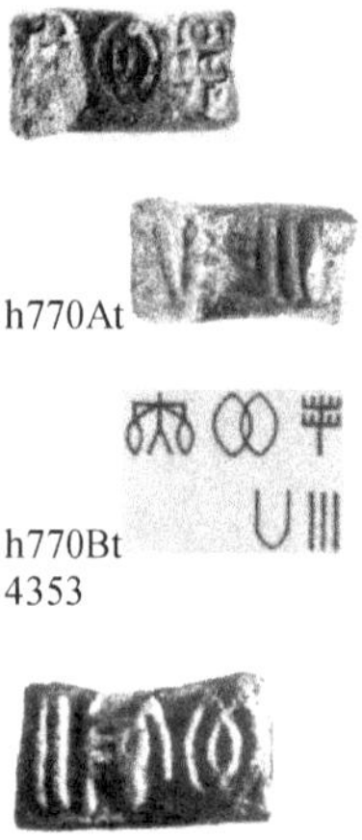
h770At

h770Bt
4353

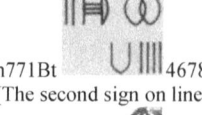

h771At

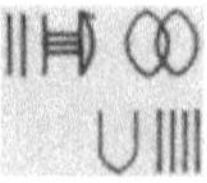
h771Bt 4678
[The second sign on line 1 is a squirrel ].

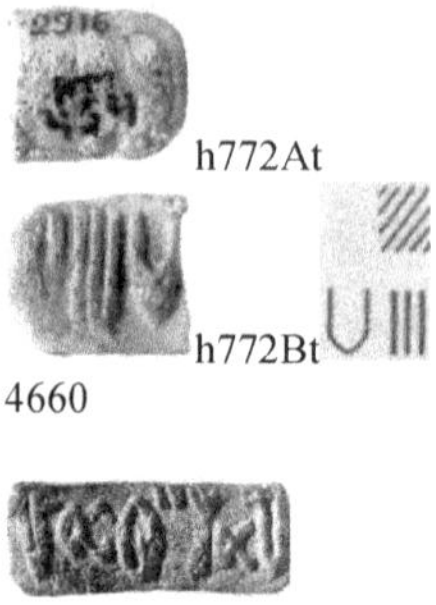
h772At

h772Bt
4660

h773At

h773Bt
4351

h774At

h774Bt
4672

h775At

h776At

h776Bt
4350

h777At

h777Bt

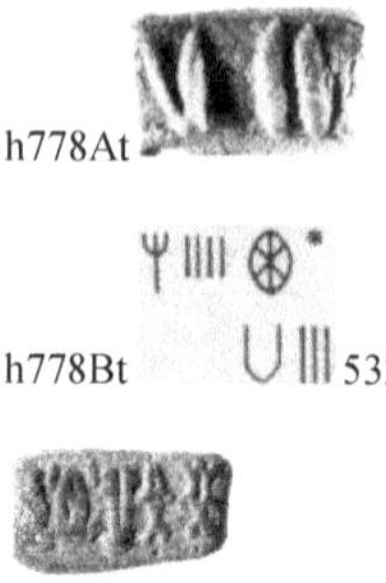
h778At

h778Bt 5322

h779At

h779Bt

h780At

h780Bt
4361

h781At

h781Bt 4670

h782At

h782Bt
5328

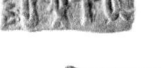

h783At

h783Bt

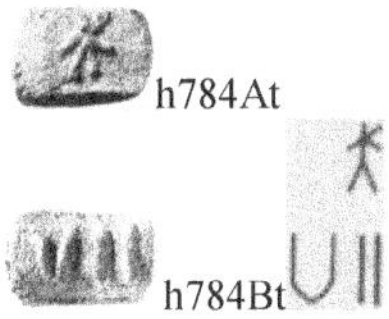

h784At

h784Bt

4364

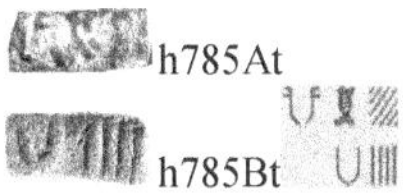

h785At

h785Bt

4681

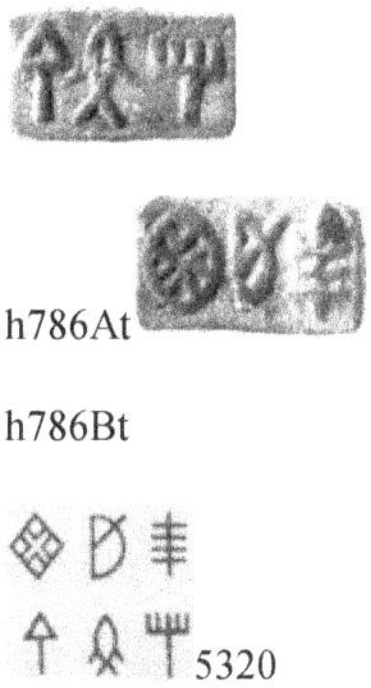

h786At

h786Bt

5320

h787At

h787Bt

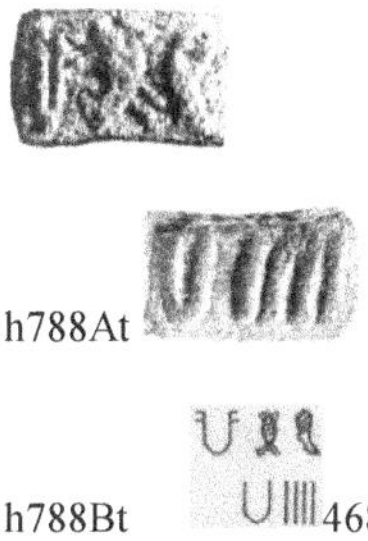

h788At

h788Bt

4683

h789At

h789Bt

4604

h790At

h790Bt

4605

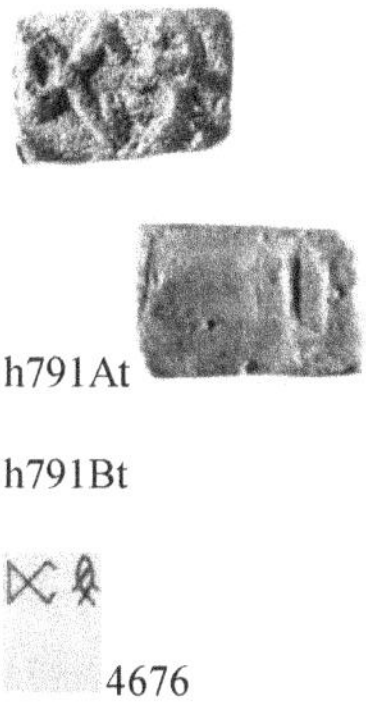

h791At

h791Bt

4676

h792At

h792Bt

4692

h793At

h793Bt

4680

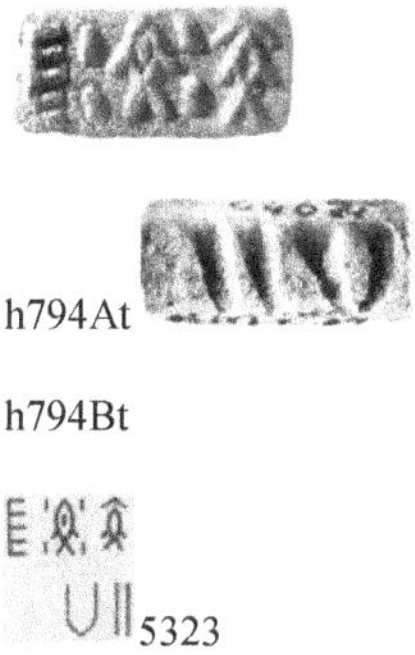

h794At

h794Bt

5323

h795At

h795Bt

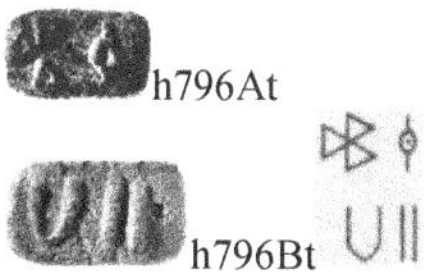

h796At

h796Bt

5327

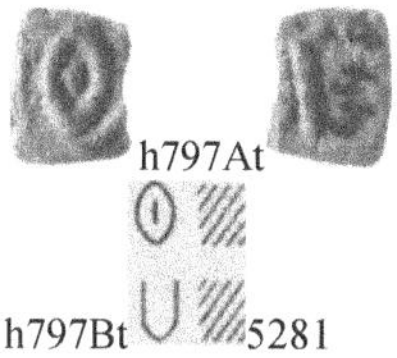

h797At

h797Bt

5281

h798At

h798Bt

4607

h799At

h799Bt

4603

h800At

h800Bt

4689

h801At

h801Bt

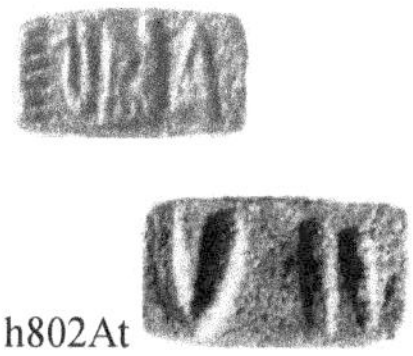

h802At

h802Bt

4679

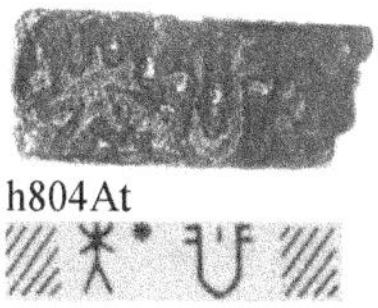

h804At

5233

h806At

h806Bt

5237

h807At

h807Bt

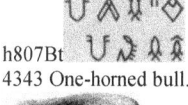

4343 One-horned bull.

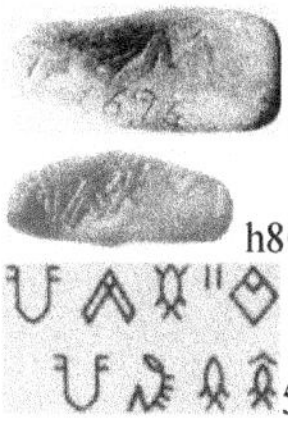

h808At

h808Bt

5238

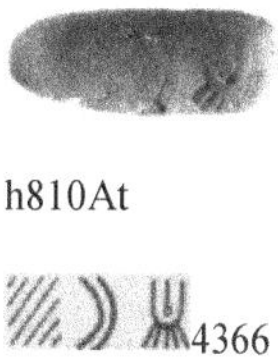

h810At

4366

h811At

h811Bt

4349

h812At 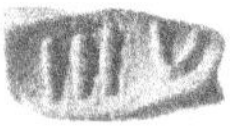

h812Bt

4686

h813At

h813Bt

4682

h814At

h814Bt 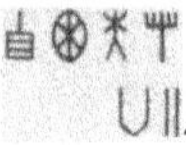
4606

h815At

h815Bt

h816At

h816Bt 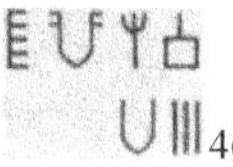
4602

h817At

h817Bt Inscribed object in the shape of a double-axe.One or more dotted circles.

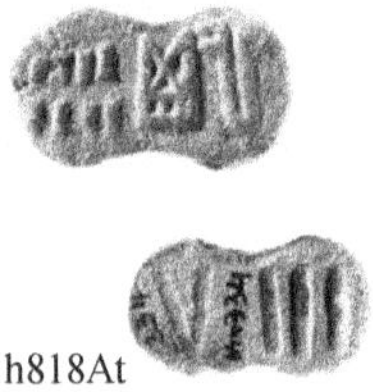

h818At

h818Bt Inscribed object in the shape of a double-axe. 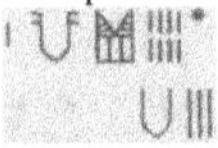
4376

h819At

h819Bt Shape of object: Blade of a weapon?

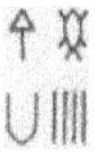
5302

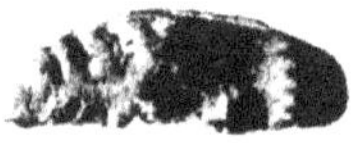

h820At

h820Bt

h821At

h821Bt Shape of object: axe.

h822At 

h822Bt Shape of object: axe.

5319

h823At

h823Bt 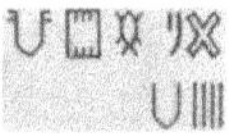
4346

h824At

h824Bt 
5278

h825At

h825Bt Shape of object: sickle? 
5324

h827At

h827Bt Shape of object: axe?

h829At

h829Bt 
5303

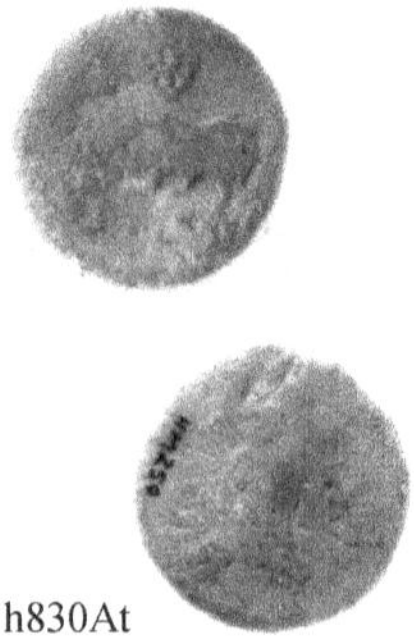

h830At

h830Bt Tablet in bas-relief. Bovid. 
4311

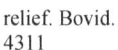

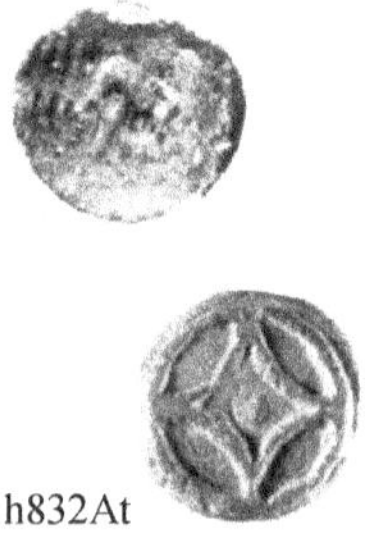

h832At

h832Bt Tablet in bas-relief
Pict-121: Lozenge within a circle with a dot in the center.

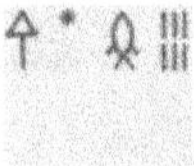
4377

h833At

h833Bt

4370

h834At

h834Bt

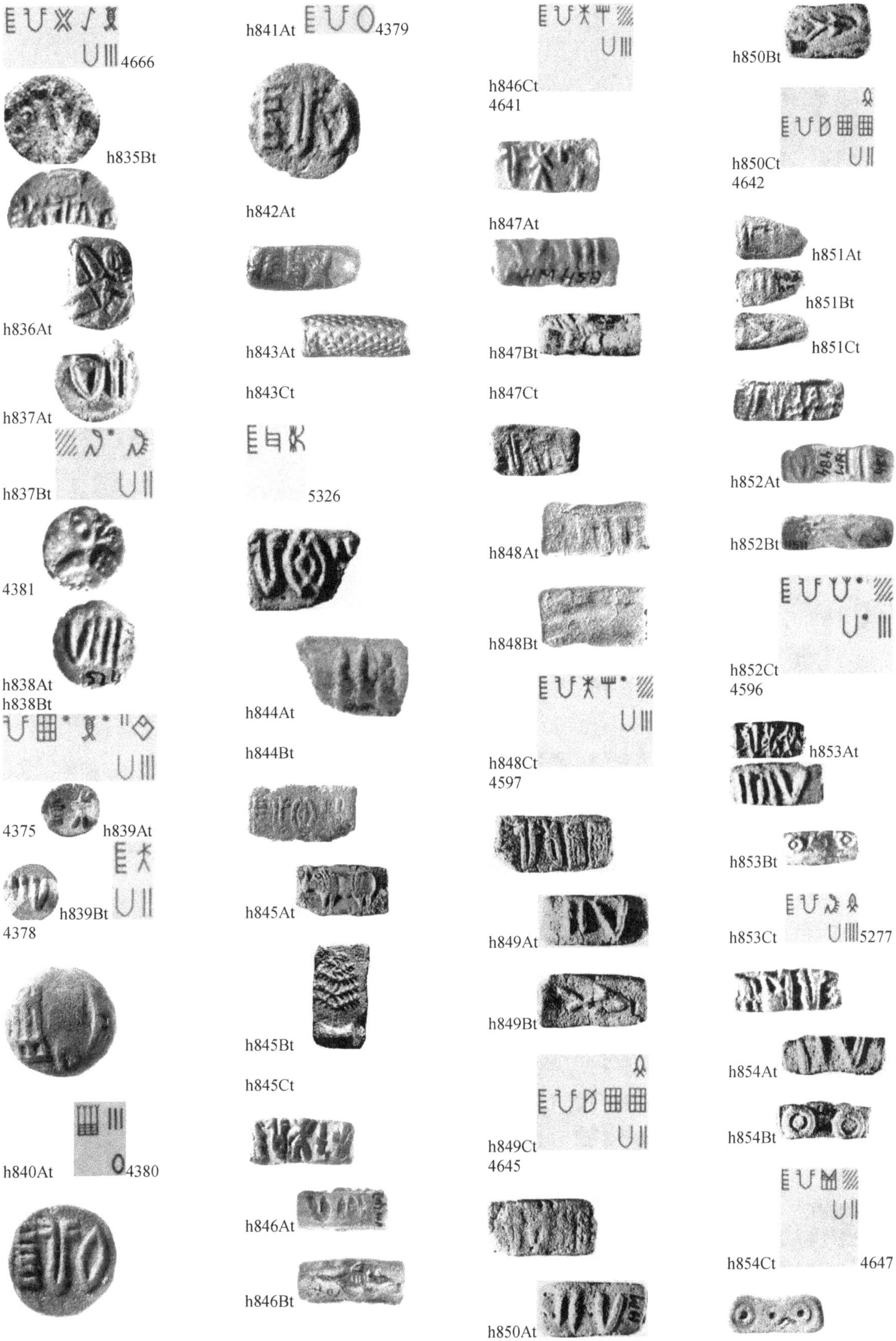

4666
h835Bt
h836At
h837At
h837Bt
4381
h838At
h838Bt
4375 h839At
h839Bt
4378
h840At 4380

h841At 4379
h842At
h843At
h843Ct
5326
h844At
h844Bt
h845At
h845Bt
h845Ct
h846At
h846Bt

h846Ct
4641
h847At
h847Bt
h847Ct
h848At
h848Bt
h848Ct
4597
h849At
h849Bt
h849Ct
4645
h850At

h850Bt
h850Ct
4642
h851At
h851Bt
h851Ct
h852At
h852Bt
h852Ct
4596
h853At
h853Bt
h853Ct 5277
h854At
h854Bt
h854Ct 4647

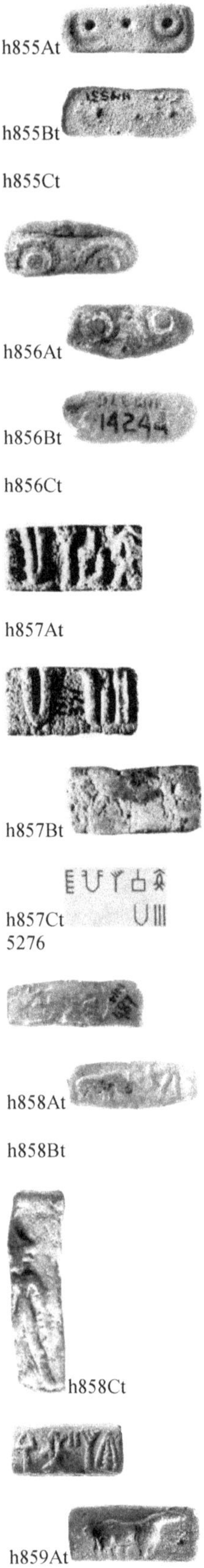

h855At

h855Bt

h855Ct

h856At

h856Bt

h856Ct

h857At

h857Bt

h857Ct
5276

h858At

h858Bt

h858Ct

h859At

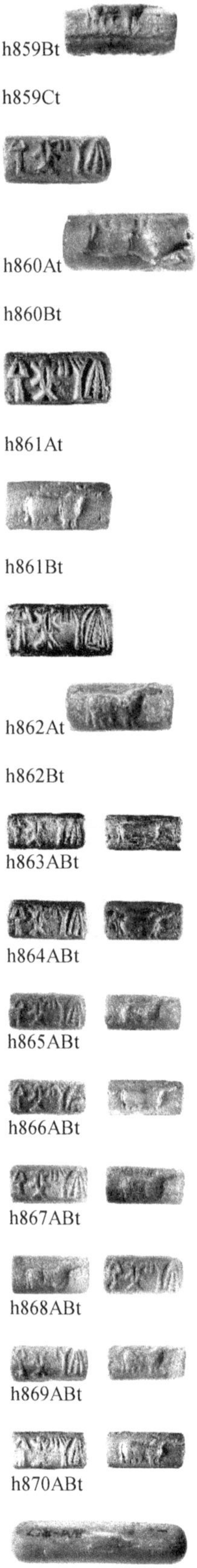

h859Bt

h859Ct

h860At

h860Bt

h861At

h861Bt

h862At

h862Bt

h863ABt

h864ABt

h865ABt

h866ABt

h867ABt

h868ABt

h869ABt

h870ABt

h871Bt

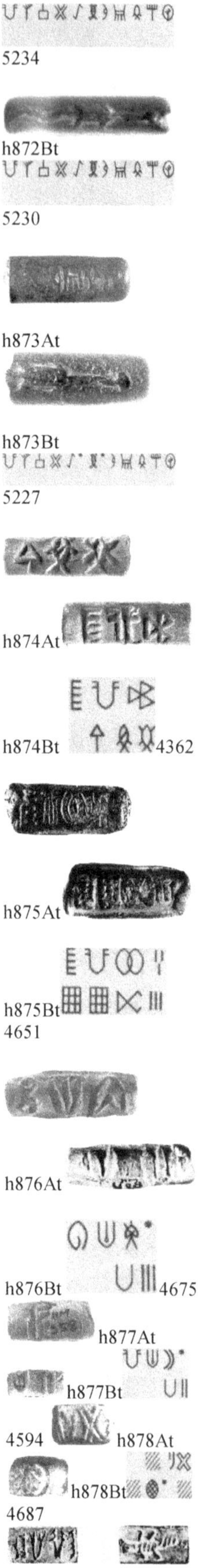

5234

h872Bt

5230

h873At

h873Bt

5227

h874At

h874Bt 4362

h875At

h875Bt
4651

h876At

h876Bt 4675

h877At

h877Bt

4594 h878At

h878Bt
4687

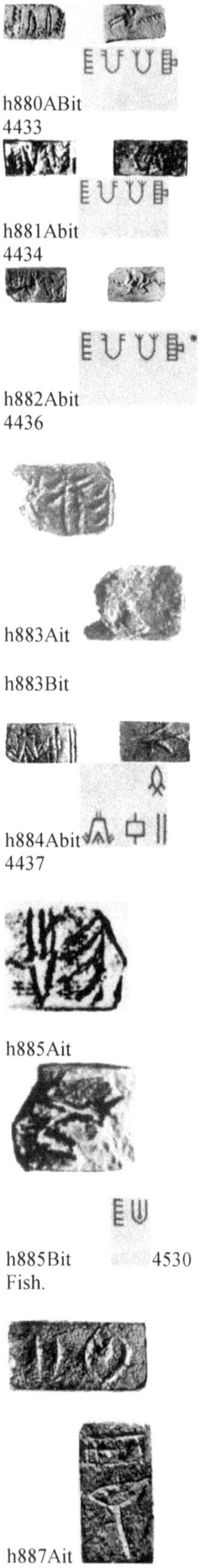

h879Abit

h880ABit
4433

h881Abit
4434

h882Abit
4436

h883Ait

h883Bit

h884Abit
4437

h885Ait

h885Bit 4530
Fish.

h887Ait
h887Bit

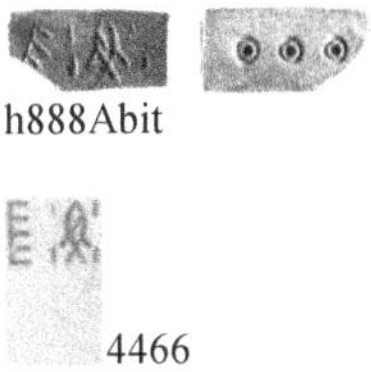
h888Abit
4466

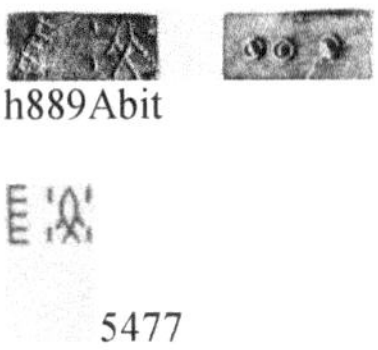
h889Abit
5477

h890ABit
4446

h891ABit

h892ABit 4451

h893Ait
h893Bit
4522

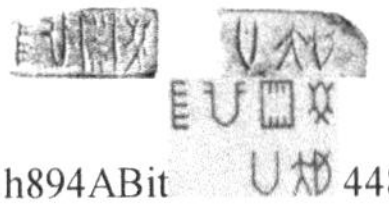
h894ABit 4487

h895Ait
h895Bit

h896ABit
4480

h897ABit

h898ABit
4506

h899Ait

h899Bit 4471

h900Ait
h900Bit 4455

h901Ait
h901Bit 4460

h902Ait
h902Bit 4535

h903Ait
h903Bit
4485

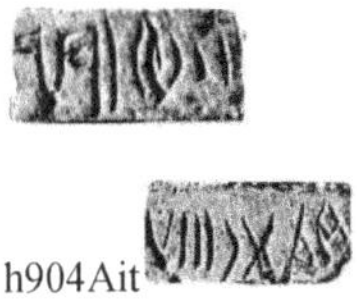
h904Ait
h904Bit

4477

h905ABit
4449

h906Ait
h906Bit

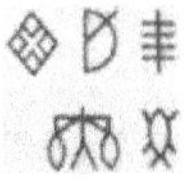
5494

h907Ait
h907Bit
4537 The second sign on h907Ait may be a ligatured fish?

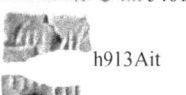

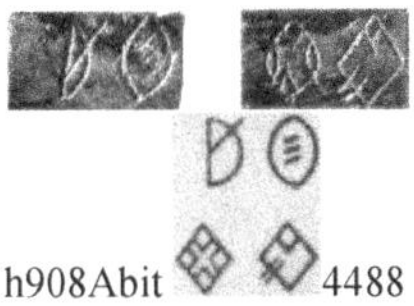
h908Abit 4488

h909ABit
5325

h910ABit 4470

h911Ait
h911Bit

4486
h912Abit

5461

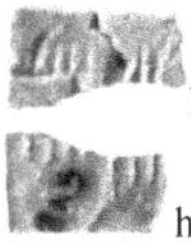
h913Ait
h913Bit

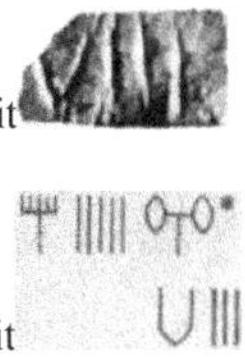
h914Ait
h914Bit
4483

h916Ait
h916Bit 4456

h917Ait
h917Bit 4472

h918Ait

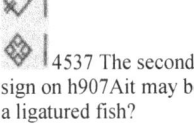

h918Bit 4481

h919Ait

h919Bit

h920Ait

h920Bit

4527

h921ABit 4514

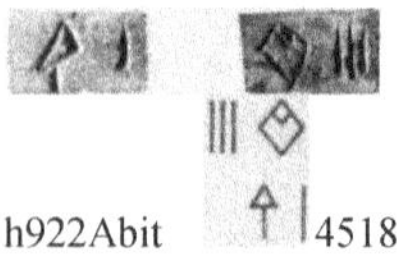
h922Abit 4518

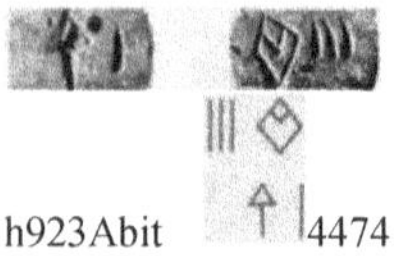
h923Abit 4474

h924ABit

h925Abit 4512

h926Abit
4519

h927Ait

h927Bit 4502

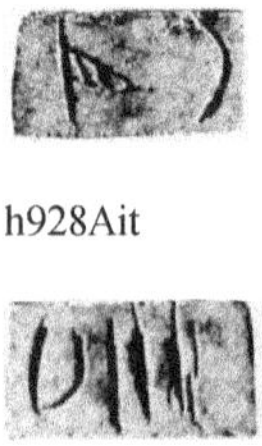
h928Ait

h928Bit

h929Ait

h929Bit

h930Ait

h930Bit 4520

h931Ait

h931Bit

4511

h932Ait

h932Bit

4403

h933Ait

h933Bit

4516

h934Ait

h934Bit

h935Ait

h935Bit

4509

h936Ait

h936Bit 5405

h937Ait

h937Bit 5458

h938Ait

h938Bit

h939Ait

h939Bit

h940Ait

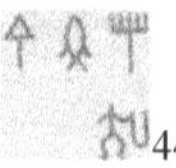
h940Bit 4453

h941Ait

h941Bit 4464

h942Ait

h942Bit

4490

h943Ait

h943Bit 4461

h944Ait

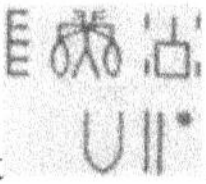
h944Bit 4475

h945Ait

h945Bit 4503

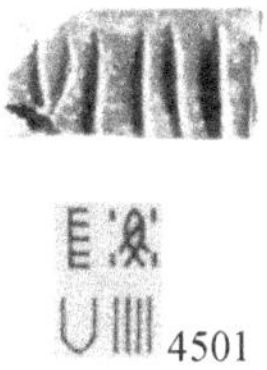
h946Ait
h946Bit 4501

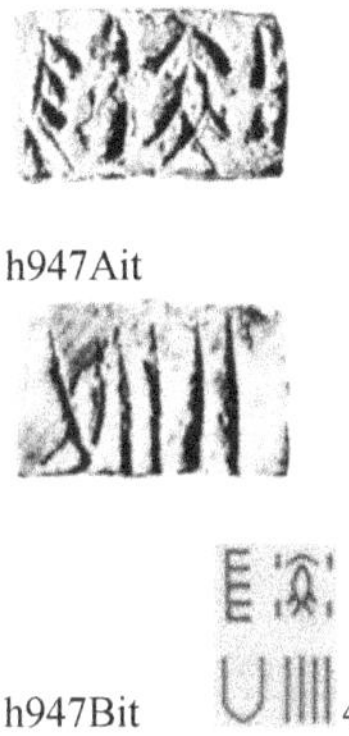
h947Ait
h947Bit 4493

h948Abit 4489

h949Abit 4479

h950ABit
4463

h951Ait
h951Bit
4498

h952Ait
h952Bit 4469

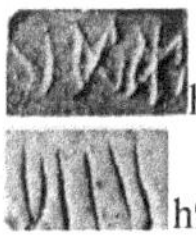
h953Ait
h953Bit

h954Ait
4467

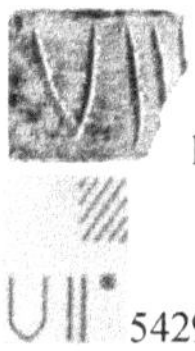
h955Bit
5429

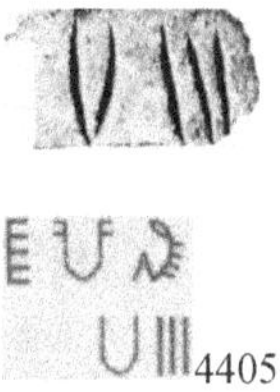
h959Ait
h959Bit 4405

h960Ait
h960Bit 4543

h961Ait
h961Bit
5449

h962Ait
h962Bit

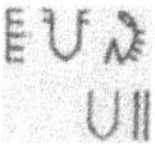
4548

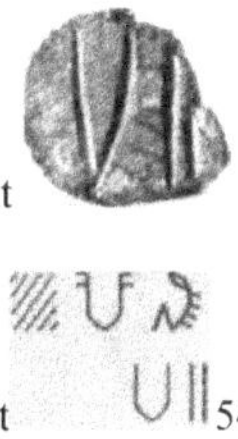
h963Ait
h963Bit 5420

h964Ait
h964Bit 5456

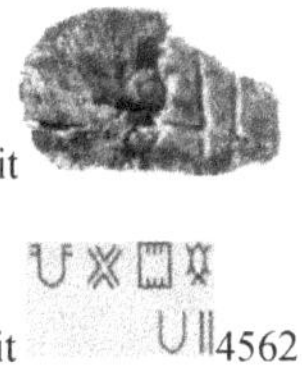
h965Ait
h965Bit 4562

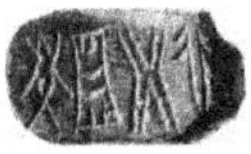

h966Ait
h966Bit 5479

h967Ait
4563

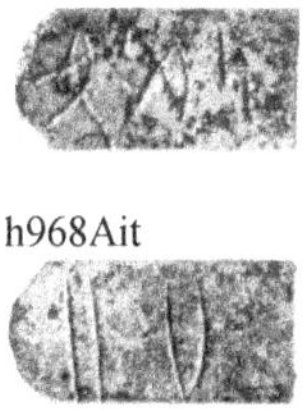
h968Ait
h968Bit

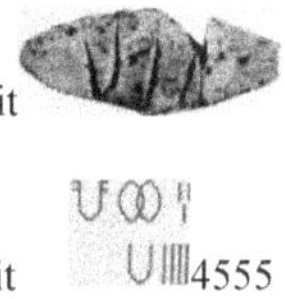
h969Ait
h969Bit 4555

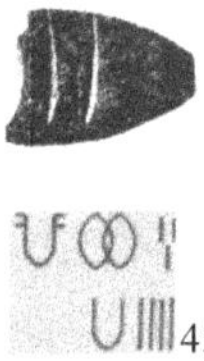
h970Ait
h970Bit 4553

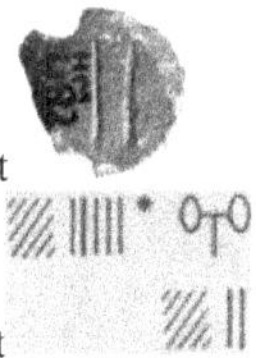
h971Ait
h971Bit
4557 Shape of object: double-axe?

h972Ait
h972Bit
4418 Pict-128: Inscribed object in the shape of a leaf? Dotted circles on obverse.

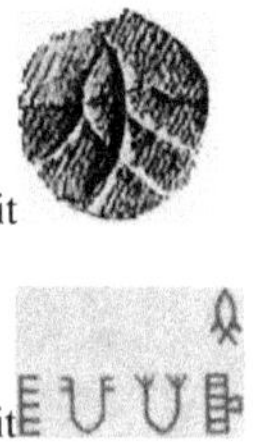
h973Ait
h973Bit
4411

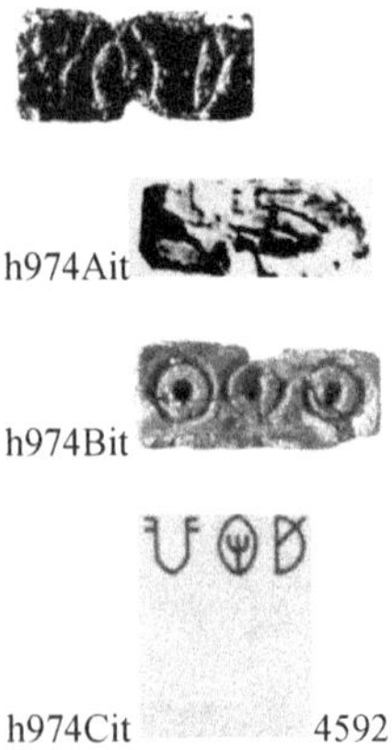
h974Ait
h974Bit
h974Cit 4592

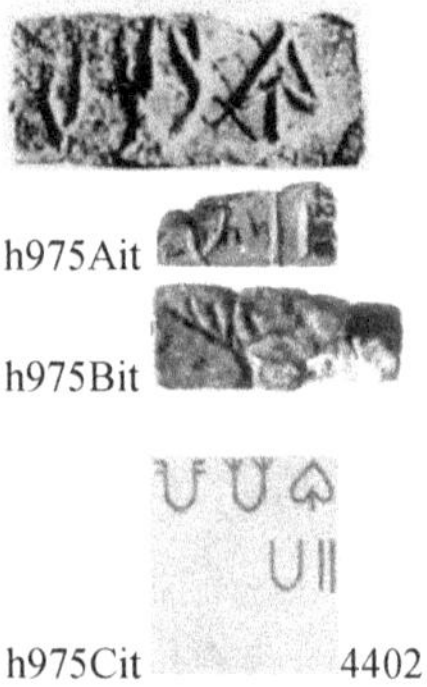
h975Ait
h975Bit
h975Cit 4402

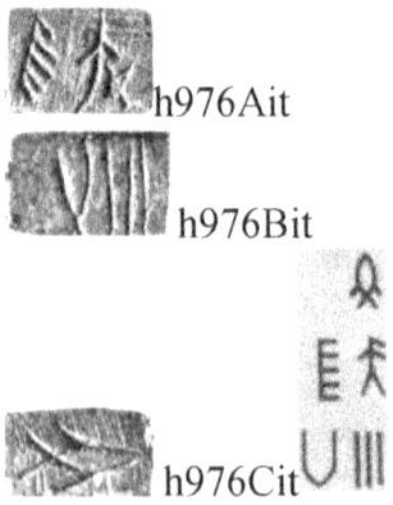
h976Ait
h976Bit
h976Cit
4588

h977Ait
h977Bit

h977Cit 4591

h978Ait
h978Bit
h978Cit 5412

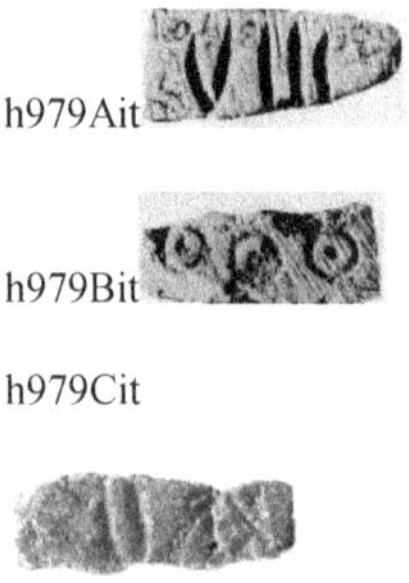
h979Ait
h979Bit
h979Cit

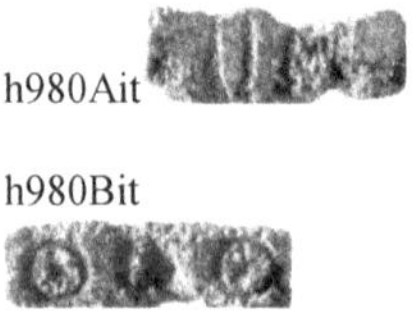
h980Ait
h980Bit
h980Cit

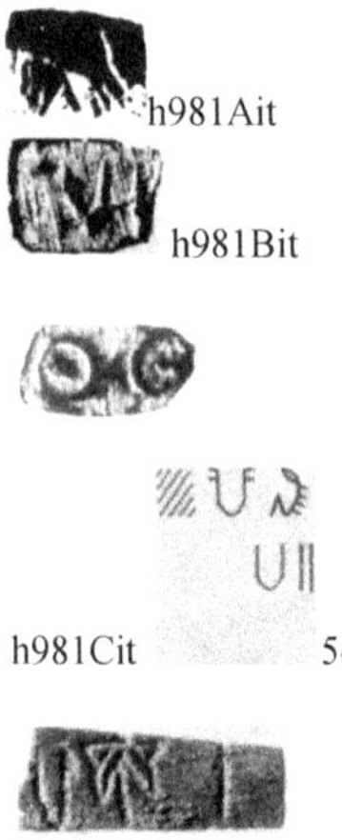
h981Ait
h981Bit
h981Cit 5415

h982Ait

h982Bit
h982Cit
m4574

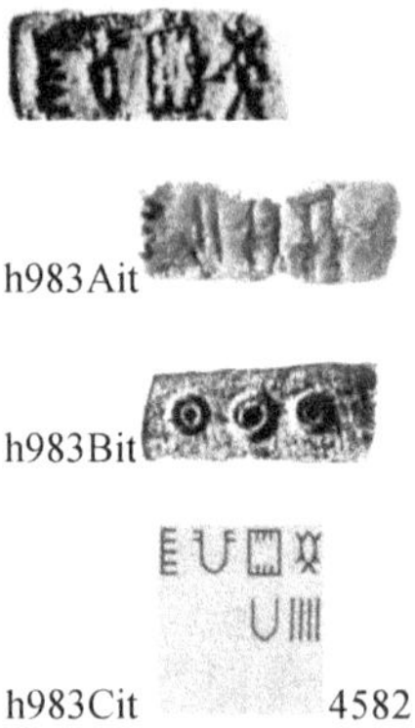
h983Ait
h983Bit
h983Cit 4582

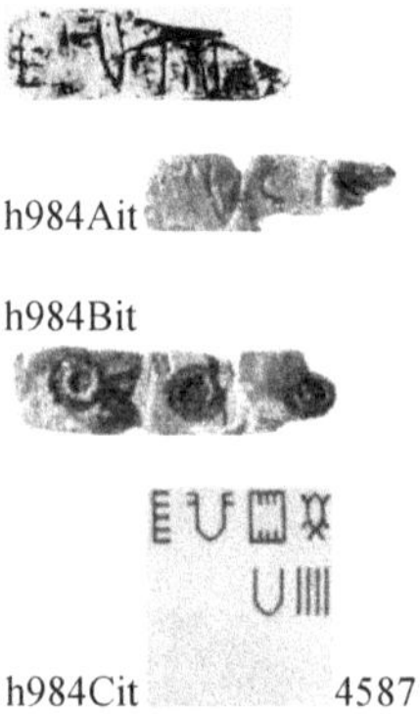
h984Ait
h984Bit
h984Cit 4587

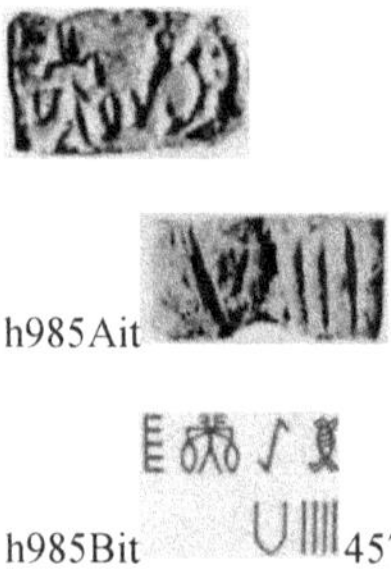
h985Ait
h985Bit 4577

h987Ait
h987Bit

h987Cit 4586

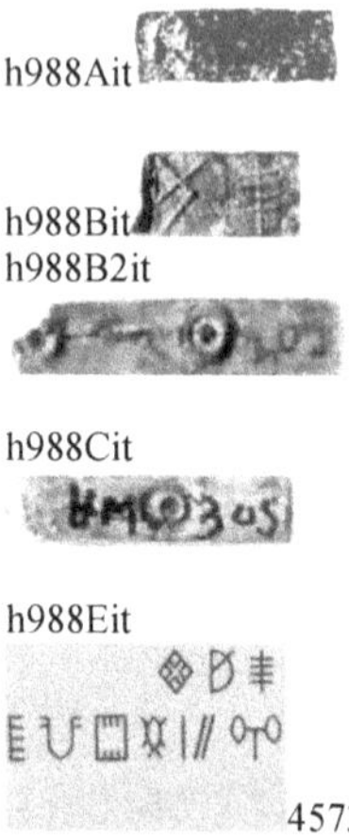
h988Ait
h988Bit
h988B2it
h988Cit
h988Eit 4573

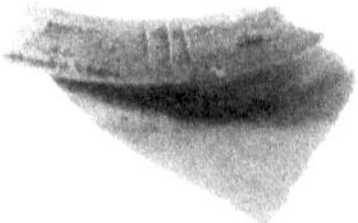
h990

h992

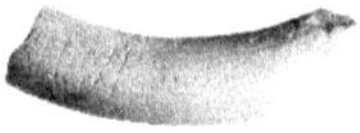
h994

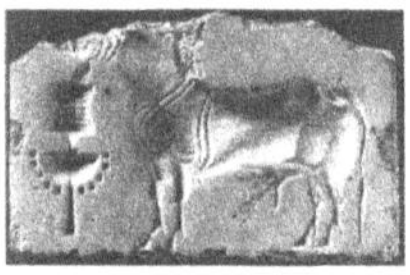
h1020

h1021

h1022

h1023

h1024

h1025a

h1027a

h1028

h1029a

h1030a

h1031

h1032a

h1033a

h1035

h1036

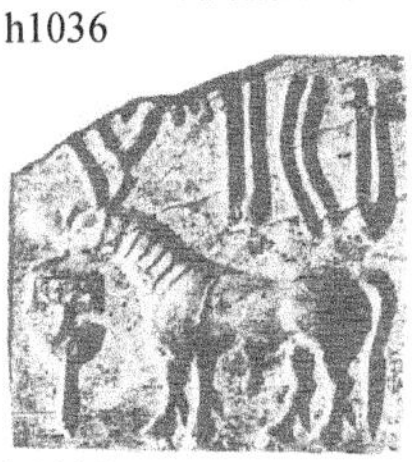

h1037

h1038

h1042a

h1043a

h1044a

h1045a

h1046

h1047a

h1048

h1049a

h1050

h1051

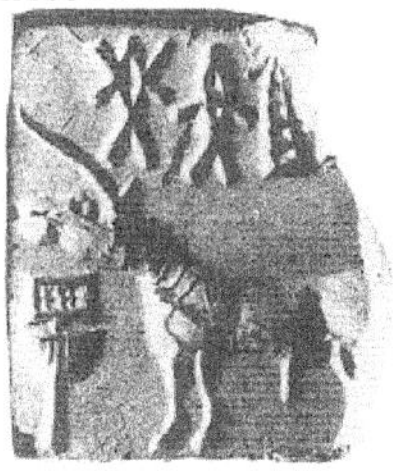

h1052

h1053a

h1056a

h1058a

h1059

h1064

h1065

h1066a

h1067a

h1068

h1071

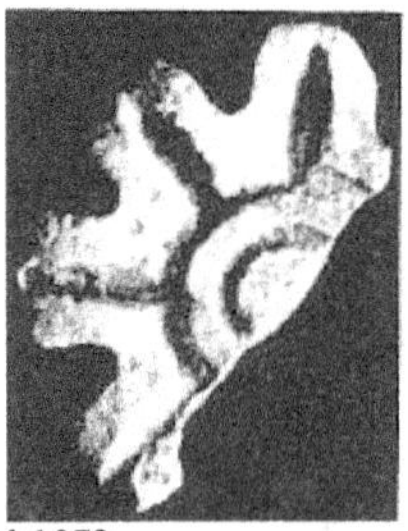
h1072

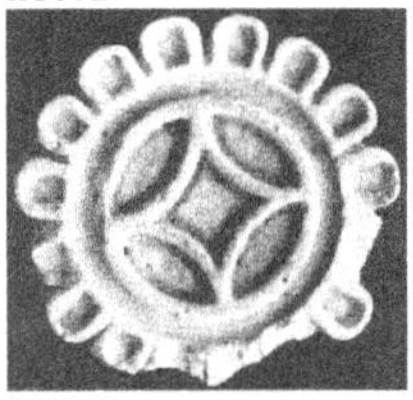
h1073

h1075

h1076

h1077

h1079a

h1080a

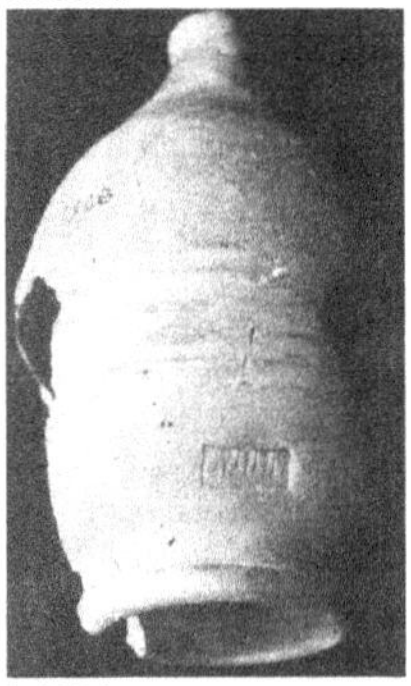
h1081

h1082

h1083

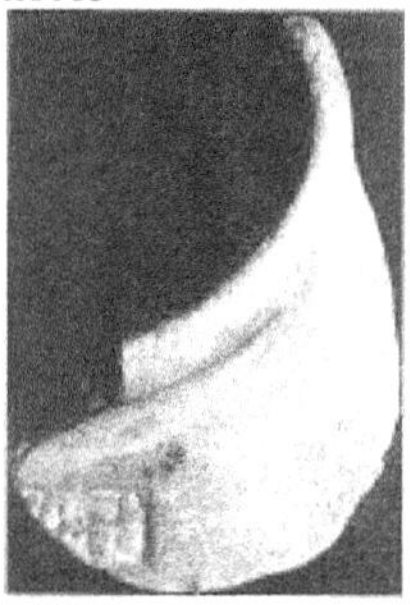
h1084

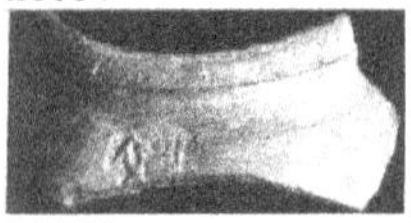
h1085

h1086

h1087

h1091

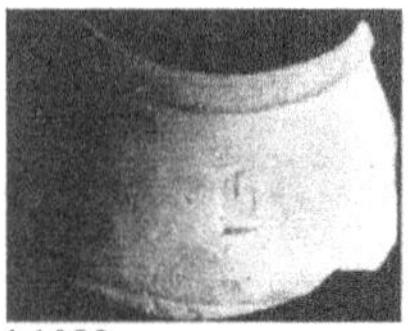
h1092

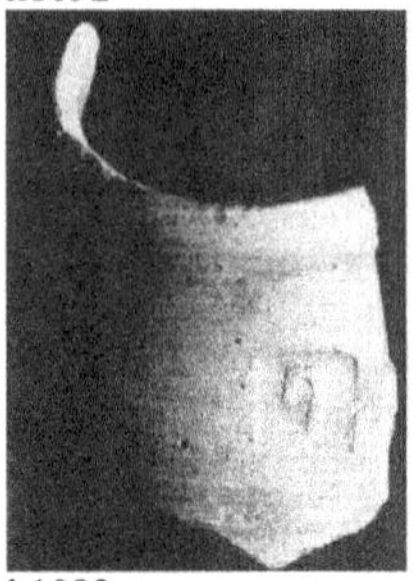
h1093

h1094

h1097

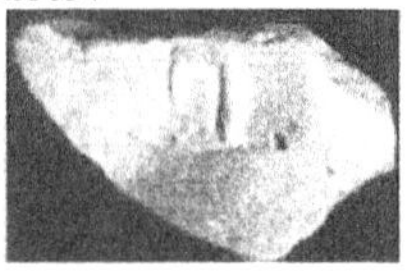
h1098

h1100A

h1100B

h1101A

h1101B

h1102A
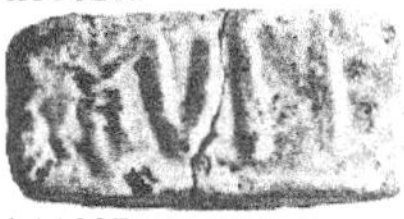
h1102B

h1103A

h1103B

h1104A

h1104B

h1105A

h1105B

h1107A

h1107B

h1108A
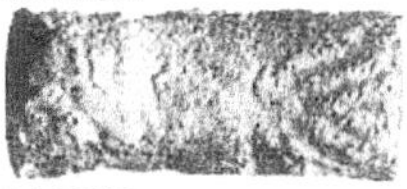
h1108B

h1109A

h1109B

h1113A

h1113B

h1114B

h1115A

h1116A

h1116B

h1117A

h1121A

h1121B
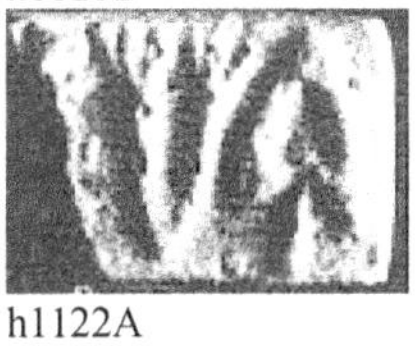
h1122A

h1122B

h1123A
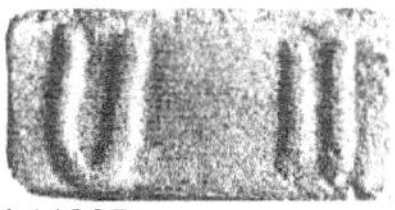
h1123B

h1124A

h1124B

h1126A
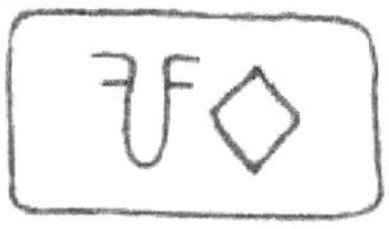
h1126B

h1130A

h1130B

h1131A

h1131B

h1133A

h1133B

h1134A

h1134B

h1137A

h1138A

h1138B
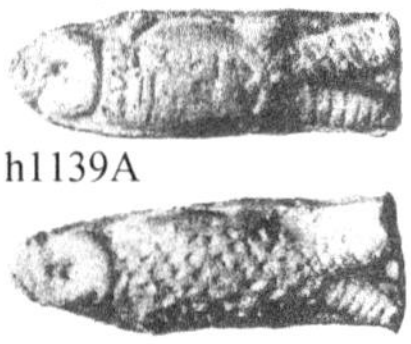
h1139A
h1139B

h1140A
h1140B

h1141A
h1141B

h1142A
h1142B

h1144A

h1144B

h1146A

h1146B

h1148A

h1149A

h1150ABC

h1151A

h1151C

h1152A

h1152B

h1155A

h1155B

h1158A

h1158B

h1159A

h1159B

h1160A

h1160B

h1165A
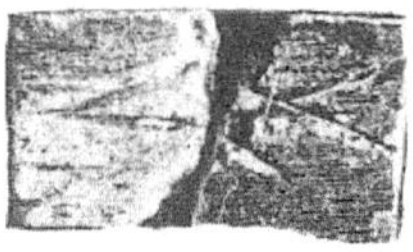
h1165B

h1166A
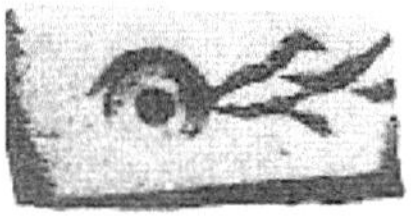
h1166B

h1178a

h1178b

h1181A

h1182A
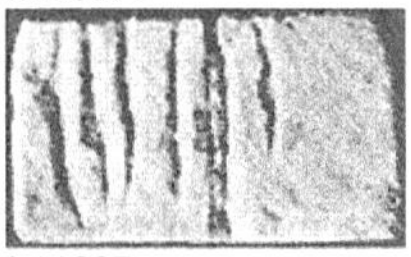
h1182B

h1184A

h1184B

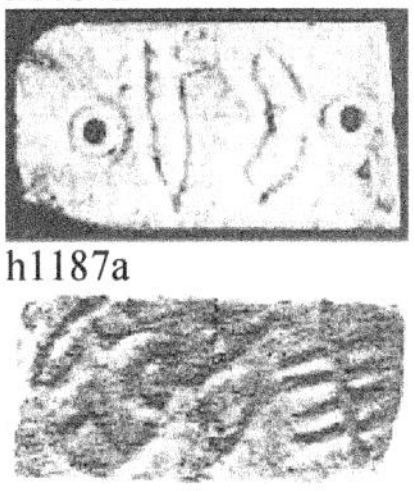
h1187a

h1187b

h1188A

h1188B

h1189A

h1189B

h1190A

h1190B

h1191A

h1191B

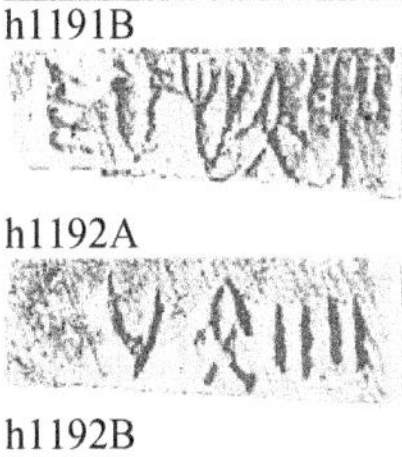
h1192A

h1192B

h1198A

h1198B

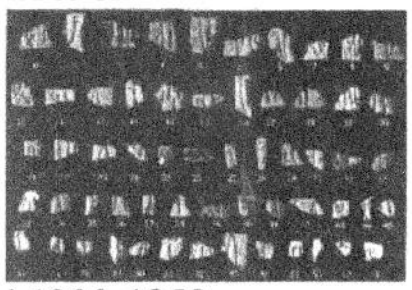
h1200-1258

h1261A

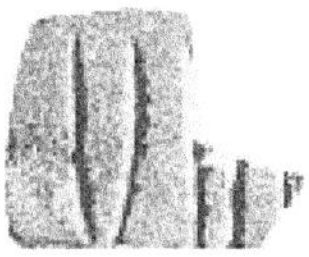
h1261B

h1272A

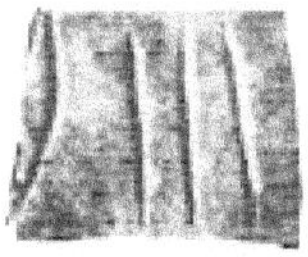
h1272B

h1273A

h1273B

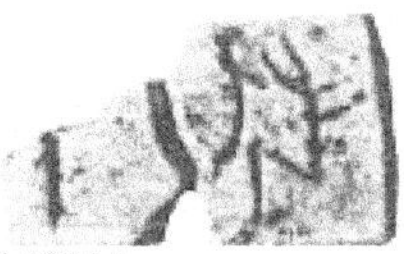
h1274A

h1274B

h1275A

h1275B

h1284A

h1284B

h1287A

h1287B

h1289A

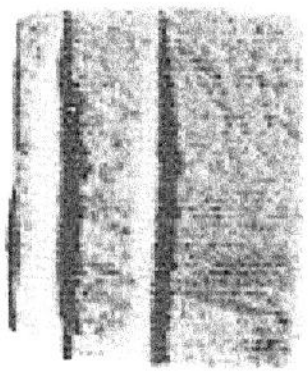
h1289B

h1293A

h1293B

h1294A

h1294B

h1302A

h1302B

h1303A

h1303B

h1304A

h1304B

h1308a

h1308b

h1309A

h1309B

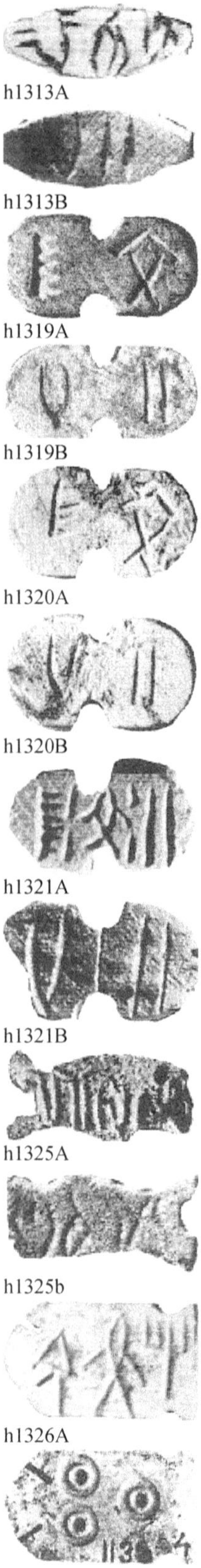

h1313A
h1313B
h1319A
h1319B
h1320A
h1320B
h1321A
h1321B
h1325A
h1325b
h1326A

h1326B
h1329A
h1329B
h1330A
h1330B
h1331A
h1331B

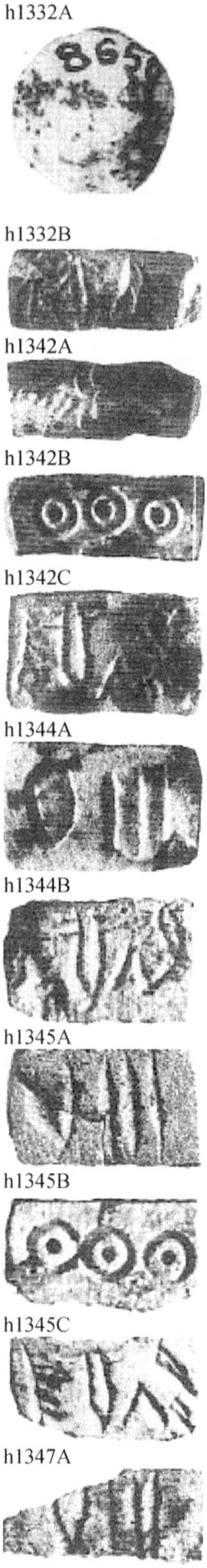

h1332A
h1332B
h1342A
h1342B
h1342C
h1344A
h1344B
h1345A
h1345B
h1345C
h1347A
h1347B

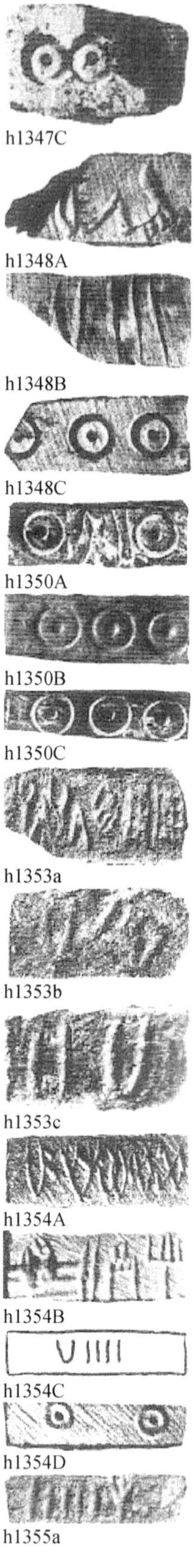

h1347C
h1348A
h1348B
h1348C
h1350A
h1350B
h1350C
h1353a
h1353b
h1353c
h1354A
h1354B
h1354C
h1354D
h1355a

h1355B

h1355c

h1355d

h1344C

h1357A

h1357b

h1366a

h1388A

h1392A

h1393A

h1394A
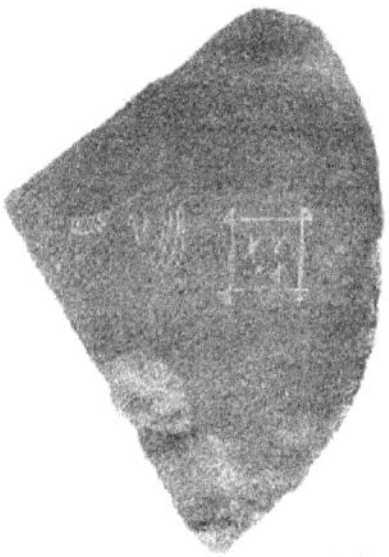
h1397A

h1403A

h1404A

h1406A

h1407A

h1408A

h1410A
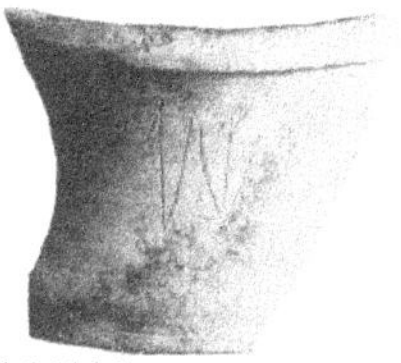
h1411A

h1412A

h1421A

h1422A

h1424A

h1431A

h1433A

h1434A

h1441A

h1444A

h1461A
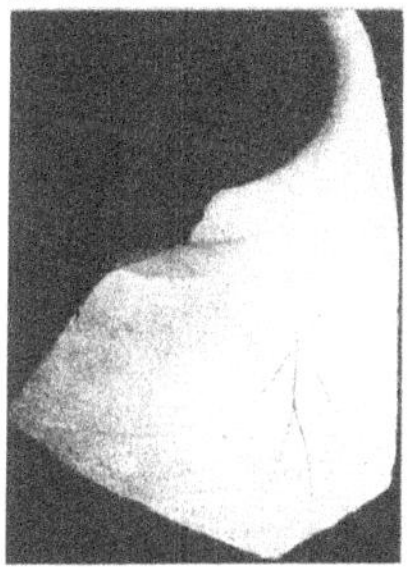
h1462A

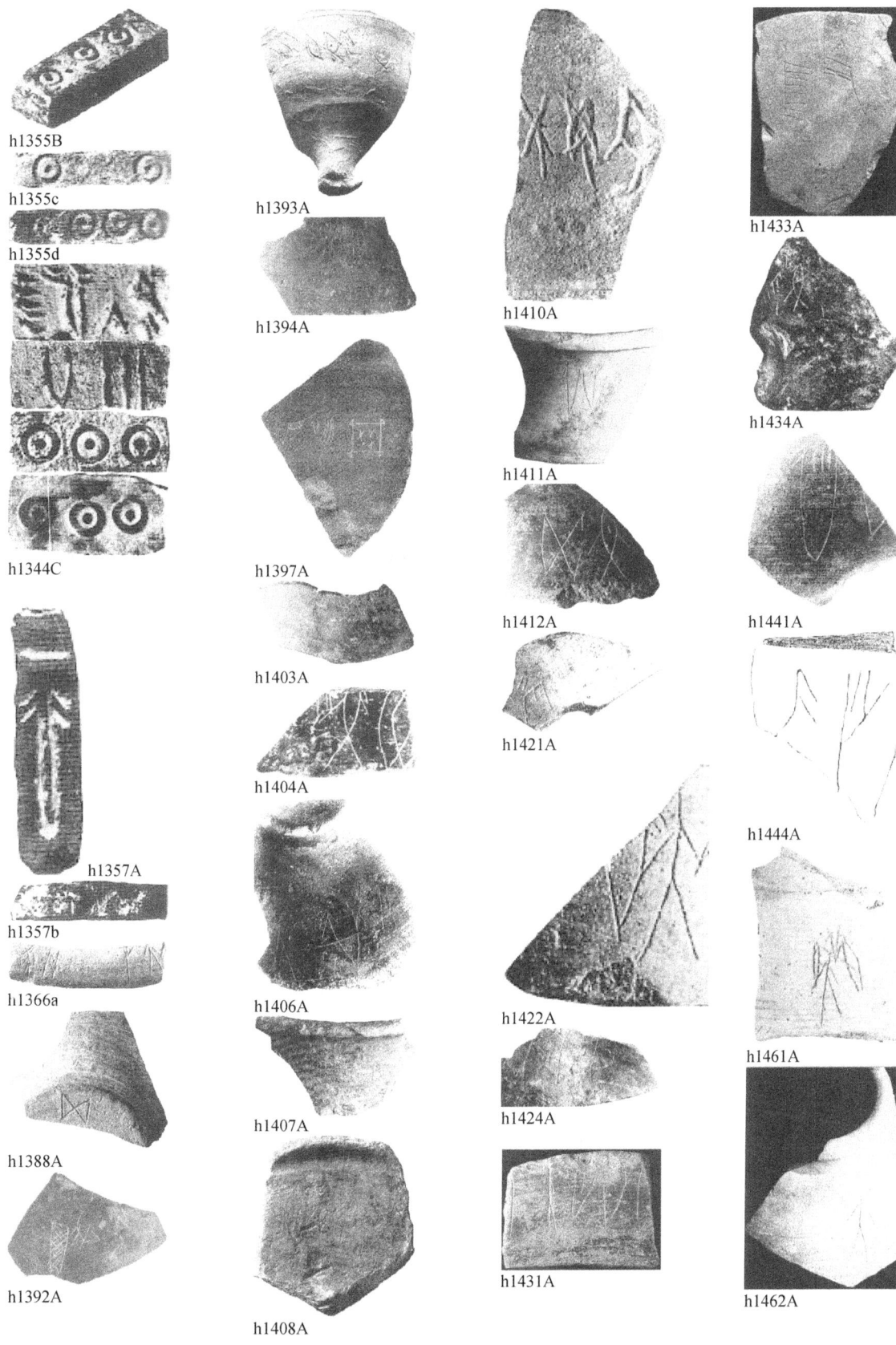

h1464A

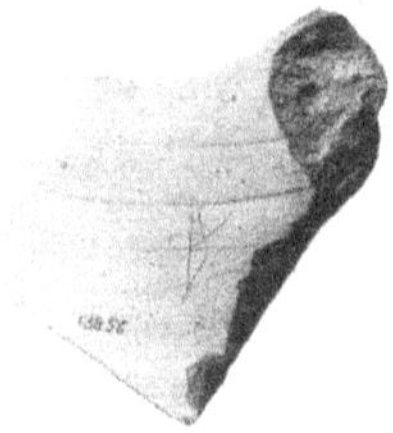

h1467A

h1468A

h1471A

h1481A

h1487A

h1491A

h1501A

h1506A

h1507A

h1513A

h1516A

h1517A

h1518copperaxe

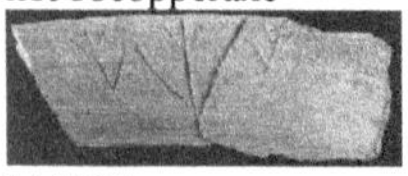

h1522A

h15533A

h1534A

h1535A

h1536A

h1537A

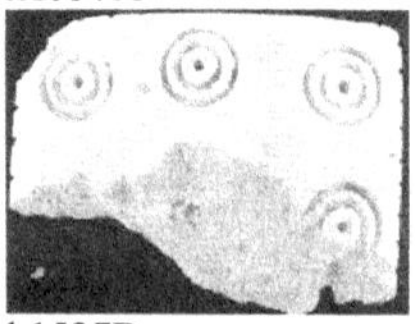

h1537B

h1538A

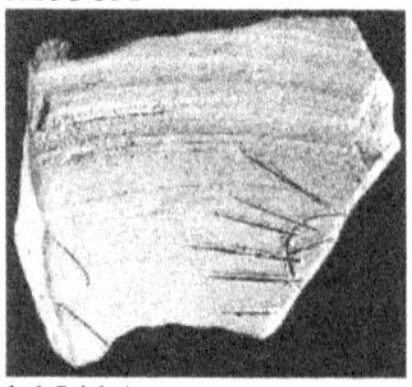

h1541A

h1544A

h1545A

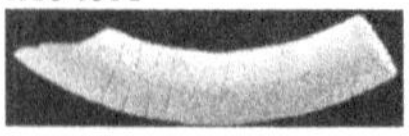

h1547A

h1559A

h1586A

h1587A

h1657A

h1662A

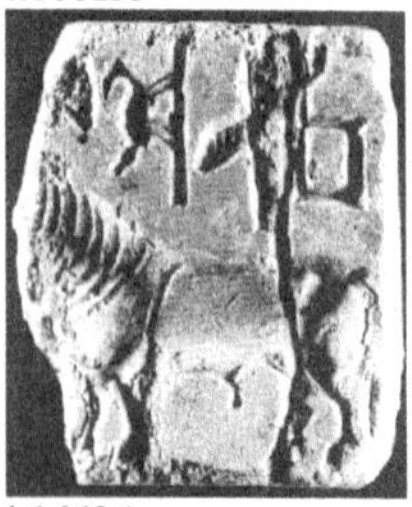

h1663A

h1664A

h1666A

h1667A
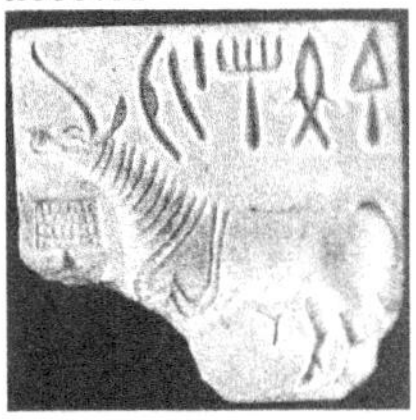
h1669A

h1670A

h1671A

h1672A

h1673A

h1676A
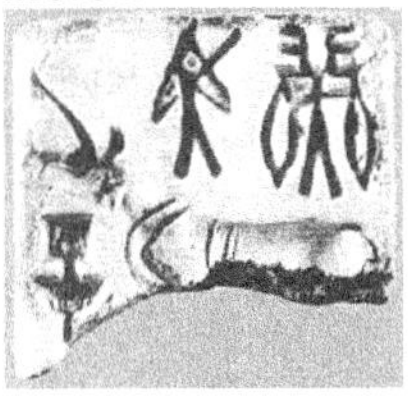
h1677A

h1678A

h1679A

h1680A

h1681A
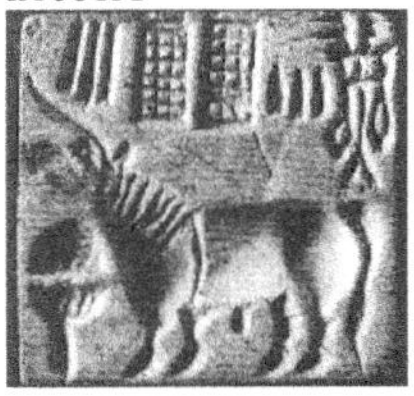
h1682A

h1684A

h1685A

h1687A
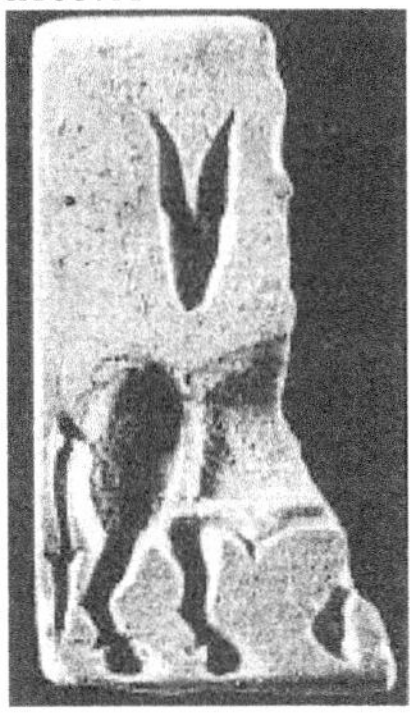
h1688A

h1690A

h1691A

h1692A

h1694A

h1695

h1696

h1697

h1698

h1699

h1700

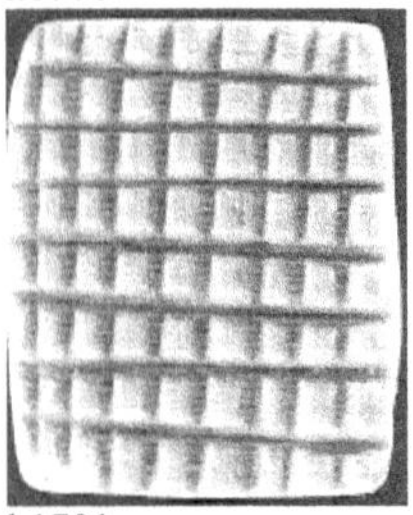
h1701

h1702

h1703

h1704

h1705

h1706

h1707

h1708

h1709

h1710

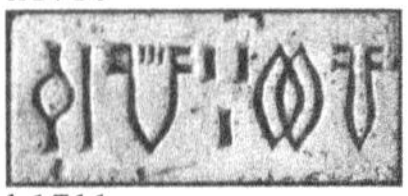
h1711

h1712

h1713

h1714

h1715

h1716

h1719A

h1719B

h1720A

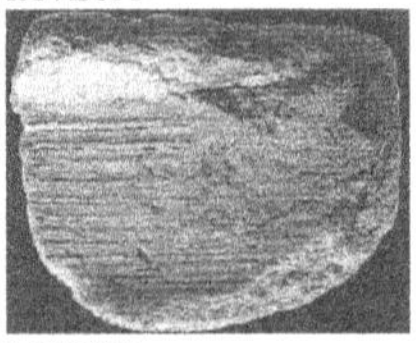
h1720B

h1721A

h1721B

h1722A

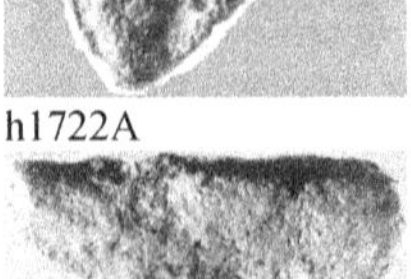
h1723

h1724

h1725A

h1726

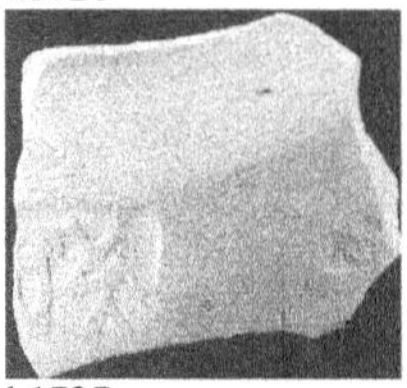
h1727

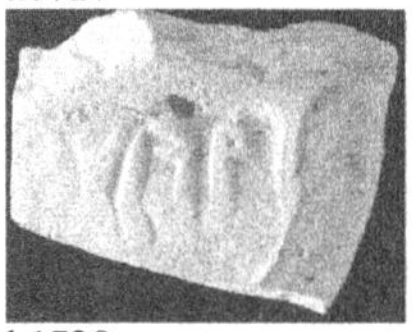
h1728

h1729

h1731

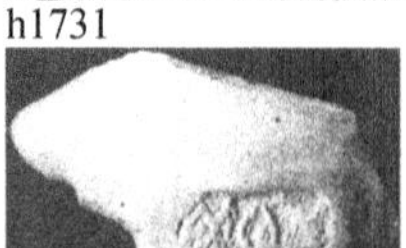
h1732

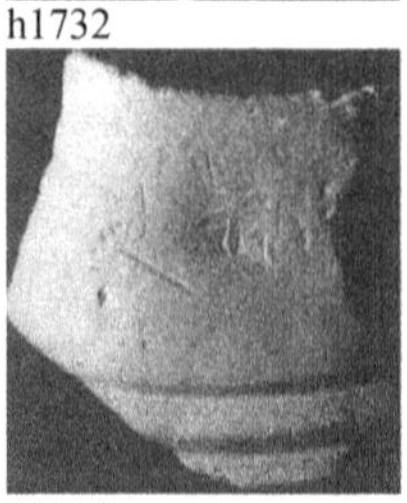
h1733

h1734

h1735

h1736

h1737

h1739

h1740

h1742

h1743

h1744

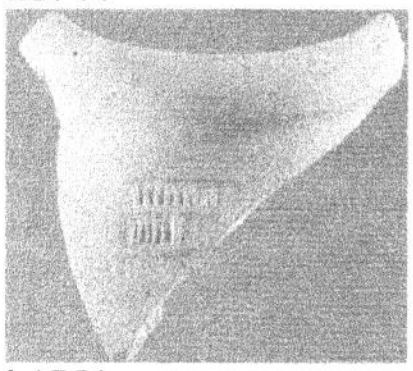
h1751

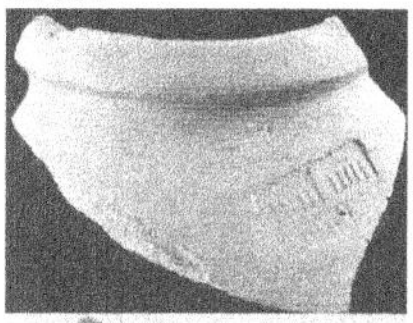

h1753

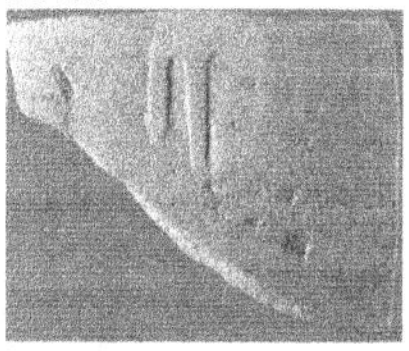
h1756

h1757

h1758

h1759

h1760

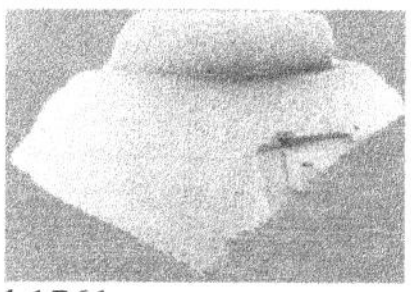
h1761

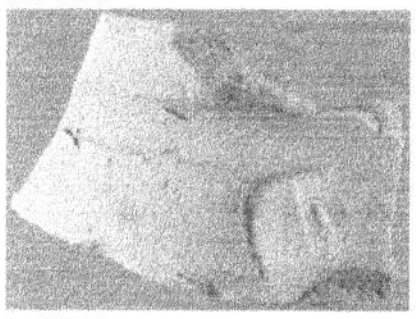
h1762

h1767A

h1767B

h1768A

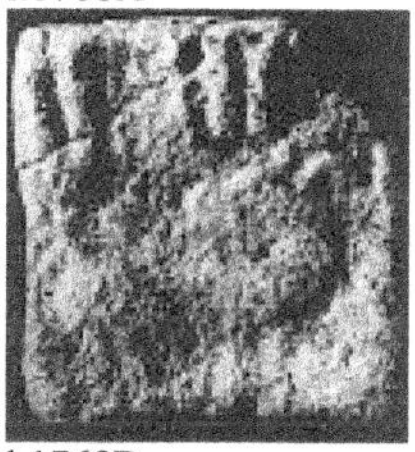
h1768B

h1770A

h1770B

h1771A

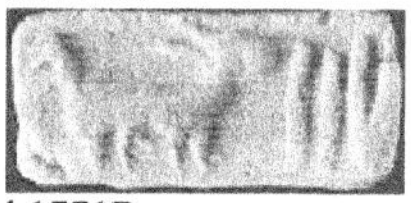
h1771B

h1772A

h1772B

h1773A

h1773B

h1774A

h1774B

h1775A

h1775B

h1776A

h1776B

h1777A

h1777B

h1778A

h1778B

h1779A

h1779B

h1780A

h1780B

h1781A

h1781B

h1782A

h1783A
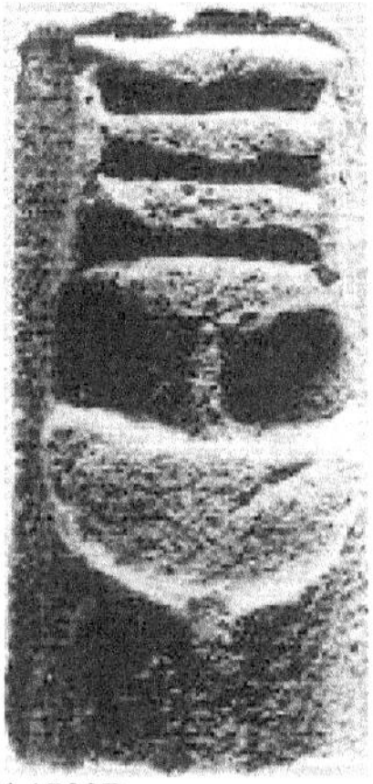
h1783B

h1785A

h1785B
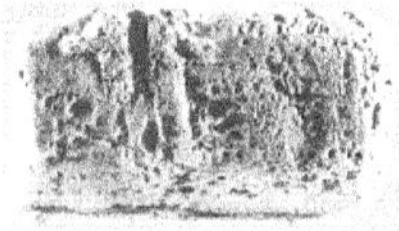
h1786A

h1786B
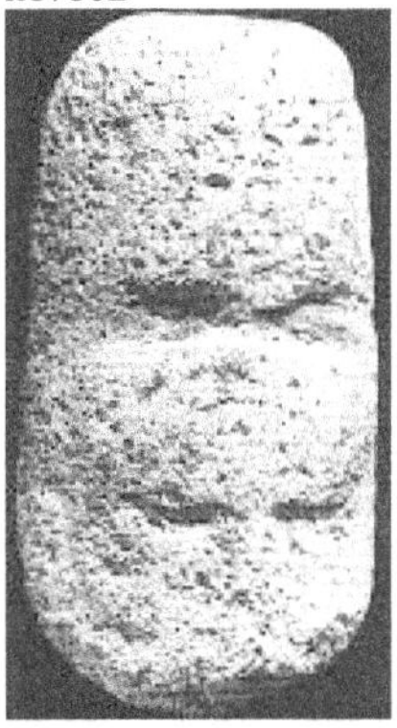
h1787A

h1787B

h1788A

h1788B

h1791A

h1791B

h1792A

h1792B
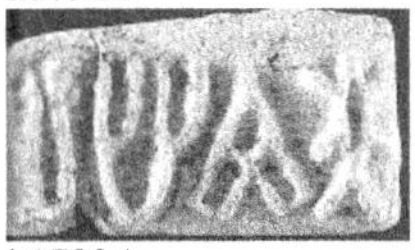
h1793A

h1793B

h1796A

h1796B

h1797A

h1797B

h1799A

h1800A

h1800B

h1801A

h1801B

h1802A

h1802B

h1803A
h1803B

h1804A

h1804B
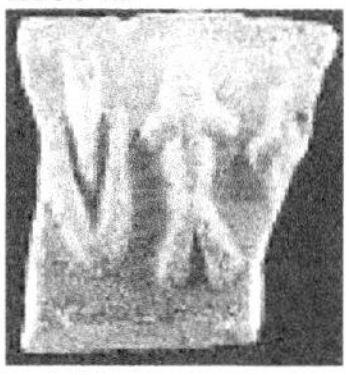
h1805A
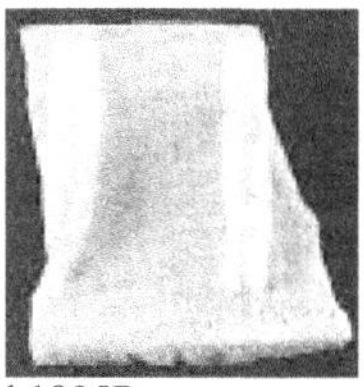
h1805B

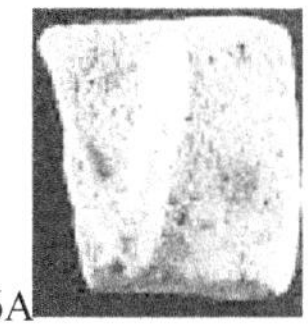
h1806A

h1806B

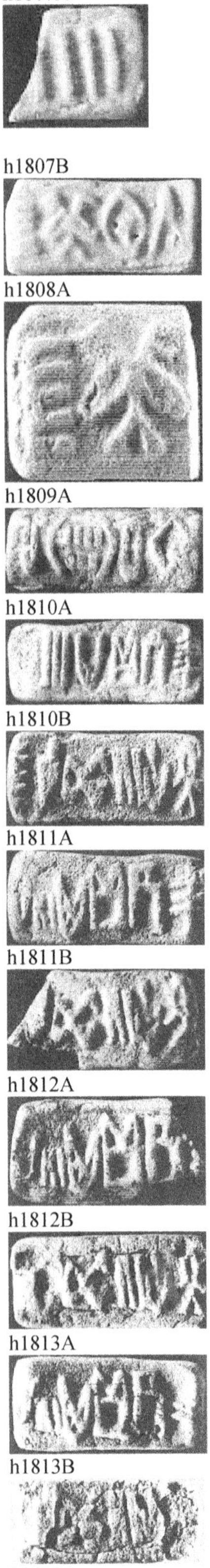

h1807A
h1807B
h1808A
h1809A
h1810A
h1810B
h1811A
h1811B
h1812A
h1812B
h1813A
h1813B
h1815A

h1815B
h1816A
h1816B
h1817A
h1817B
h1818A
h1818B
h1819A
h1819B
h1820A
h1820B
h1821A

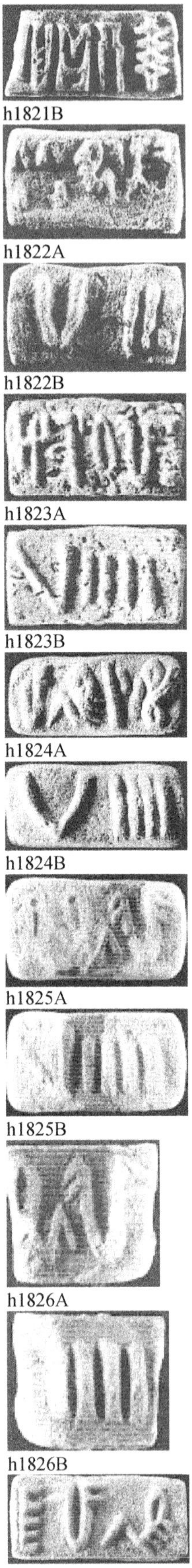

h1821B
h1822A
h1822B
h1823A
h1823B
h1824A
h1824B
h1825A
h1825B
h1826A
h1826B

h1827A
h1827B
h1829A
h1829B
h1830A
h1830B
h1831A
h1831B
h1832A
h1832B
h1833A
h1833B
h1834A

h1834B

h1835A
h1835B

h1836A
h1836B
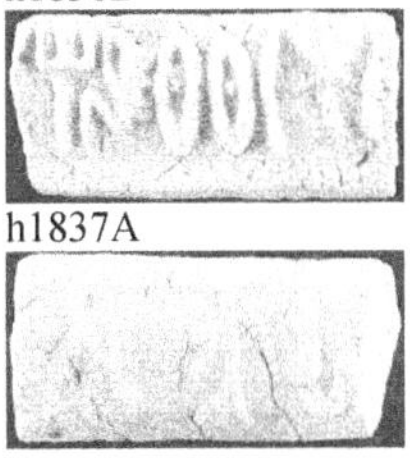
h1837A
h1837B

h1839A
h1839B

h1840A
h1840B

h1841A
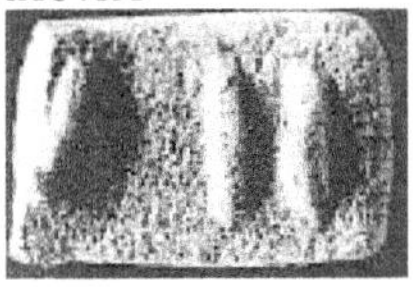
h1841B

h1842A
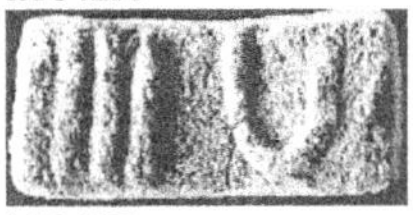
h1842B

h1843A

h1843B

h1844A

h1844B

h1845A

h1845B

h1846A

h1846B

h1848A
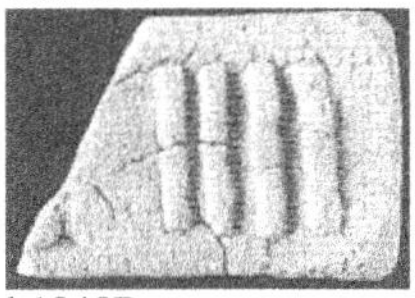
h1848B

h1849A

h1849B
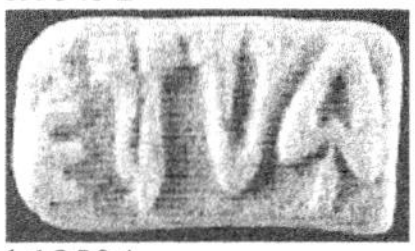
h1850A

h1850B

h1851A
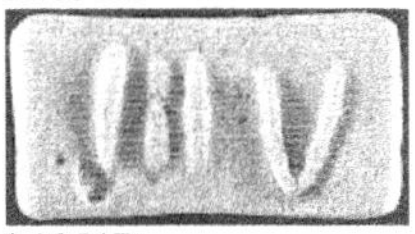
h1851B

h1853A

h1853B

h1854A

h1854B

h1856A

h1856B

h1857A

h1857B

h1858A
h1858B
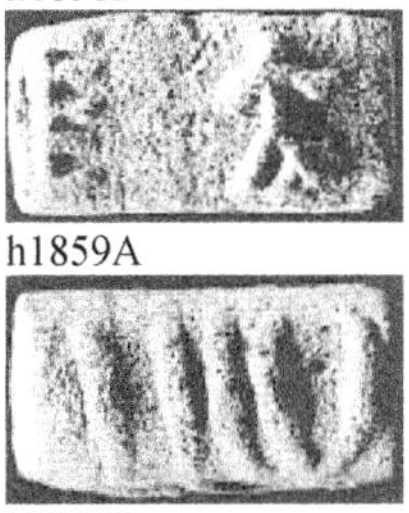
h1859A
h1859B

h1860A
h1860B

h1861A
h1861B

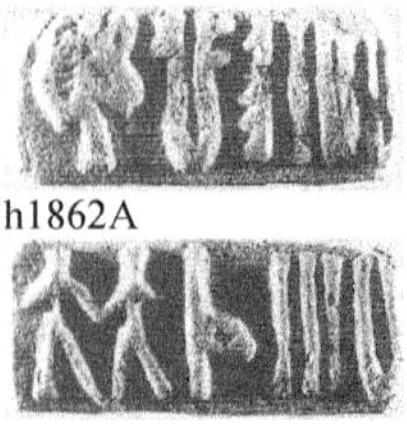
h1862A
h1862B

h1863A
h1863B

h1864A
h1864B

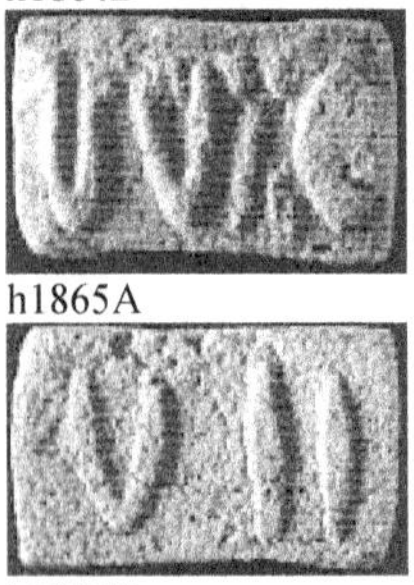
h1865A
h1865B

h1866A
h1866B

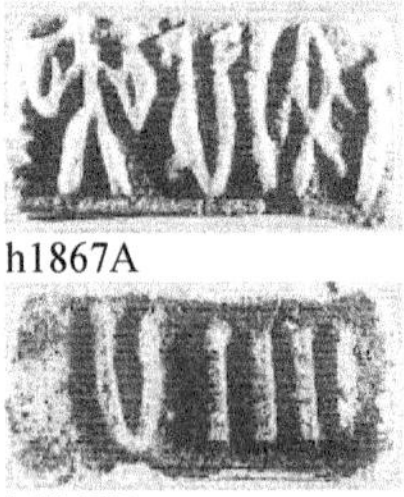
h1867A
h1867B

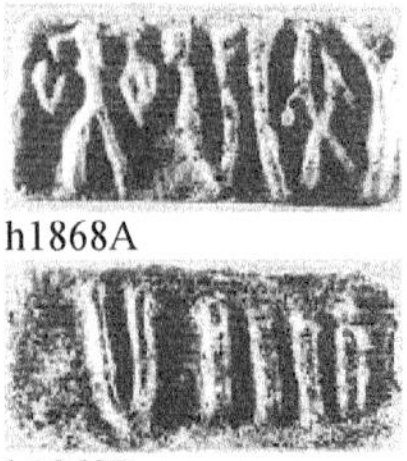
h1868A
h1868B

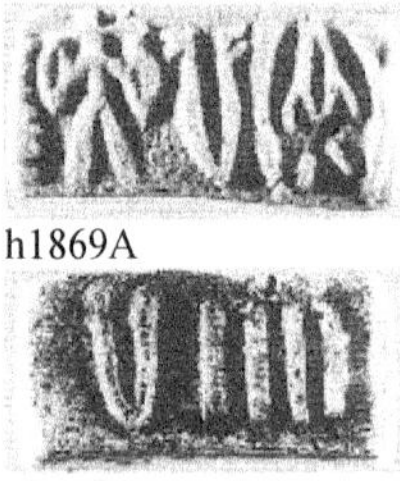
h1869A
h1869B

h1870A
h1870B

h1871A
h1871B

h1872A
h1872B

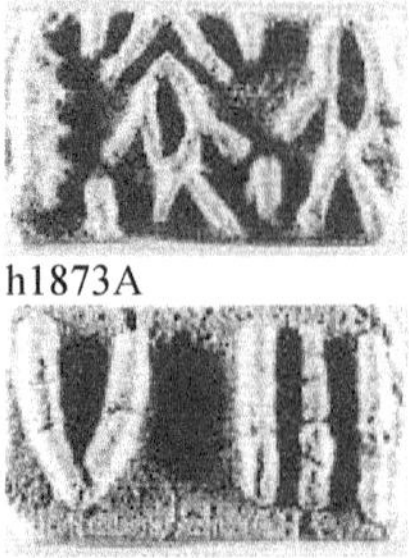
h1873A
h1873B

h1874A
h1874B

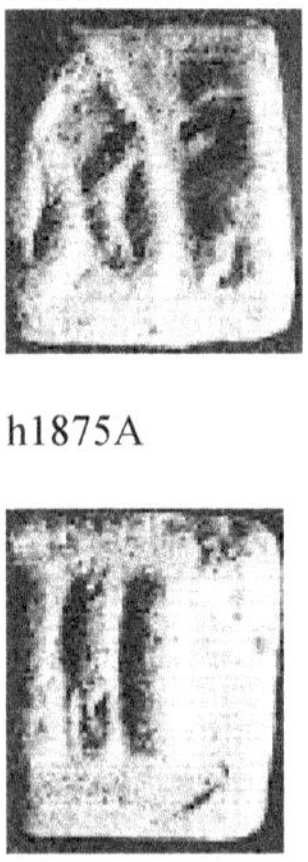
h1875A
h1875B

h1876A
h1876B

h1877A
h1877B

h1878A
h1878B

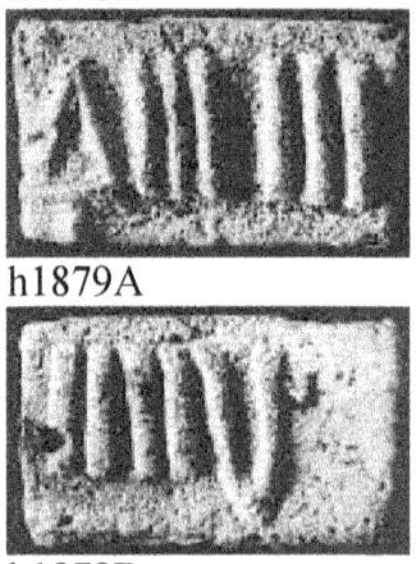
h1879A
h1879B

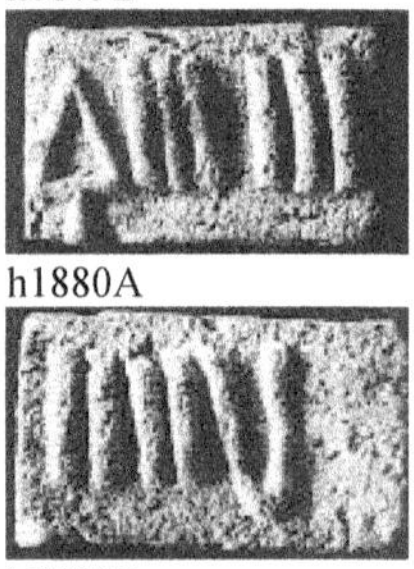
h1880A
h1880B

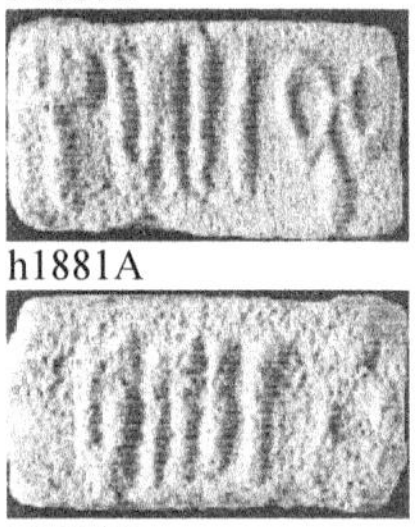
h1881A
h1881B

h1882A
h1882B

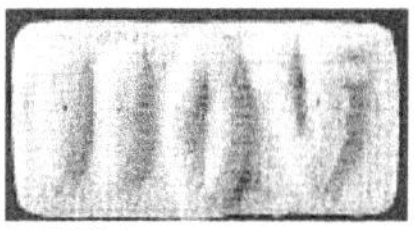
h1883A

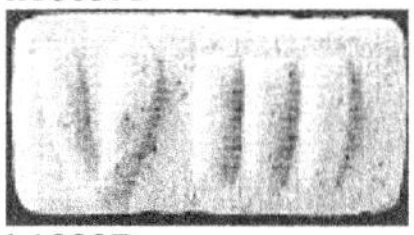
h1883B

h1886A

h1886B

h1887A

h1887B

h1892A

h1892B

h1893A

h1893B

h1895A

h1895B

h1896A

h1896B

h1897A

h1897B

h1898A

h1898B

h1899A

h1899B

h1900A

h1900B

h1901A

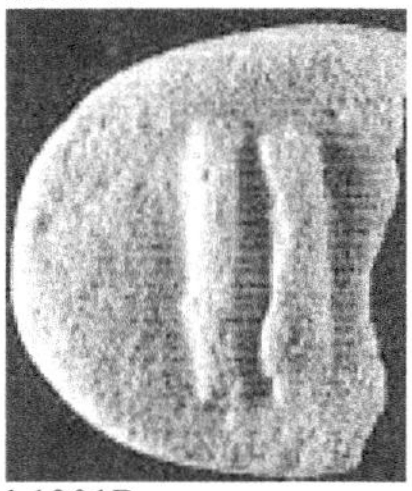
h1901B

h1902A

h1902B

h1903A

h1903B

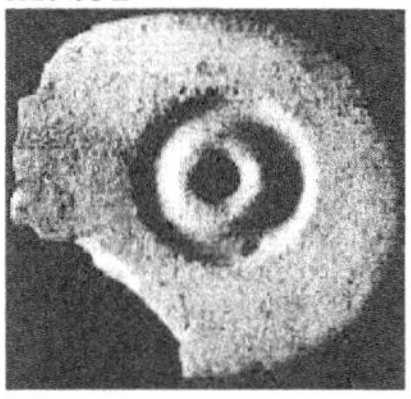
h1904A

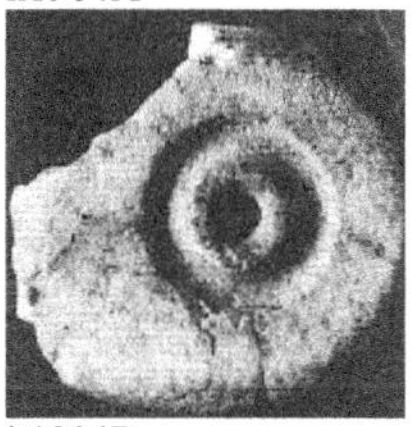
h1904B

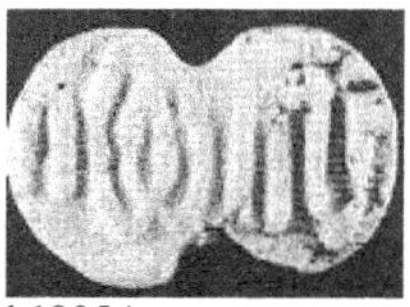
h1905A

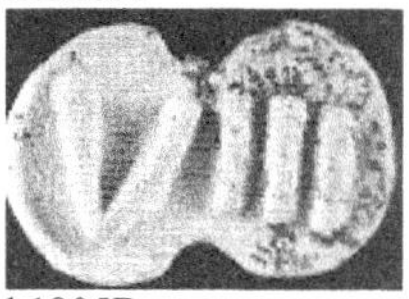
h1905B

h1906A

h1907A

h1907B

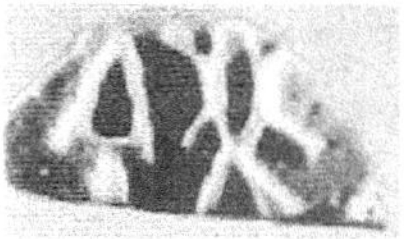
h1908A

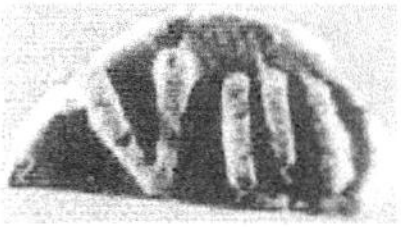
h1908B

h1909A

h1909B

h1910A

h1910B

h1911A

h1911B

h1912A
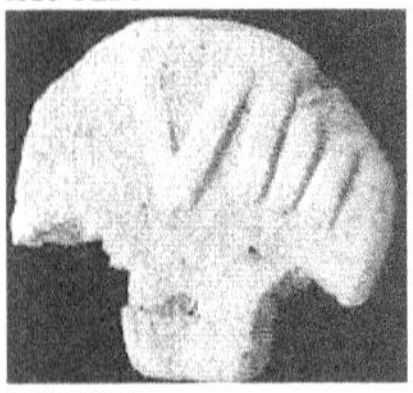
h1912B
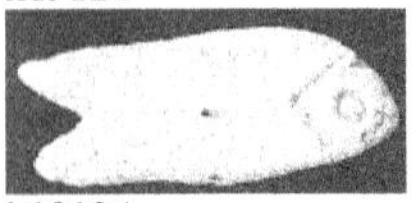
h1913A

h1913B

h1914A

h1914B

h1915A

h1915B

h1916A
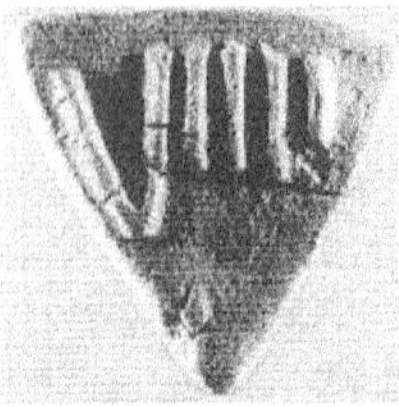
h1916B

h1917A

h1917B
h1918A
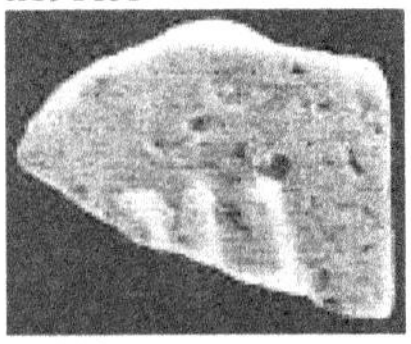
h1918B

h1920A

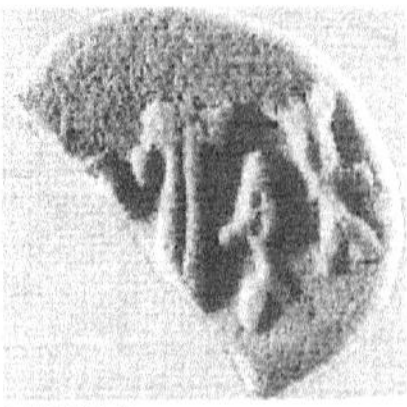
h1921A

h1921B

h1922A

h1922B

h1923A

h1923B
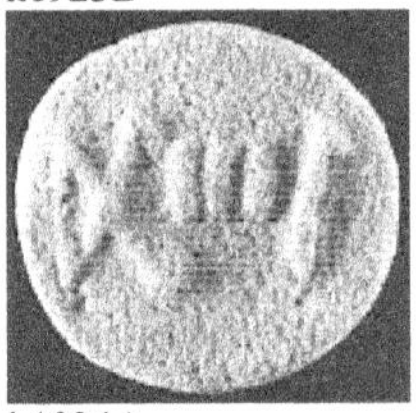
h1924A

h1924B

h1925A

h1925B

h1926A

h1927A

h1929A

h1929B
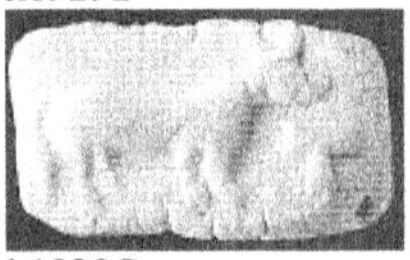
h1929C

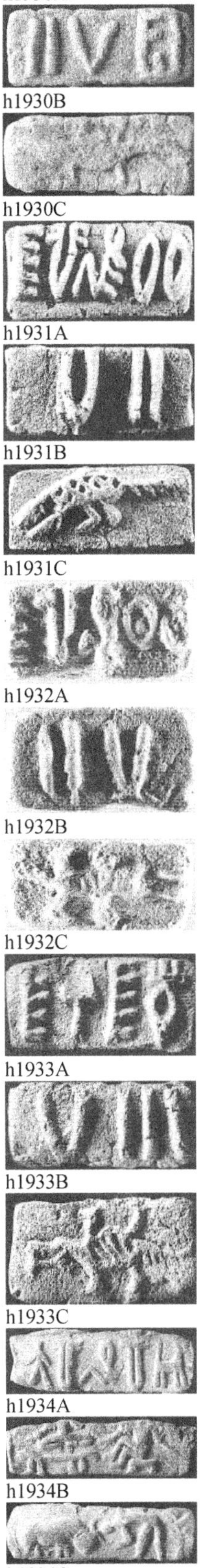
h1930A
h1930B
h1930C
h1931A
h1931B
h1931C
h1932A
h1932B
h1932C
h1933A
h1933B
h1933C
h1934A
h1934B
h1934C

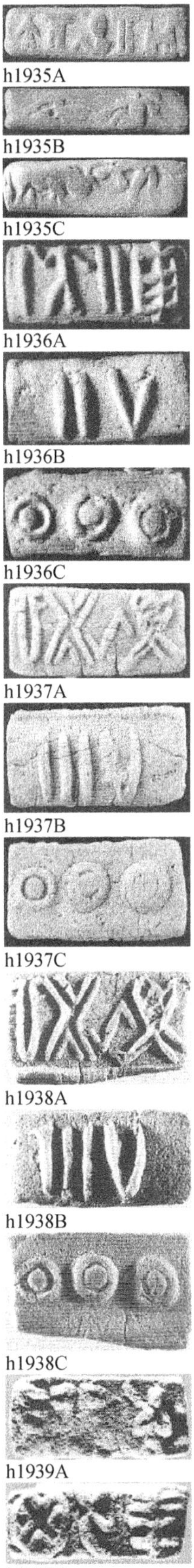
h1935A
h1935B
h1935C
h1936A
h1936B
h1936C
h1937A
h1937B
h1937C
h1938A
h1938B
h1938C
h1939A
h1939B

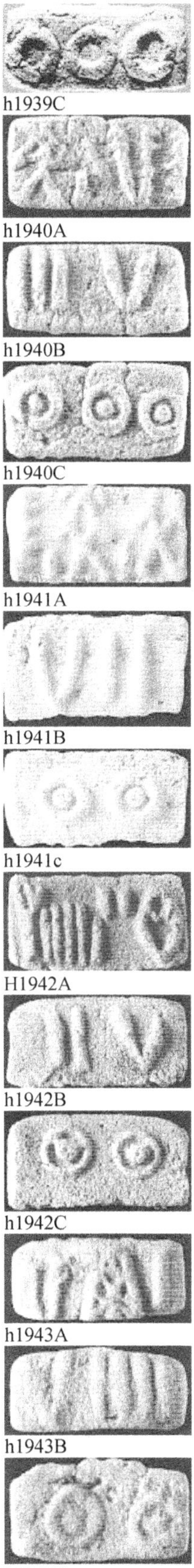
h1939C
h1940A
h1940B
h1940C
h1941A
h1941B
h1941c
H1942A
h1942B
h1942C
h1943A
h1943B
h1943C

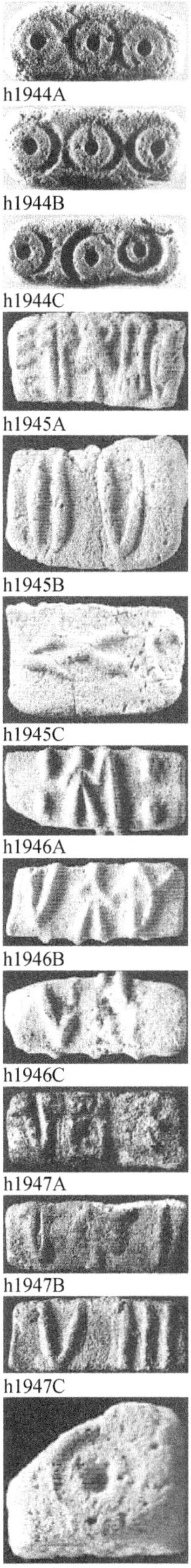
h1944A
h1944B
h1944C
h1945A
h1945B
h1945C
h1946A
h1946B
h1946C
h1947A
h1947B
h1947C
h1950A

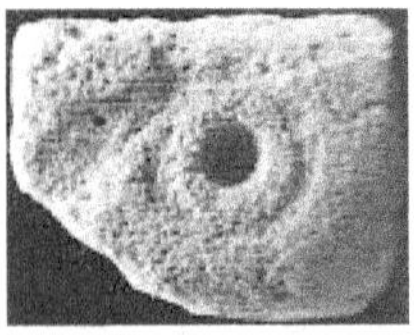

h1950B

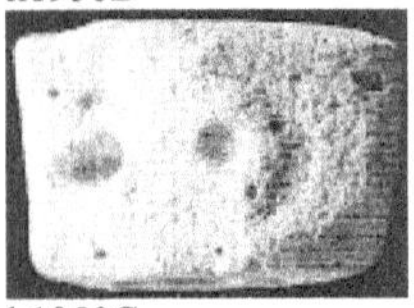

h1950C

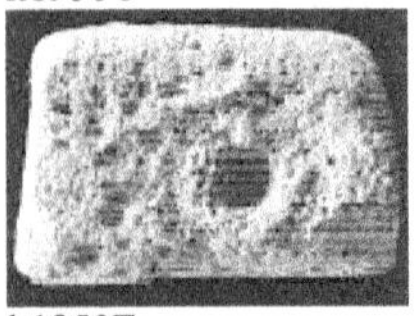

h1950E

h1951A

h1951B

h1953A

h1953B

h1955A

(bird+fish)

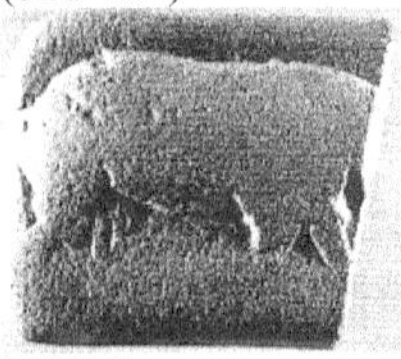

h1955B

h1958A

h1958B

h1959

h1961A

h1961B

h1962A

h1962B

h1963A

h1963B

h1964A

h1964B

h1966A

h1966B

h1967A

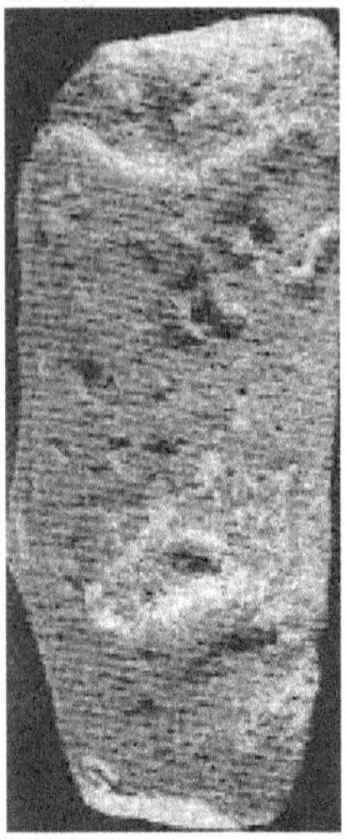

h1967B

h1968A

h1968B

h1969A

h1969B

h1970A

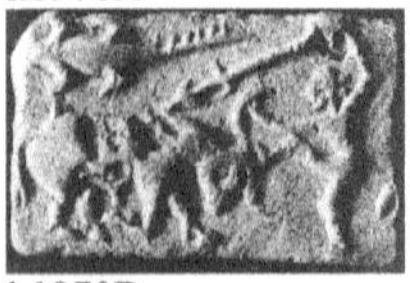

h1970B

h1971A

h1971B

h1972A

h1972B

h1973A

h1973B

h1974A

h1974B

h1975A

h1975B

h1976A

h1976B

h1977A

h1977B

h1978A

h1978B

h1979A

h1979B

h1980A

h1981B

h1981A

h1981B

h1985A

h1985B

h1987A

H1987B

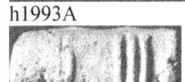

h1988A, h1989A, h1990A

h1988B, h1989B, h1990B

h1991A

h1991B

h1992B

h1993A

h1993B

h1994A
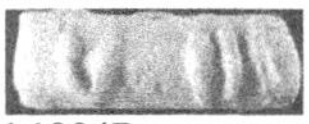
h1994B
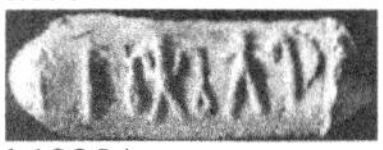
h1995A

h1995B

h1997A

h1997B

h1999A

h1999B
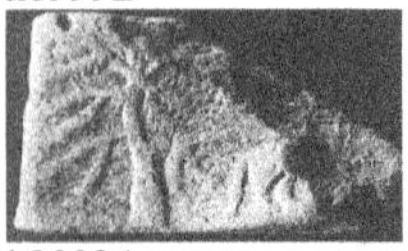
h2002A

h2003Ah

h2003B

h2005A

h2005B

h2006A

h2006B

h2010A

h2010B

h2012A

h2012B

h2013A

h2014A

h2014B
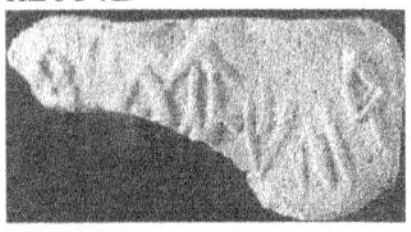
h2015A

h2015B

h2016A

h2018A

h2018B

h2019A

h2019B

h2019C

h2020A

h2020C

h2021A

h2021B

h2021C

h2022A

h2022B

h2022C
h2023A

h2023B

h2023C

h2024A

h2024B

h2024C

h2025C

h2026A

h2026B

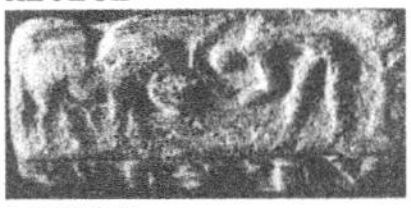

h2026C

h2026D

h2027A

h2027B

h2027C

h2027D

h2028A

h2028B

h2028C

h2028A

h2029A

h2029B

h2029C

h2029D

h2030A

h2030C

h2030D

h2031A

h2031B

h2031C

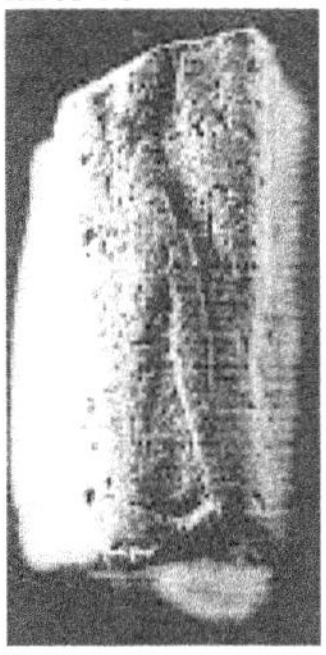

h2031D

h2032B

h2032D

h2033A

h2033B

h2034A

h2034B

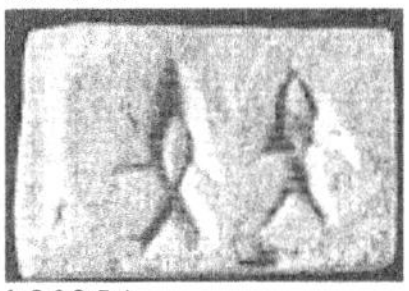

h2035A

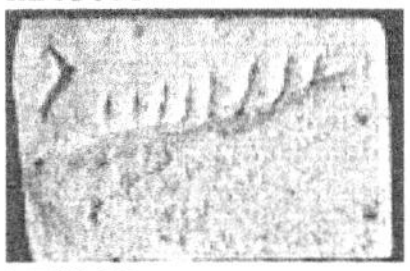

h2035B

h2036A

h2036B

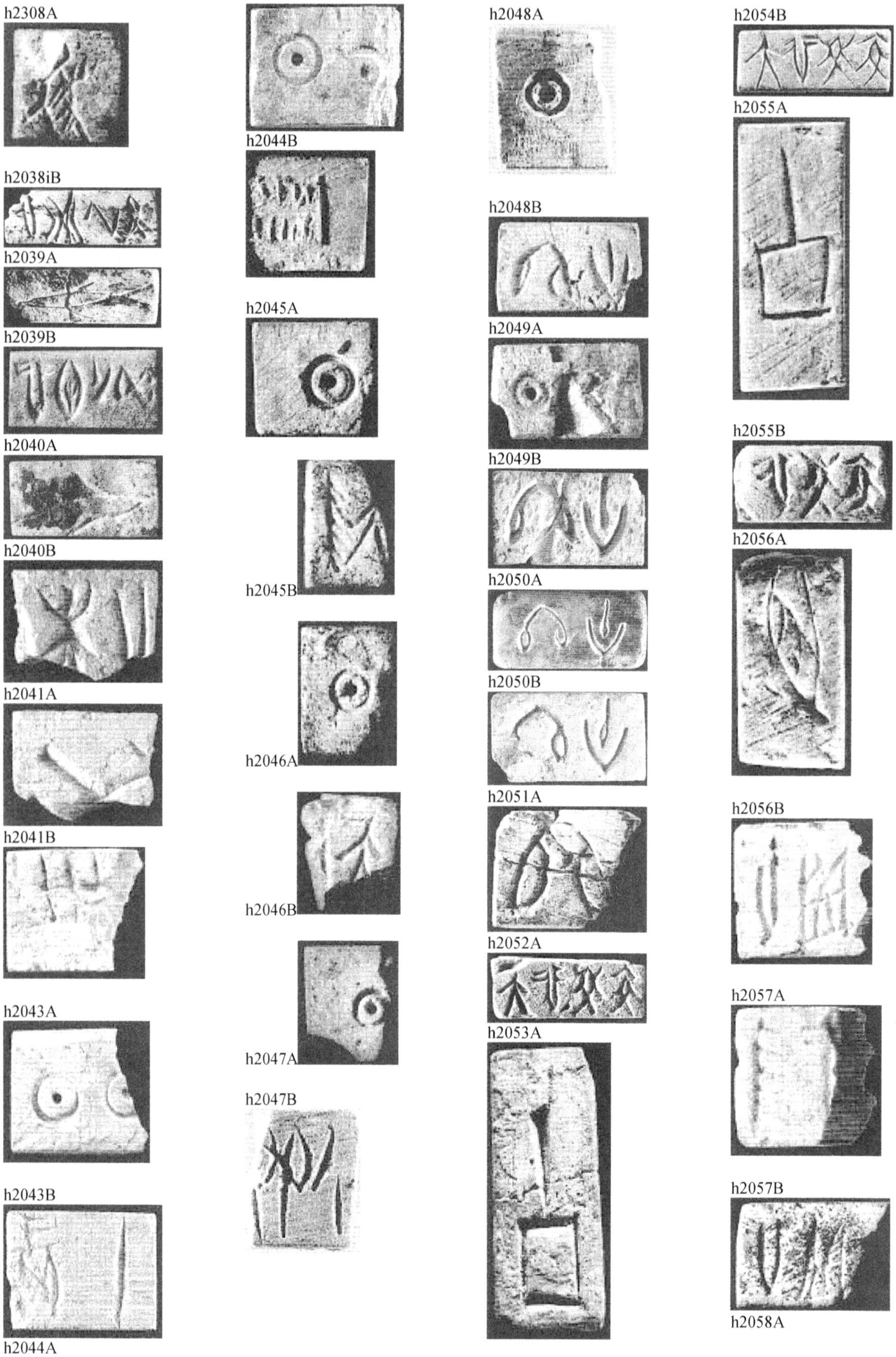

h2308A
h2038iB
h2039A
h2039B
h2040A
h2040B
h2041A
h2041B
h2043A
h2043B
h2044A
h2044B
h2045A
h2045B
h2046A
h2046B
h2047A
h2047B
h2048A
h2048B
h2049A
h2049B
h2050A
h2050B
h2051A
h2052A
h2053A
h2054B
h2055A
h2055B
h2056A
h2056B
h2057A
h2057B
h2058A

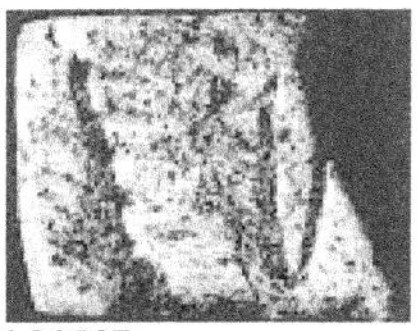

h2058B

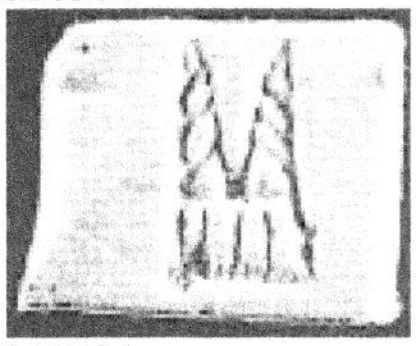

h2059A

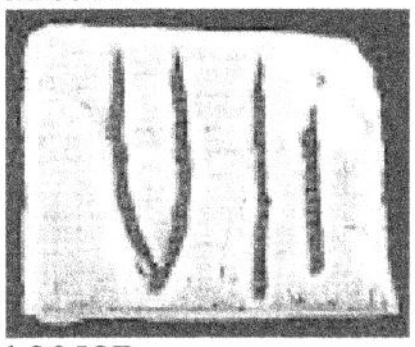

h2059B

h2062A

h2062B

h2063A

h2063B

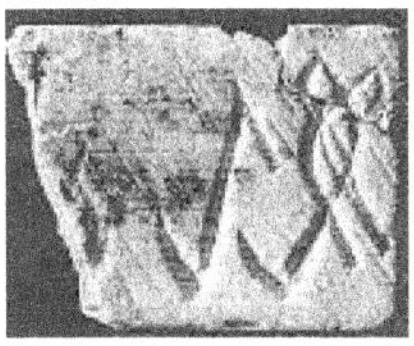

h2064A

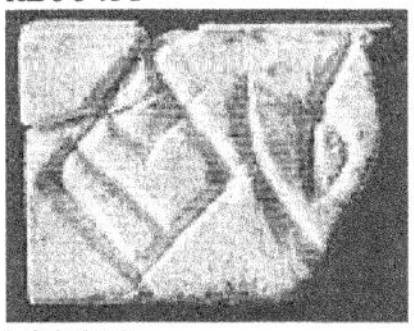

h2064B

h2065A

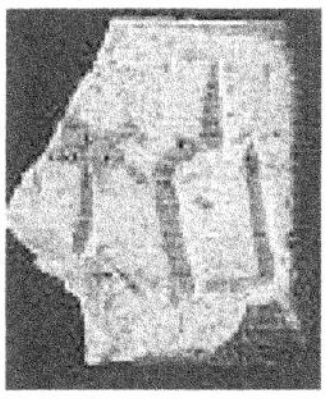

h2065B

h2066A

h2066B

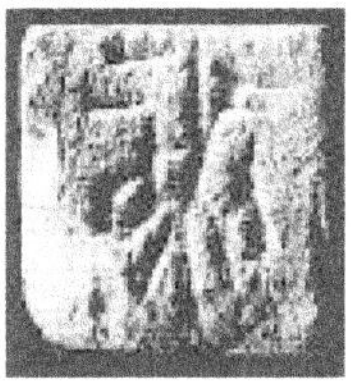

h2067A

h2067B

h2068A

h2068B

h2069A

h2069B

h2070A

h2070B

h2071A

h2071B

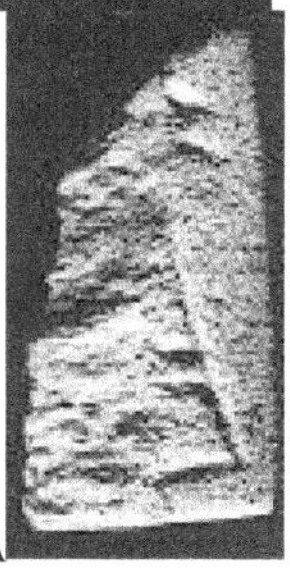

h2072A

h2072B

h2073A

h2073B

h2074A

h2074B

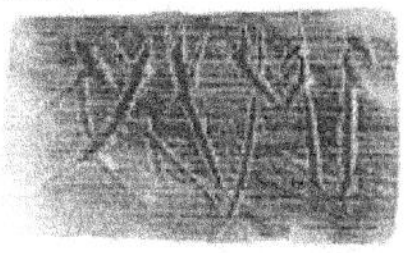

h2076A

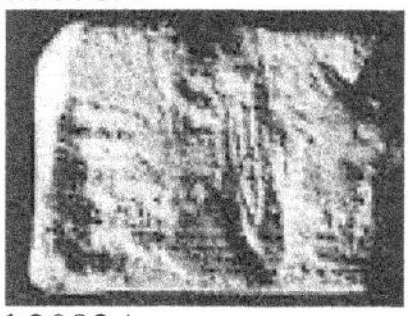

h2082A

h2082B

h2083A

h2083B

h2084A

h2084B

h2085A

h2085B

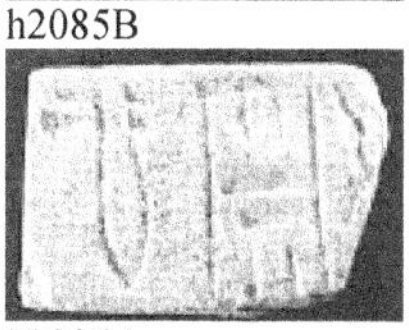

h2086A

h2086B

h2089A

h2089B

h2090A

h2090B

h2091A

h2091B

h2092A

h2092B

h2093A

h2093B

h2094A

h2094B

h2095A

h2095B

h2096A

h2096B

h2097A

h2097B

h2098A

h2098B

h2099A

h2099B

h2102A
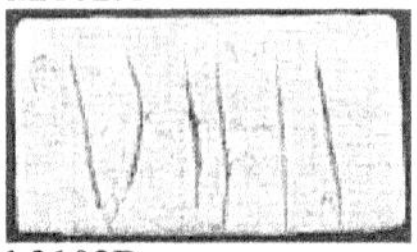
h2102B

h2104A

h2104B

h2105A

h2105B

h2106A
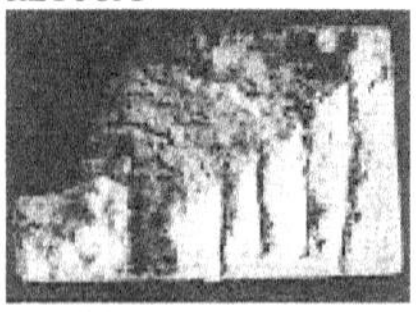
h2106B

h2107A

h2107B

h2108A

h2108B

h2109A1B2

h2109B1A2

h2110A

h2110B

h2111B

h2112A

h2112B

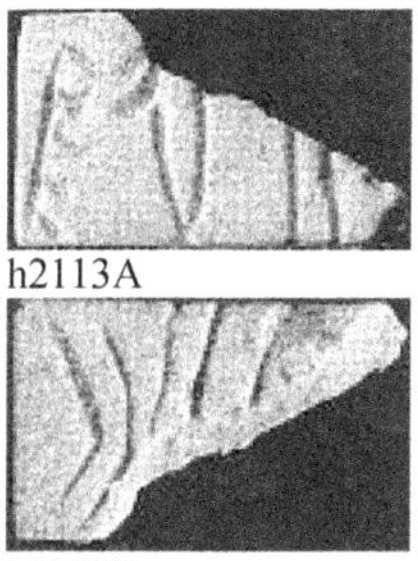
h2113A
h2113B
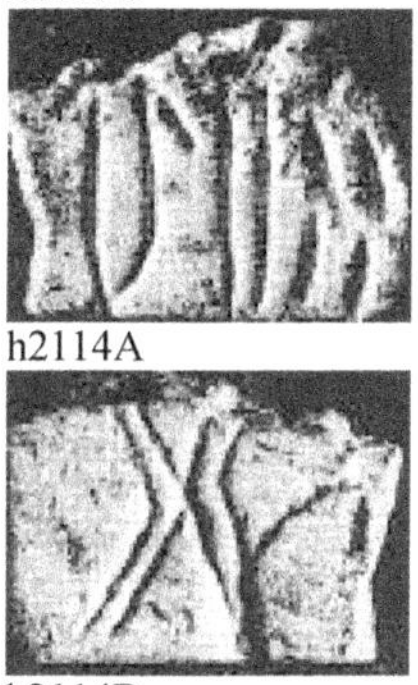
h2114A
h2114B

h2115A
h2115B

h2119A

h2119b
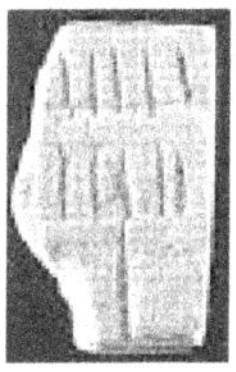
h2120A

h2120B

h2121A

h2121B

h2123A

h2123B
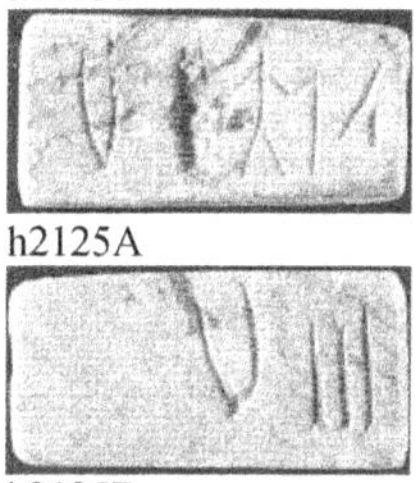
h2125A
h2125B

h2127A

h2127B

h2128A

h2128B

h2129A

h2129B

h2130A

h2130B

h2131A

h2131B

h2132A
h2132B

h2133A

h2133B

h2134A

h2134B

h2135A

h2136A

h2137A
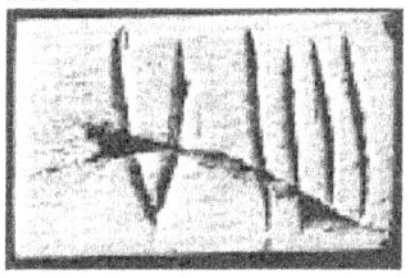
h2137B

h2138A

h2138B

h2139A

h2139B

h2140A

h2140B

h2141A

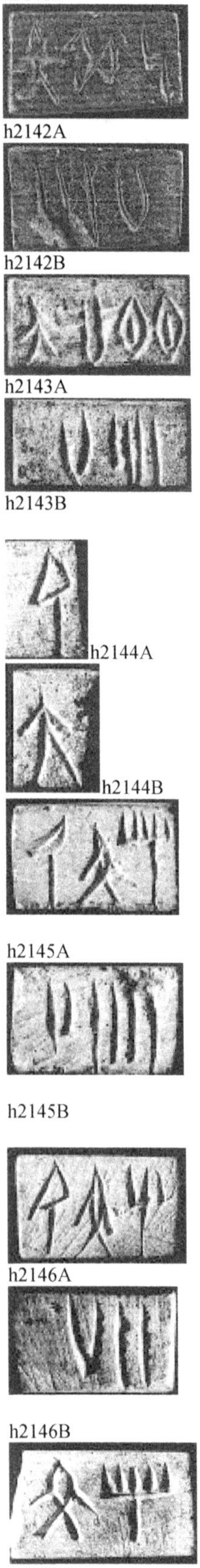

h2141B
h2142A
h2142B
h2143A
h2143B
h2144A
h2144B
h2145A
h2145B
h2146A
h2146B
h2147A

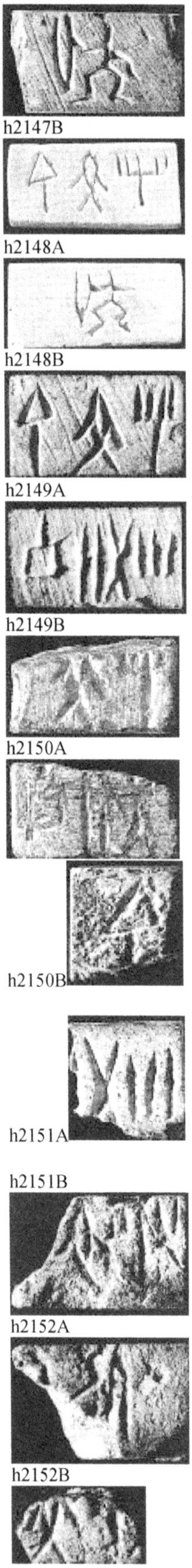

h2147B
h2148A
h2148B
h2149A
h2149B
h2150A
h2150B
h2151A
h2151B
h2152A
h2152B

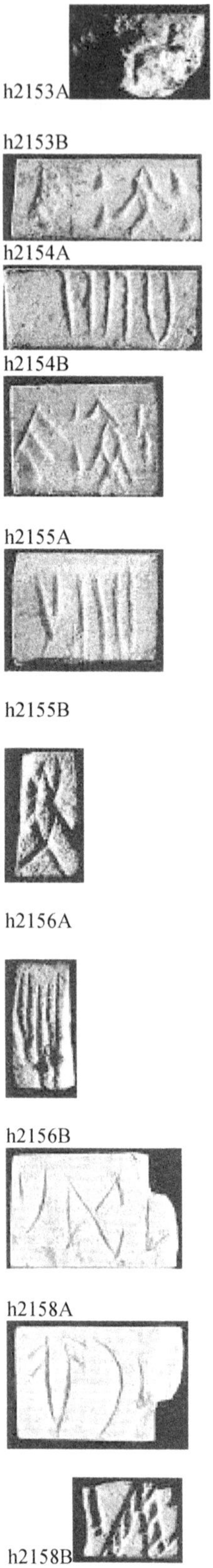

h2153A
h2153B
h2154A
h2154B
h2155A
h2155B
h2156A
h2156B
h2158A
h2158B
h2159A

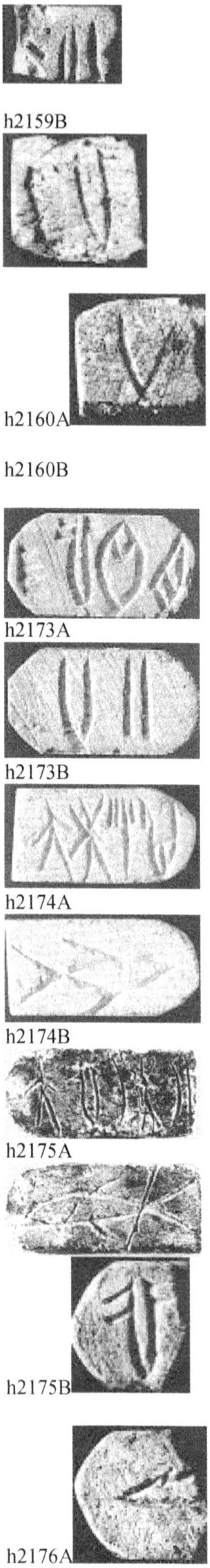

h2159B
h2160A
h2160B
h2173A
h2173B
h2174A
h2174B
h2175A
h2175B
h2176A

h2176B
h2177A
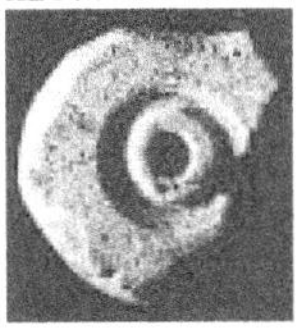
h2177B
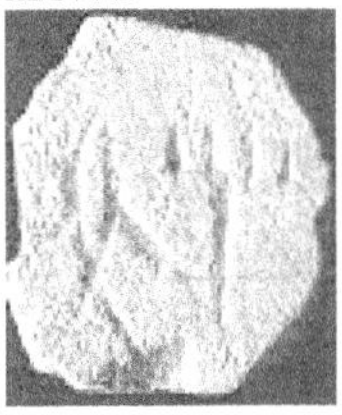
h2178A
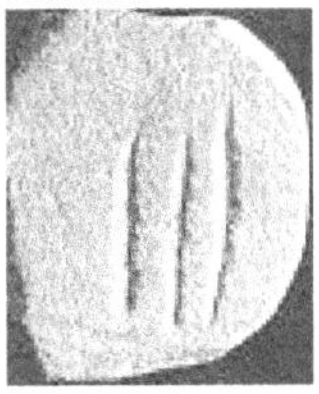
h2178B
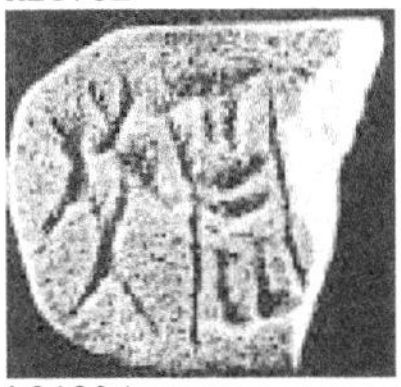
h2180A
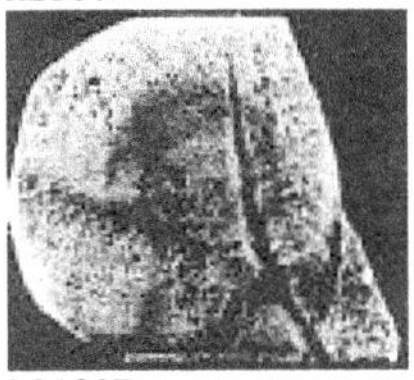
h2180B

h2181A

h2181B
h2182A
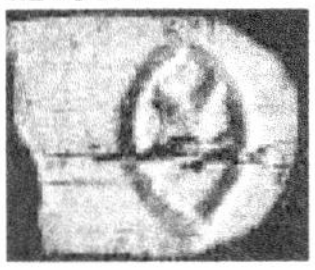
h2182B

h2183A
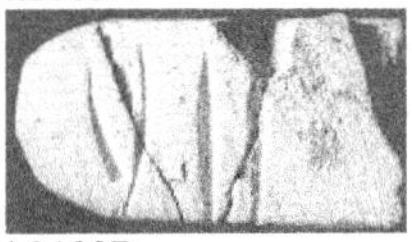
h2183B

h2184A
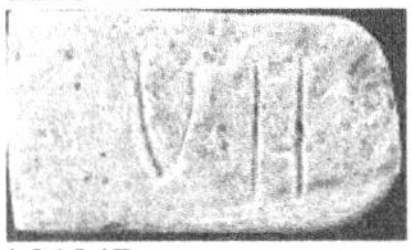
h2184B

h2185A

h2185B

h2186A

h2186B

h2187A

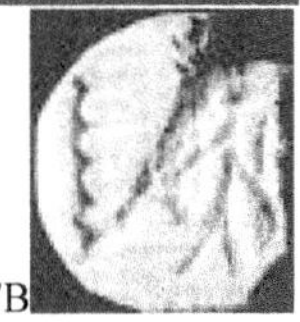
h2187B
h2188A
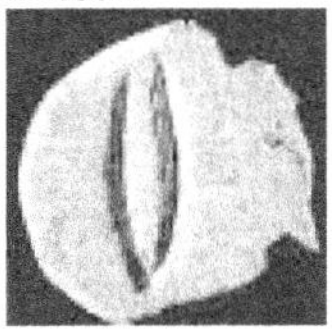
h2188B
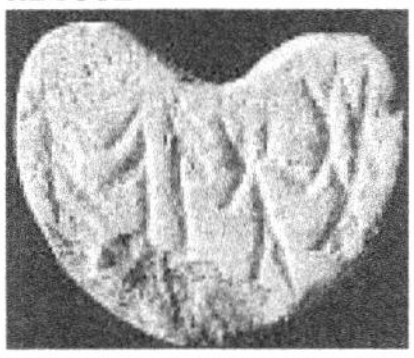
h2189A

h2189B

h2190A

h2190B

h2192A

h2192B

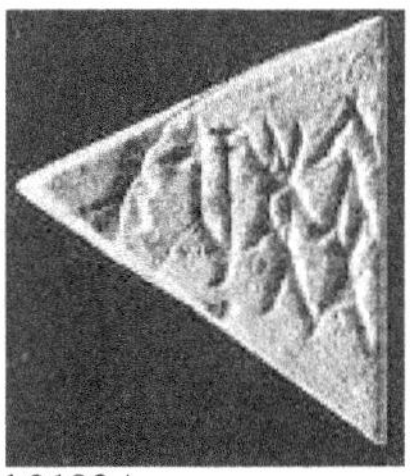
h2193A

h2193B
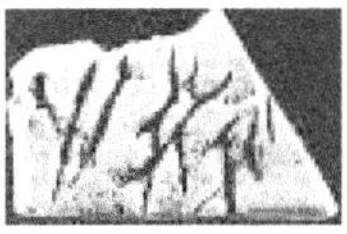
h2194A

h2194B
h2195A

H2195B

h2197A
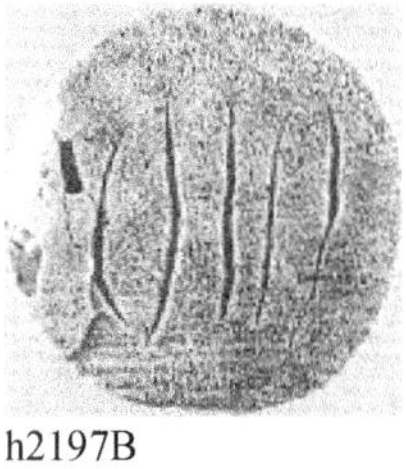
h2197B

h2198A
h2198B
h2200A
h2200B
h2200C
h2201A
h2201B
h2201C
h2204A
h2204B
h2204c
h2205A

h2205B
h2205C
h2207A
h2207B
h2207C
h2208A
h2208B
h2208C
h2209A
h2209B
h2209C
h2210A
h2210B

h2210C
h2211A
h2211B
h2211C
h2212A
h2212B
h2212C
h2213A
h2213B
h2213C
h2214A
h2214N
h2214C

h2215B
h2215C
h2217A
h2217B
h2217C
h2218A
h2218B
h2218C
h2219A
h2219B
h2219C
h2220A

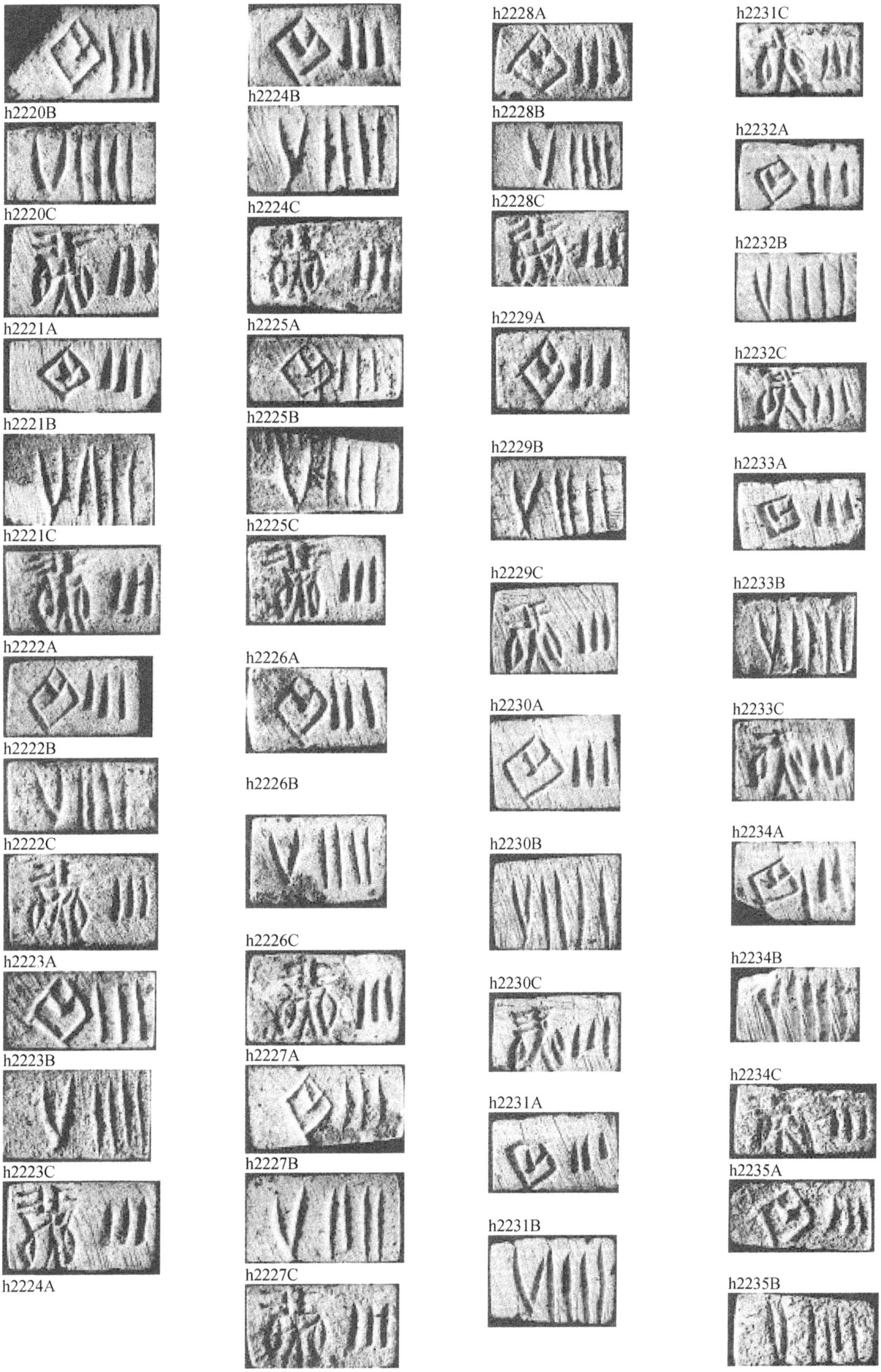

h2220B
h2220C
h2221A
h2221B
h2221C
h2222A
h2222B
h2222C
h2223A
h2223B
h2223C
h2224A
h2224B
h2224C
h2225A
h2225B
h2225C
h2226A
h2226B
h2226C
h2227A
h2227B
h2227C
h2228A
h2228B
h2228C
h2229A
h2229B
h2229C
h2230A
h2230B
h2230C
h2231A
h2231B
h2231C
h2232A
h2232B
h2232C
h2233A
h2233B
h2233C
h2234A
h2234B
h2234C
h2235A
h2235B

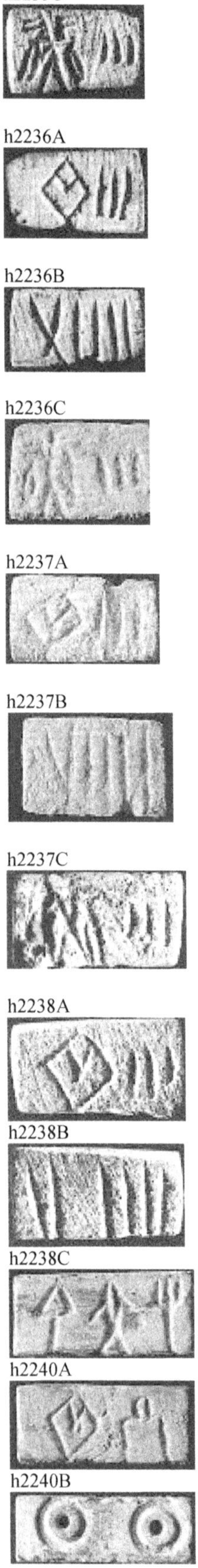

h2235C
h2236A
h2236B
h2236C
h2237A
h2237B
h2237C
h2238A
h2238B
h2238C
h2240A
h2240B

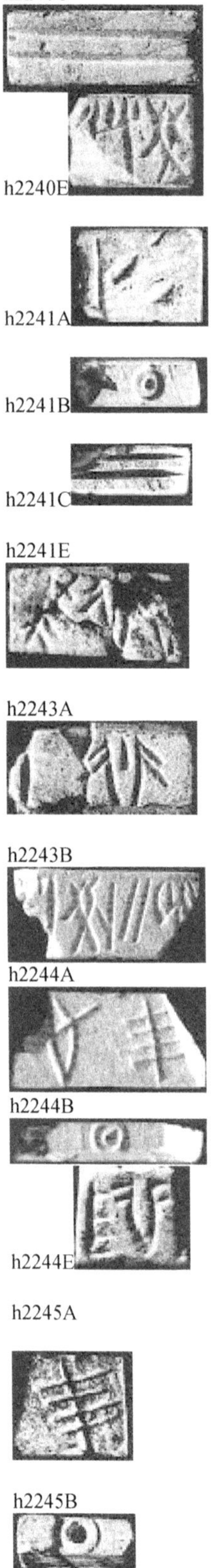

h2240C
h2240E
h2241A
h2241B
h2241C
h2241E
h2243A
h2243B
h2244A
h2244B
h2244E
h2245A
h2245B

h2245E
h2246A
h2246B
h2246C
h2246E
h2247A
h2247B
h2249Acopper
h2250Acopper
h2251Acopper
h2252Acopper
h2253Acopper
h2254Acopper

h2255Acopper
h2256Acopper
h2257Acopper
h2264A
h2265A
h2270A
h2334A
h2339A
h2340A
h2341A
h2345A

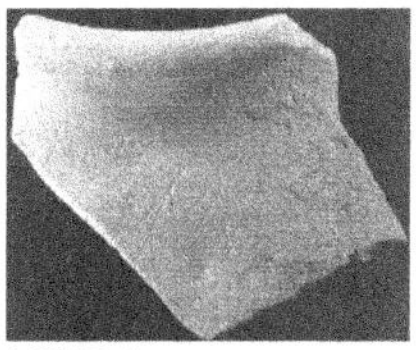
h2353A

h2354A

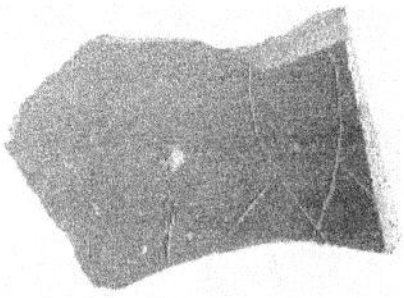
h2357A

h2358A

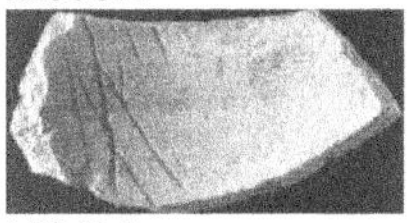
h2360A

h2367A

h2368A

h2373A

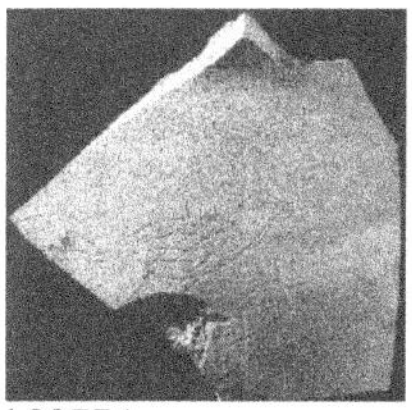
h2377A

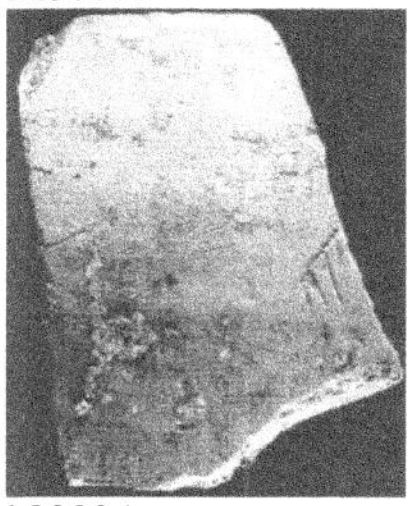
h2380A

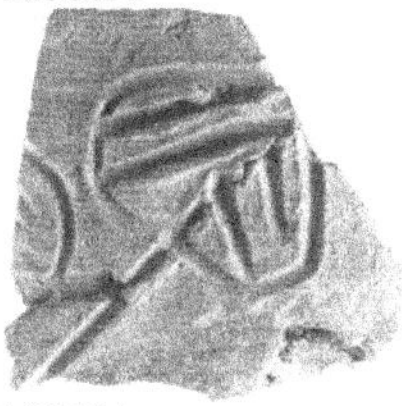
h2383A

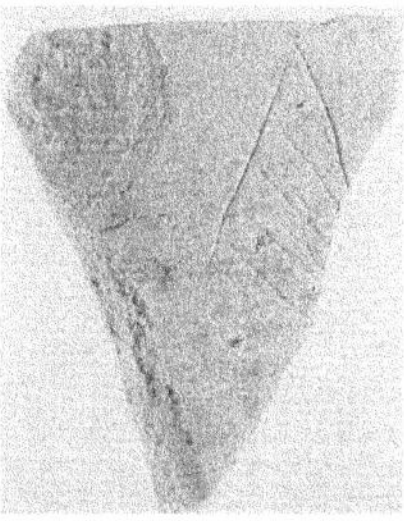
h2384A

h2390A

h2397A

j2398A

h2399A

h2400A

h2403A

h2405A

h2548A

h2549A

h2586A

h2569Alead

h2570Abone

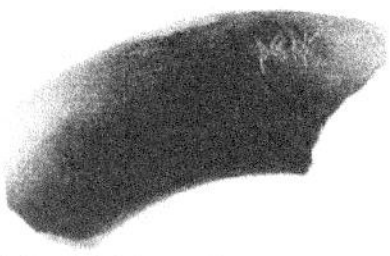
h2576Abangle

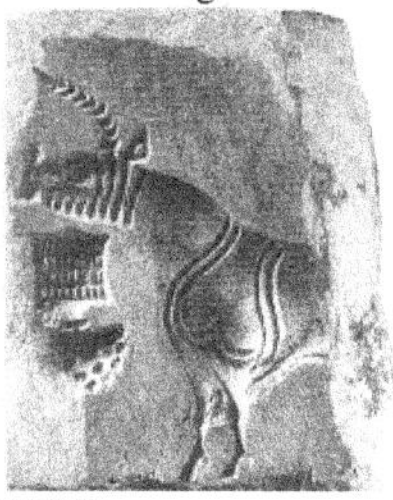
h2590

Harappa Texts (Either unmatched with inscribed objects or objects not illustrated)

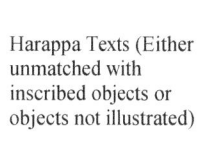

4015

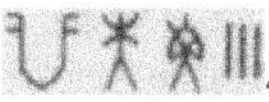
4033

4035

4046

4067

4073

4101

4108

4114

4117

4119

4134

4136

4140

4158

4164

4292

4296

4305

Pict-90: Standing person with horns and bovine features holding a staff or mace on his shoulder.

4324

4330

4334

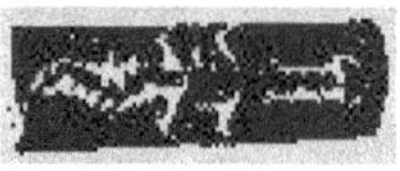

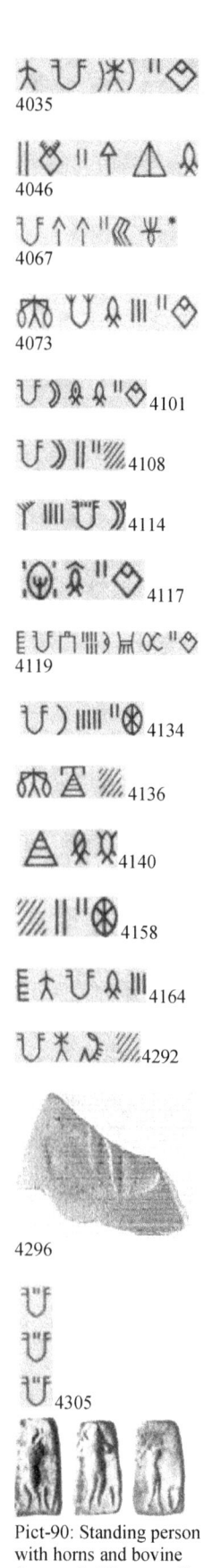

Pict-63: Gharial, sometimes with a fish held in its jaw and/or surrounded by a school of fish.

4343

Tablet in bas-relief One-horned bull

4344

4347

4356

4360

4363

4369

4373

4384

4404

4406

4407

Pict-129: Inscribed object in the shape of a double-axe or double-shield?

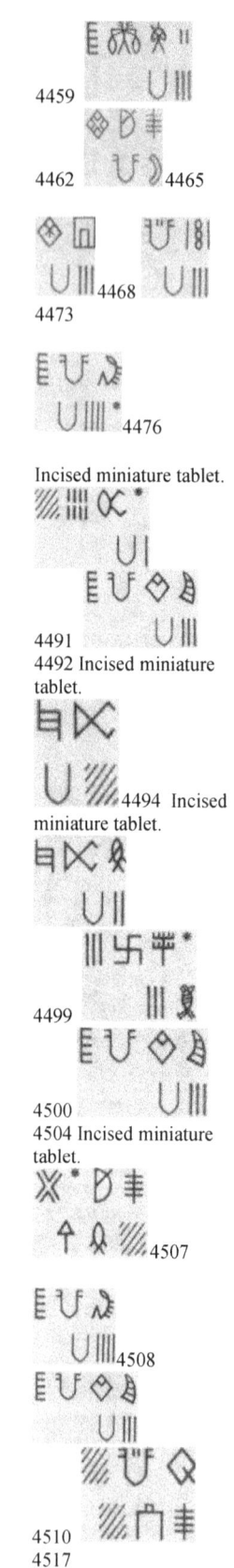

4408

4422

4427

4428

4432

4435

4438

4439

4447

4448

4452

4458

4459

4462

4465

4468

4473

4476

Incised miniature tablet.

4491

4492 Incised miniature tablet.

4494 Incised miniature tablet.

4499

4500

4504 Incised miniature tablet.

4507

4508

4510

4517

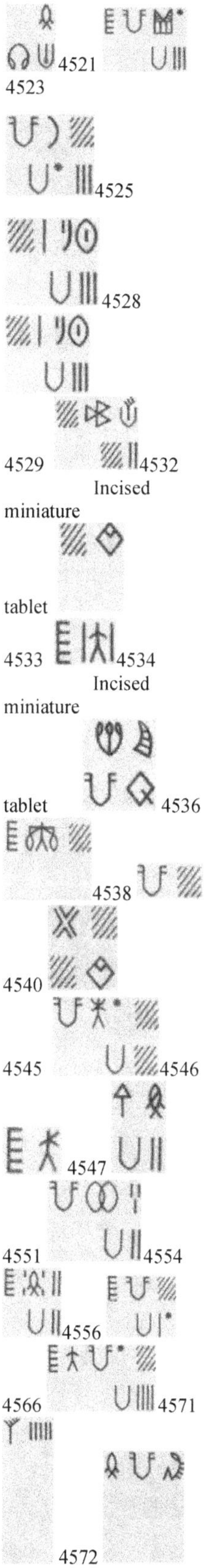

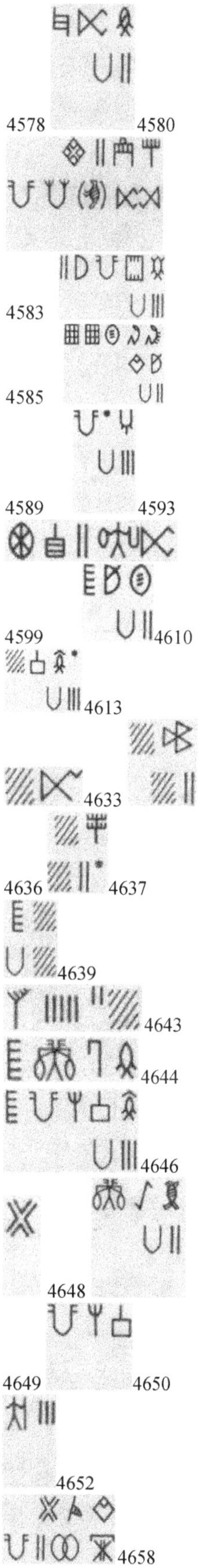

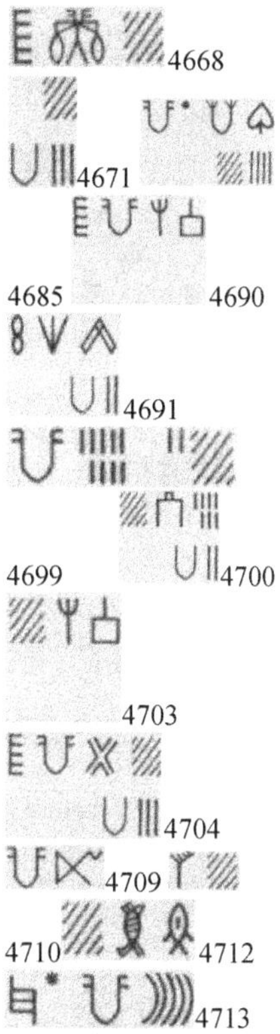

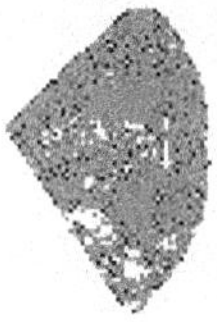

Pict-134: Motif on a pottery graffiti showing a rectangular enclosure with four marks within; the marks looks like X and V.

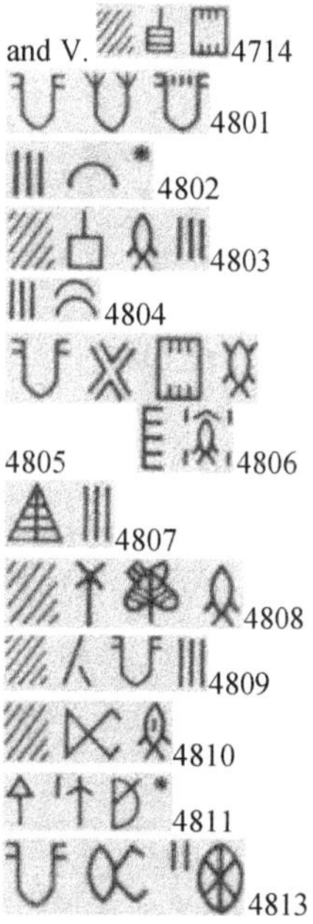

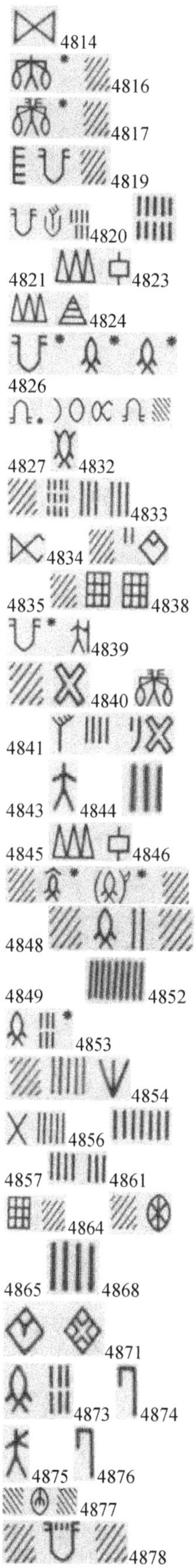

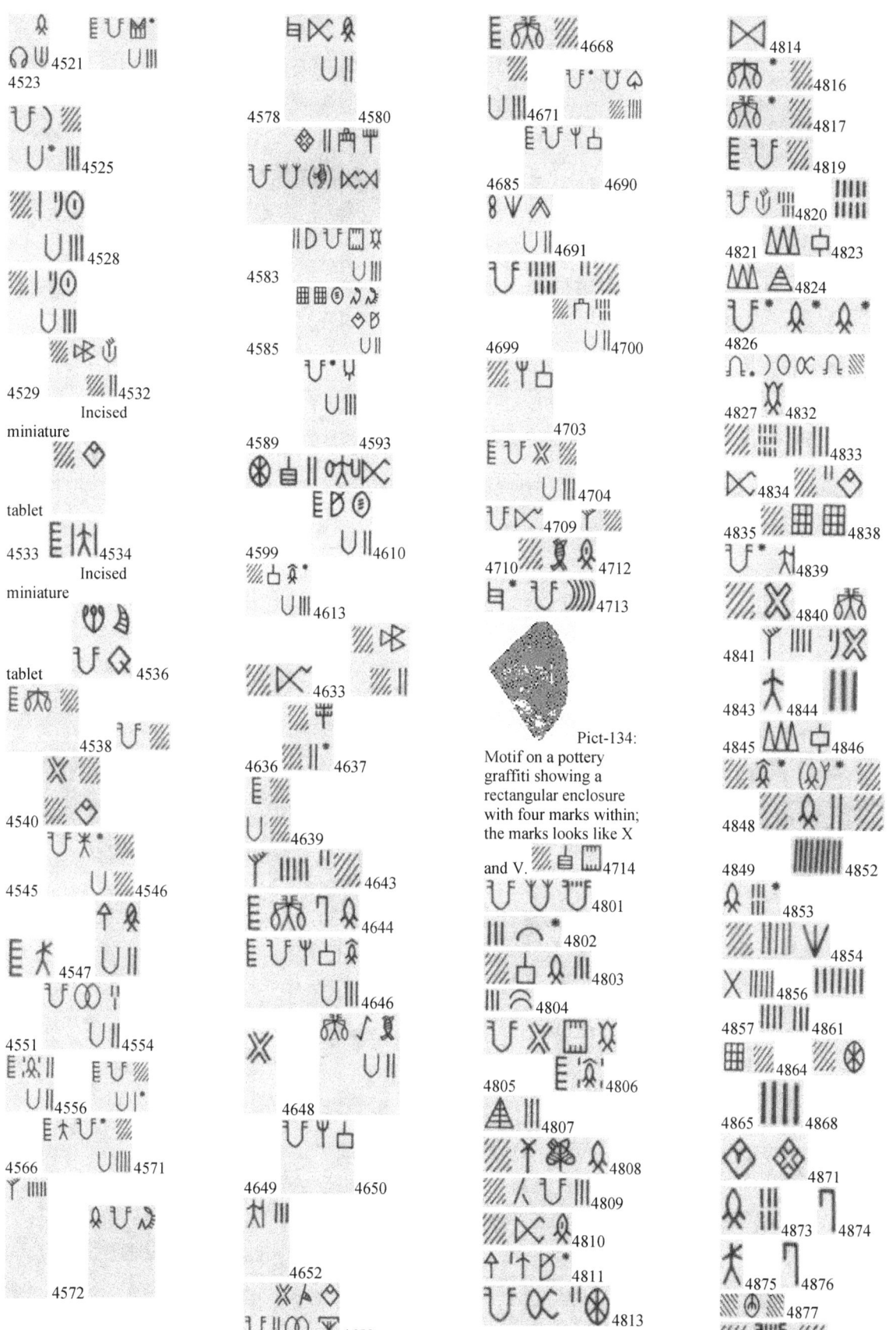

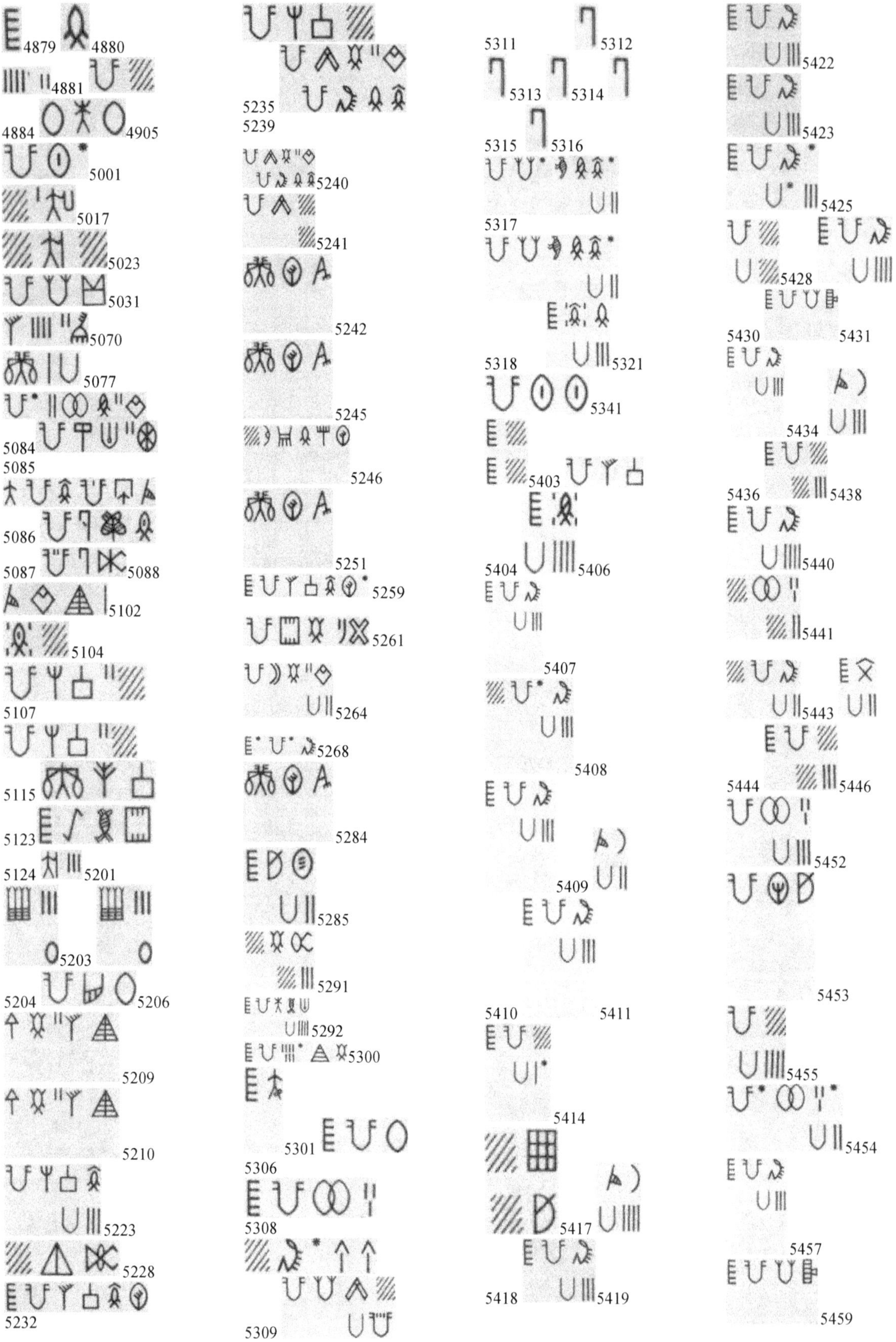
4879 4880
4881
4884 4905
5001
5017
5023
5031
5070
5077
5084
5085
5086
5087 5088
5102
5104
5107
5115
5123
5124 5201
5203
5204 5206
5209
5210
5223
5228
5232
5235
5239
5240
5241
5242
5245
5246
5251
5259
5261
5264
5268
5284
5285
5291
5292
5300
5301
5306
5308
5309
5311 5312
5313 5314
5315 5316
5317
5318 5321
5341
5403
5404 5406
5407
5408
5409
5410 5411
5414
5417
5418 5419
5422
5423
5425
5428
5430 5431
5434
5436 5438
5440
5441
5443
5444 5446
5452
5453
5455
5454
5457
5459

5469
5470
5478
5480
5481
5482
5484
5485 5486
5487
5488
5489
5490
5492
5495
5501
5504 5505
5506
5507
5508 5510

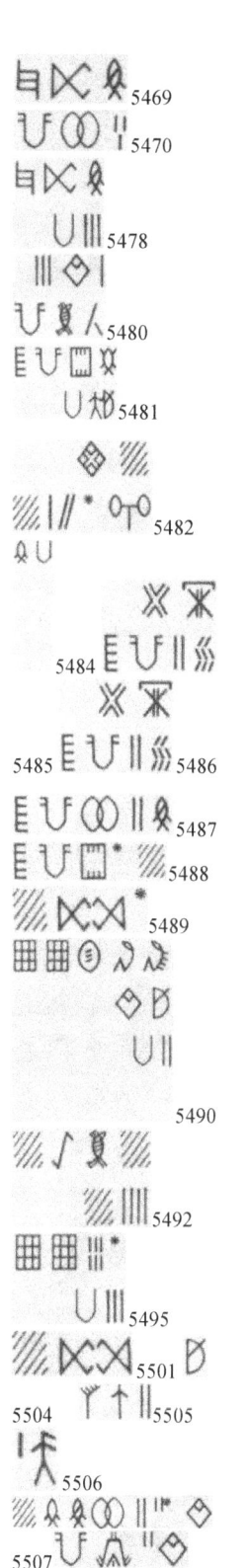

5511 5513
5514 5515
5516
5517 5518
5519
5601
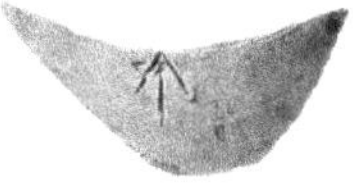
hd06

hulas

jhukar1
jhukar2

9001
jhukar3
8009 8036
8048
8060 8061
8062
8202 8214
8215
8217

Kalibangan002

8019
Kalibangan003
8030

Kalibangan004
8026

Kalibangan005
8017

Kalibangan006
8020

Kalibangan007
8043

Kalibangan008
8041

Kalibangan009
8021

Kalibangan010
8006

Kalibangan011
8034
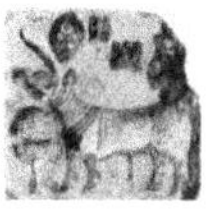
Kalibangan012

Kalibangan013
8051

Kalibangan014
8012

Kalibangan015
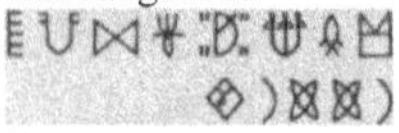
8056

Kalibangan016

8044

Kalibangan017
8027

Kalibangan018
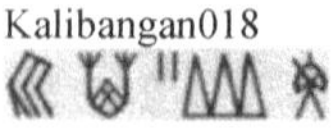

8040

Kalibangan019

8058

Kalibangan020
8047

Kalibangan021

Kalibangan022

8008

Kalibangan023
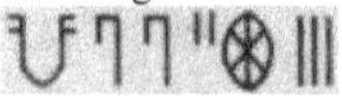

8029

Kalibangan024

Kalibangan025

8037

Kalibangan026
8071

Kalibangan027
8022 'Unicorn' with two horns! "Bull with two long horns (otherwise resembling the 'unicorn')", generally facing the standard. That it is the typical 'one-horned bull' is surmised from two ligatures: the pannier on the shoulder and the ring on the neck.

Kalibangan028
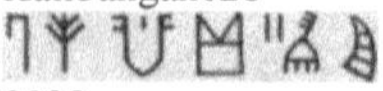
8038

Kalibangan029
8018

Kalibangan030
8002

Kalibangan031a

8007

Kalibangan032a

Kalibangan033
8025

Kalibangan034 8052

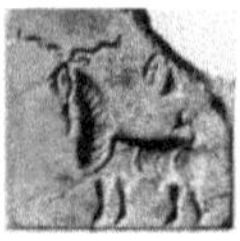

Kalibangan035

Kalibangan036

Kalibangan037 8042

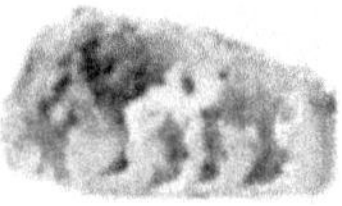

Kalibangan038

Kalibangan039
8011

Kalibangan040

8072

Kalibangan041

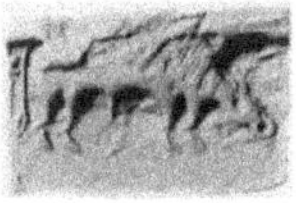

Kalibangan042a

Kalibangan043

8039 Pict-59:Composite motif: body of an ox and three heads: of a one-horned bull (looking forward), of antelope (looking backward), and of short-horned bull (bison) (looking downward).

Kalibangan044

8045

Kalibangan045

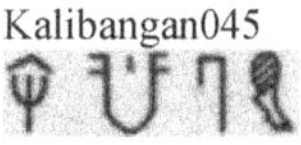

8054

Kalibangan046

 8053

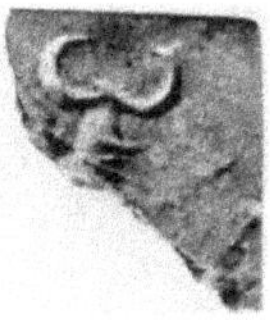

Kalibangan047

Kalibangan048

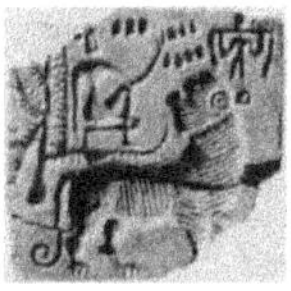

Kalibangan049

 8013

Kalibangan050c

8031 Pict-53: Composition: body of a tiger, a human body with bangles on arm, a pig-tail, horns of an antelope crowned by a twig.

Kalibangan051

8003

Kalibangan052

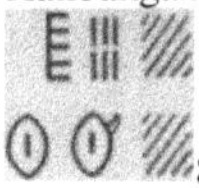 8015

Kalibangan053

Kalibangan054

8033

Kalibangan055a

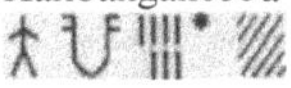

8035

Kalibangan056

8004

Kalibangan057

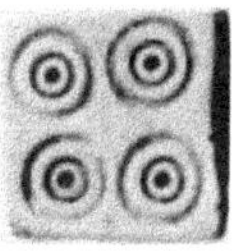

Kalibangan058

Kalibangan059

 8016

Kalibangan060

 8059

Kalibangan061

8001

Kalibangan062

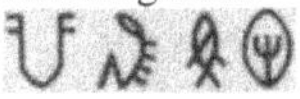

8023

Kalibangan063

8055

Kalibangan064

Kalibangan065a

Kalibangan065A6

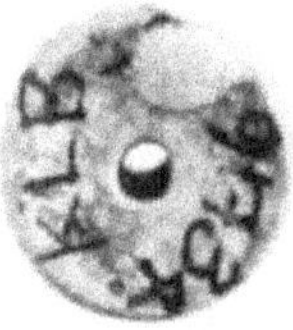

Kalibangan065E
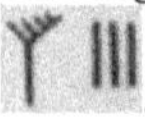
8024 Pict-104: Composition: A tree; a person with a composite body of a human (female?) in the upper half and body of a tiger in the lower half, having horns, and a trident-like head-dress, facing a group of three persons consisting of a woman (?) in the middle flanked by two men on either side throwing a spear at each other (fencing?) over her head.

Kalibangan066

8102

Kalibangan067 8121
Ox-antelope with a long tail; sometimes with a trough in front.

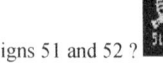

Kalibangan068A

Kalibangan068B

8117 [Is it a bird or an India River Otter? Could be a scorpion, a model for

Signs 51 and 52 ?

See variant in Text 9845 West Asia find]

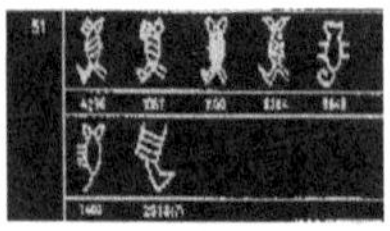

Kalibangan069A

8109

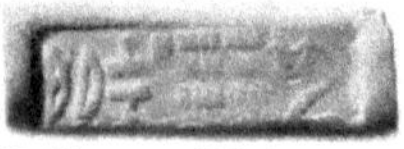

Kalibangan070A

8108

Kalibangan071

8110

Kalibangan072

8111

Kalibangan073

8112

Kalibangan074

8115

Kalibangan075

8113

Kalibangan076A

Kalibangan076B

Kalibangan077A

Kalibangan077B

8118

Kalibangan078A

Kalibangan078B

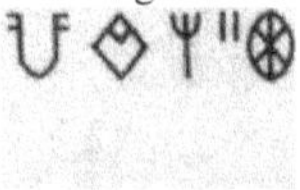

8104

Kalibangan

079AB

Kalibangan080A

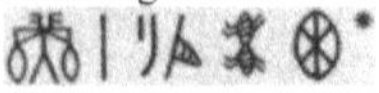
8120

Kalibangan081A

8105

Kalibangan082A

8122

Kalibangan

083A12

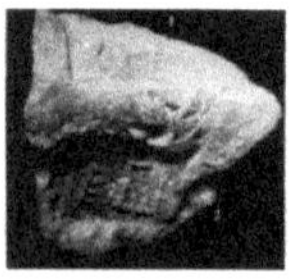

Kalibangan

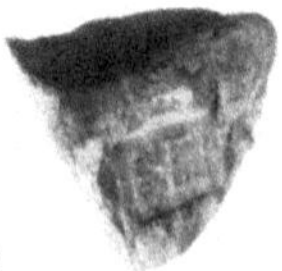

084A12

Kalibangan

084A2

8103

Kalibangan

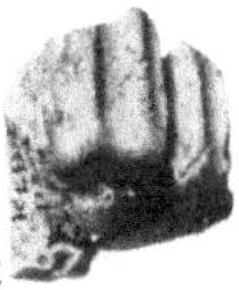

085A12

Kalibangan085B

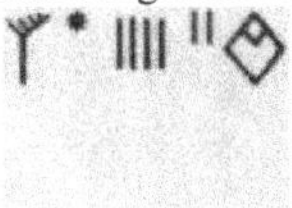

8106

Kalibangan086A14

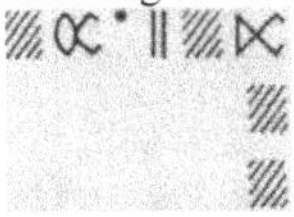

8114

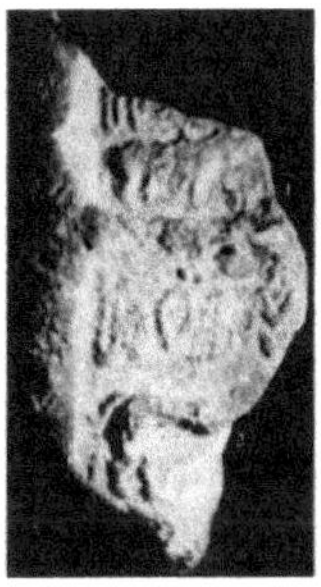

Kalibangan087A12

8116

Kalibangan

088A14

Kalibangan088B

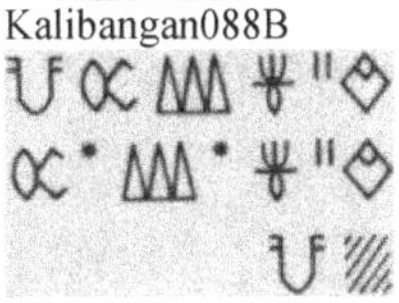

8119

Kalibangan089A14c

8101

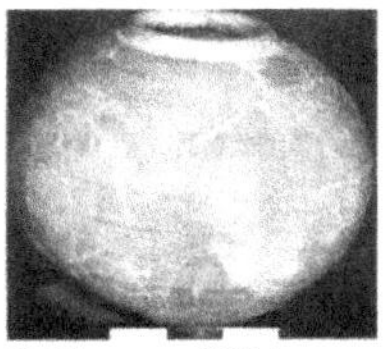

Kalibangan090A

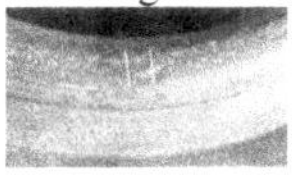

Kalibangan

090A1

Kalibangan

090A2 8202

Kalibangan091A

8212

Kalibangan092A

8210

Kalibangan093A

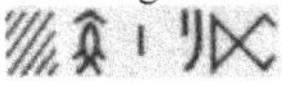

8219

Kalibangan094A

Kalibangan095A

Kalibangan096c

8221

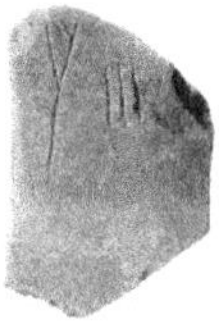

Kalibangan097A

8213

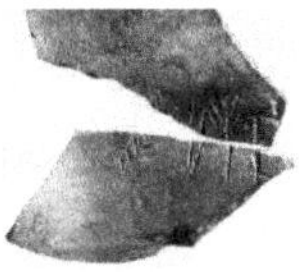

Kalibangan098A

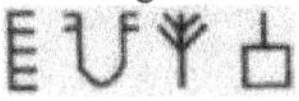

8201

Kalibangan099A

8208

Kalibangan100A

Kalibangan101A

8205

Kalibangan102A

8207

Kalibangan103A

8209

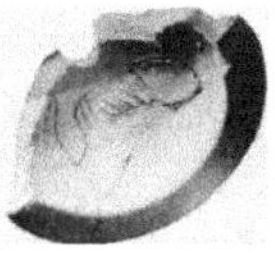

Kalibangan104A

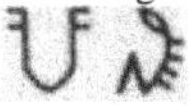

8218

Kalibangan105A

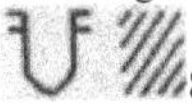

8216

Kalibangan106A

8204

Kalibangan107A

Kalibangan108A

8206

Kalibangan109A

Kalibangan110A 
8211

Kalibangan111A

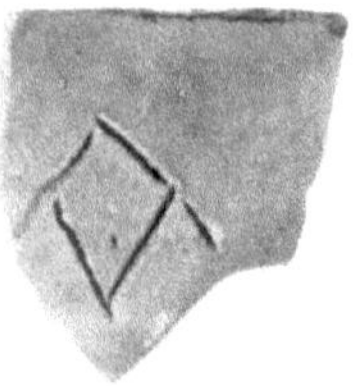

Kalibangan112A

Kalibangan118

Kalibangan119A

Kalibangan119B

Kalibangan120A

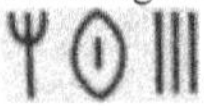
8220

Kalibangan122B

Kalibangan

122B2

Kalibangan

121A, B

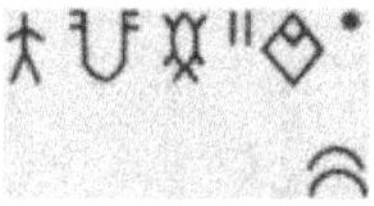

8302

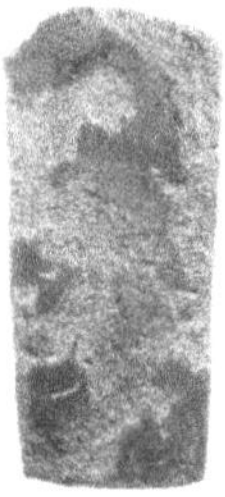

Kalibangan122A

Kalibangan

122A2
8301

Kalako-deray 01

Kalako-deray 05

Kalako-deray 06

Kalako-deray 07

Kalako-deray 08

Kalakoderay10

Khirsara1a

Khirsara2a

9051 Kot-diji

Lewandheri01

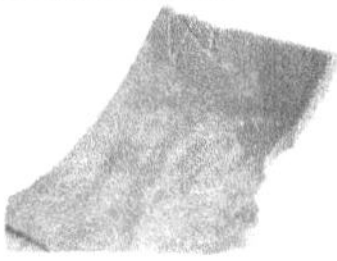

Loebanr01

Lohumjodaro1a

9011

Lothal001

7015

Lothal002

7031

Lothal003

Lothal004a

7080

Lothal005

7044

Lothal006a

7038

Lothal007a

Lothal008a

Lothal009

7022

Lothal010

7009

Lothal011

7026

Lothal012a

7089

Lothal013

7050

Lothal014a

7094

Lothal015

7086

Lothal016

7002

Lothal017

7008

Lothal018

7096

Lothal019a

7092

Lothal020

7078

Lothal021

7047

Lothal022a

7035

Lothal023a

7043

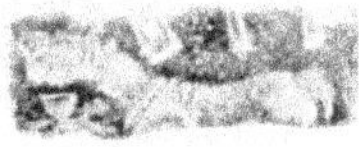

Lothal024

Lothal025

7104

Lothal026

7024

Lothal027

7036

Lothal028

7045

Lothal029

7005

Lothal030a

Lothal031

7076

Lothal032a 

Lothal033a

Lothal035

7101

Lothal036a

7081

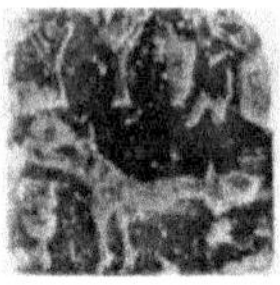

Lothal037

7034

Lothal038a 

7053

Lothal039

7102

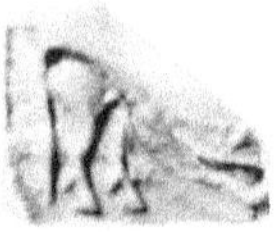

Lothal040a

Lothal041

7066

Lothal042

Lothal043 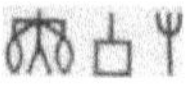

7049

Lothal044

Lothal045

7028

Lothal046

7107

Lothal047a

7074

Lothal048 

7025

Lothal049

Lothal050

Lothal051a

 7057

Pict-127: Upper register: a large device with a number of small circles in three rows with another row of short vertical lines below; the device is horned. A seed-drill? [Is this an orthographic model for  Sign 176?]

Lothal052  7011

Lothal054a 

7099

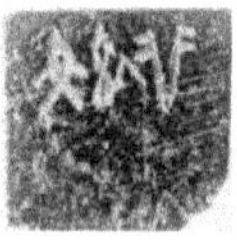

Lothal055

 7106

Lothal056
7100

Lothal057
7095

Lothal058a
7029

Lothal059
7097

Lothal060
7039

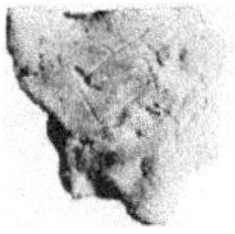

Lothal061

Lothal062
7054

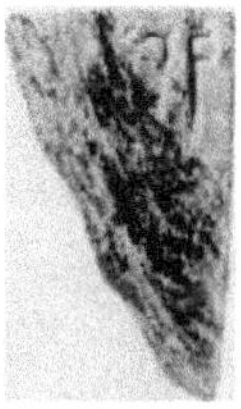

Lothal063

Lothal064
7030

Lothal065 7103

Lothal066acdef
7048

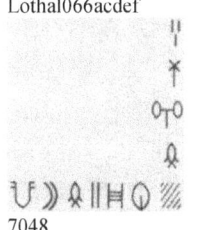

Lothal068
7070

Lothal069

Lothal070

Lothal071

Lothal072

Lothal075

Lothal076a

Lothal077

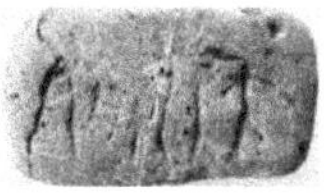

Lothal078 7077

Lothal079
7063

Lothal080a

Lothal081
7093

Lothal082 7105

Lothal083
7068

Lothal084 7112

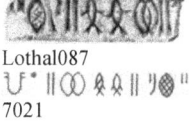

Lothal085

Lothal086 7007

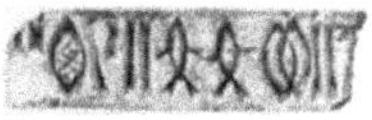

Lothal087
7021

Lothal088
7017

Lothal089
7090

Lothal090 7032

Lothal091
7111

Lothal092
7062

Lothal093
7064

Lothal094a
7073

Lothal095
7042

Lothal096
7023

Lothal097 7072

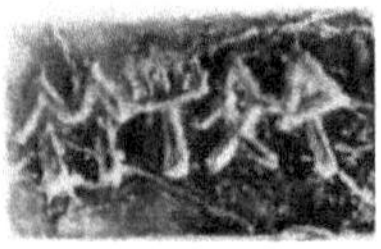

Lothal098
7082

Lothal099

Lothal100a

Lothal100B
7055

Lothal101
7001

Lothal102 7040

Lothal103
7018

Lothal104
7085

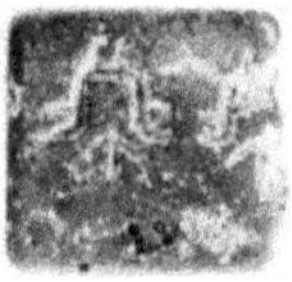

Lothal105
7016

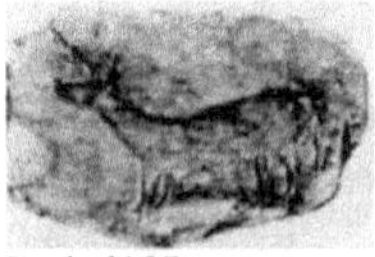

Lothal107

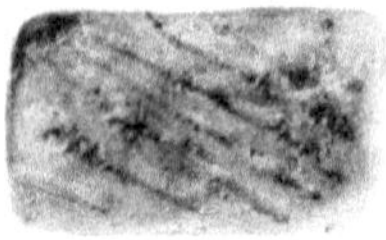

Lothal108

Lothal109a
7046

Lothal110
7006

Lothal111
7056

Lothal112
7020

Lothal113a
7004

Lothal114a
7013

Lothal115

7065

Lothal116 7027

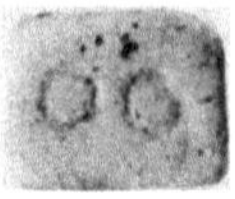

Lothal117 7075

Lothal118
7019

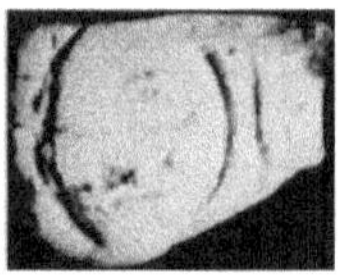

Lothal119

Lothal120

Lothal121

Lothal122
7069

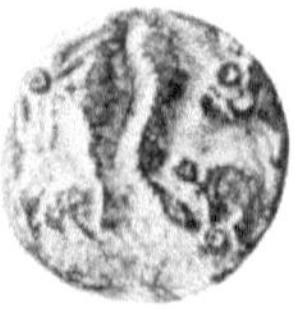

Lothal123A

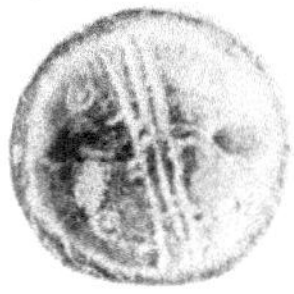

Lothal123B

Lothal124A

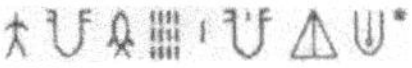

7224

Lothal125A

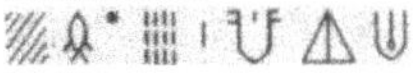

7241

Lothal126A

7242

Lothal127A

7221

Lothal128A

7239

Lothal129A

Lothal130A

Lothal131A 

7255

Lothal132A 

7213

Lothal133A 

7245

Lothal134A

7252

Lothal135A

7220

Lothal136A

 7225

Lothal137A

 7257

Lothal138A

Lothal138B

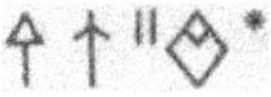 7214

Lothal139A

7223

Lothal140A

7244

Lothal141A1

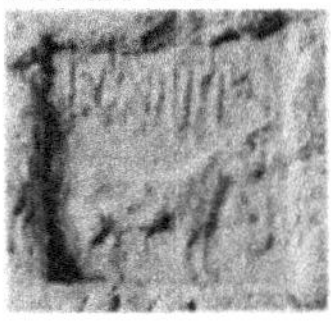

Lothal141A2

7280

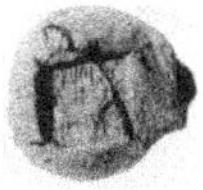

Lothal142A

Lothal142B

 7204

Lothal143A

Lothal143B

7243

Lothal144A

7274

Lothal145A

Lothal146AB

7279

Lothal147A

7260

Lothal148A

7270

Lothal149A

7272

Lothal150A
7268

Lothal151A

7266

Lothal152A
7222

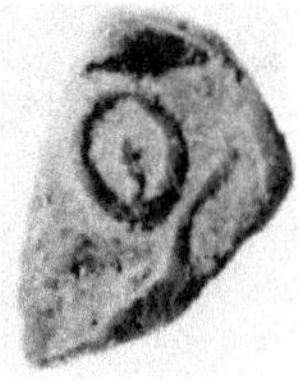

Lothal153A
7271

Lothal154A

Lothal155A

Lothal156A

Lothal157A

Lothal158A

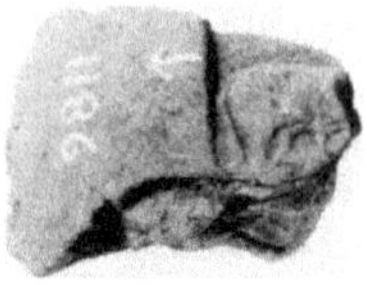

Lothal159A

Lothal160A

Lothal161A

7205

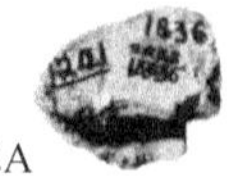

Lothal162A

Lothal162B

Lothal163A

Lothal163C

7228

Lothal164A

7230

Lothal165A

7203

Lothal166A

7206

Lothal167A

7231

Lothal168A

7234

Lothal169A

7235

Lothal170A 
7229

Lothal171A

Lothal172A

Lothal173A

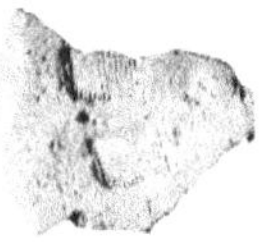

Lothal174A

Lothal175A

Lothal176A
7216

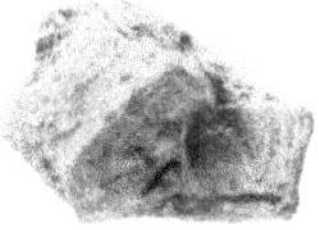

Lothal177A
7211

Lothal179A

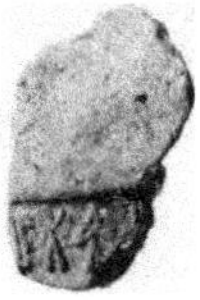

Lothal180A
7240

Lothal181A
7273

Lothal182A
7238

Lothal183A

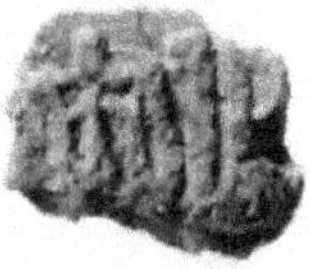

Lothal184A

Lothal185A

Lothal186A 
7259

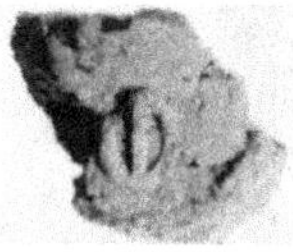

Lothal187A 
7209

Lothal188A

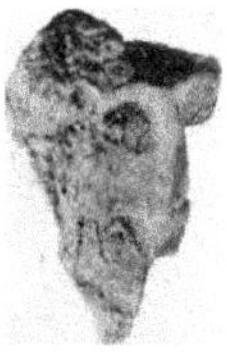

Lothal189A12

Lothal189A34
7217

Lothal190A13

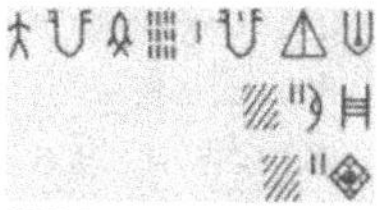

7236

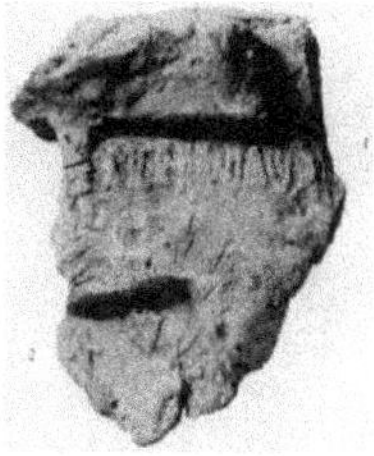

Lothal191A12

7249

Lothal192A12
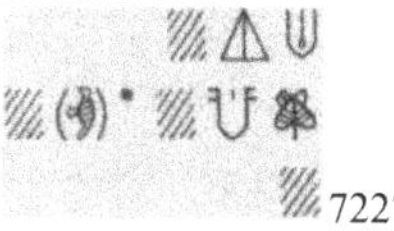7227

Lothal193A12

Lothal193A3
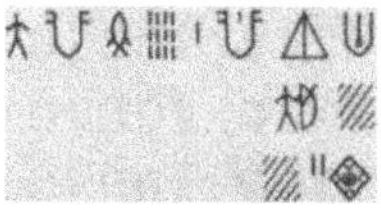
7253

Lothal194A1

Lothal194A2

7251

Lothal195A12

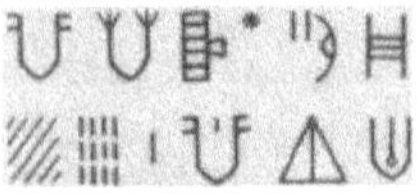
7258

Lothal196A12

7248

Lothal197A12

7237

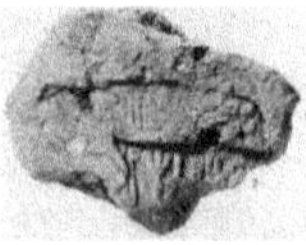

Lothal198A12

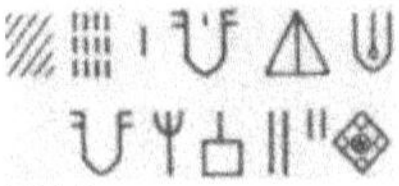
7215

Lothal199A12

7247

Lothal200A1

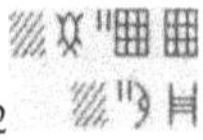
Lothal200A2
7219

Lothal201A12

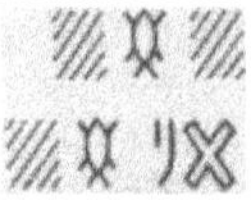
7263

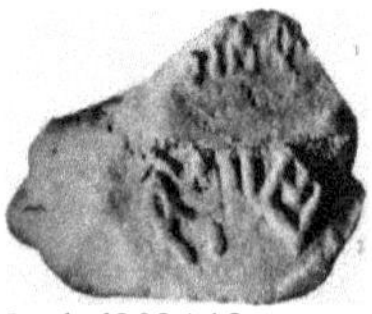

Lothal202A12

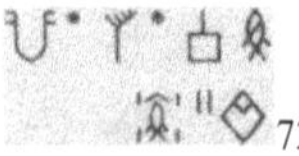
7267

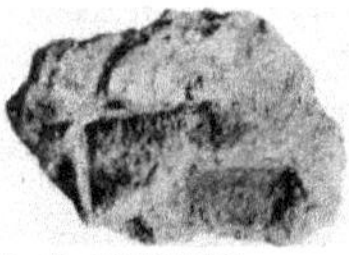

Lothal203A12

7246

Lothal204A

Lothal204F

7275

Lothal205A12

7218

Lothal206A12

7265

Lothal207A12

7281

Lothal208A12

Lothal209A12

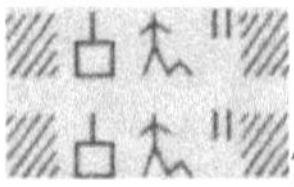
7262

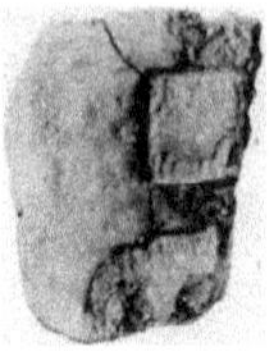

Lothal210A12
7201

Lothal211A13

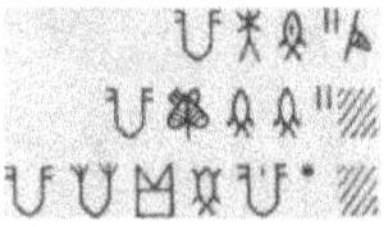
7277

Lothal212A12

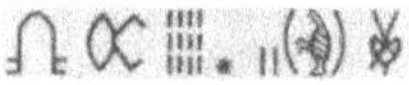
7261

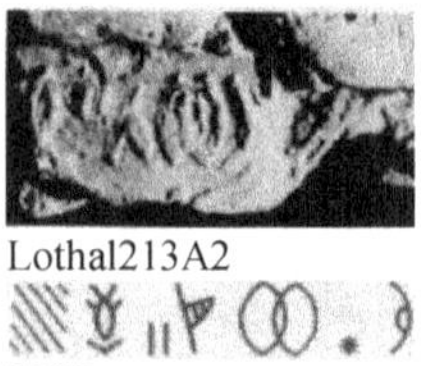

Lothal213A2
7207

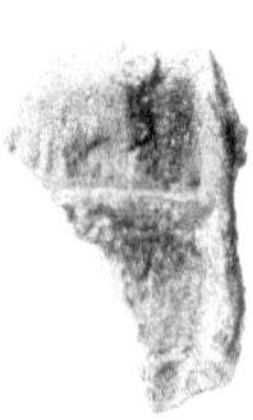

Lothal214A12

Lothal216D12

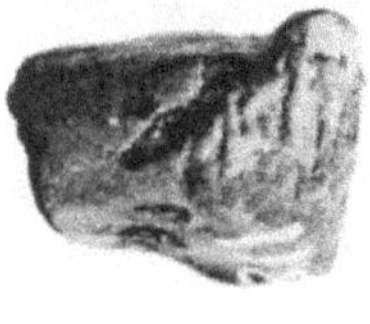
Lothal216E

7283

Lothal217A

Lothal217B

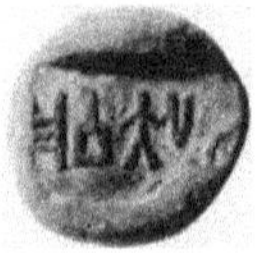
Lothal218A

7202

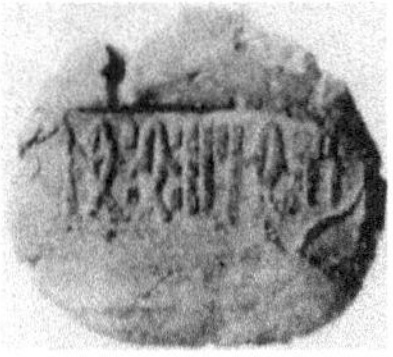
Lothal219A

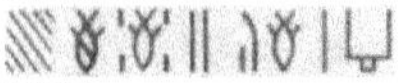
7282

Lothal220A

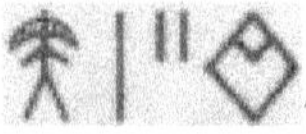
7278

Lothal221A

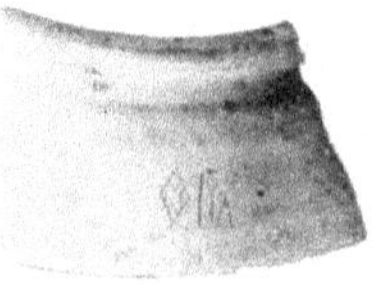
Lothal222A

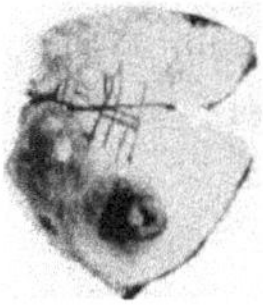
Lothal223A

Lothal224A

Lothal225A

Lothal227A

Lothal229A

Lothal230A

Lothal233A

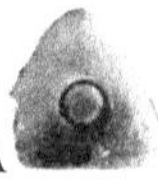
Lothal246A

Lothal269A

Lothal270A

Lothal272A

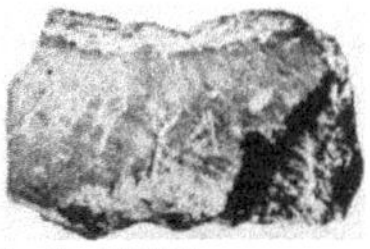
Lothal273A

7301

Lothal277A

Lothal280A

Lothal281A

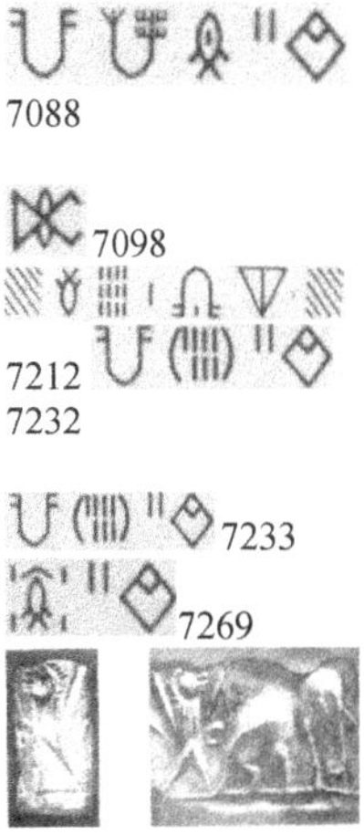
7088

7098

7212

7232

7233

7269

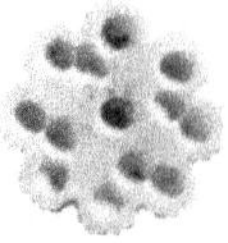
Maski
Mehi

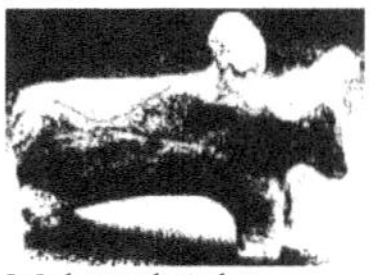
Mehrgarh zebu

Mehrgarh01

Mehrgarh04

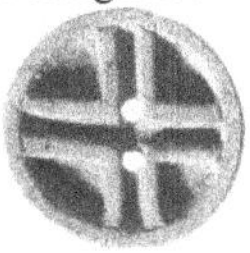
Mehrgarh05

Mehrgarh08

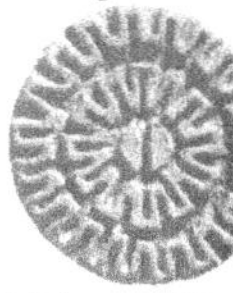
Mehrgarh10

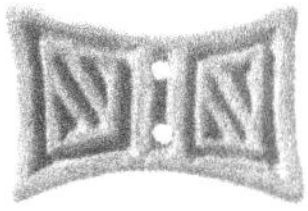
Mehrgarh11

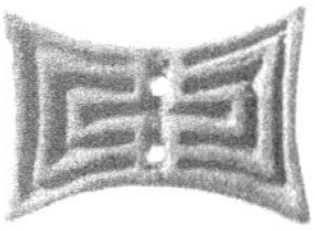
Mehrgarh12

Mehrgarh13

Mehrgarh14

Mehrgarh15

Mehrgarh16

Mehrgarh17

Mehrgarh18

m0001a

1067

m0002a

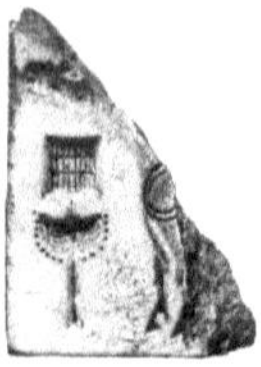

m0003a

2225

m0004a

3109

m0005

2247

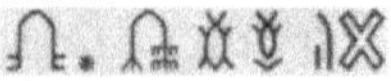

m0006a

2422

m0007

1011

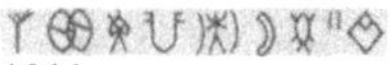

m0008a

1038

m0009a

2616

m0010

1006

m0011

m0012

3031

m0013

1069

m0014

1022

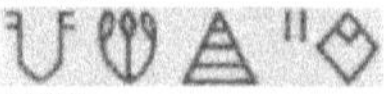

m0015

2177

m0016a

1037

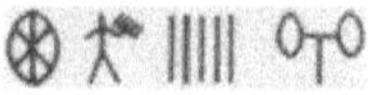

m0017

1035

m0018Ac

1548

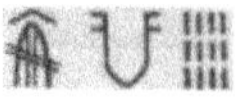

m0019a

1085

m0020a

1054

m0021a

2103

m0022a

1023

m0023a
2398

m0024
2694

m0025
1056

m0026a
2074

m0027a
2084

m0028a
2178

m0029a
2033

m0030a
2396

m0031
2576

m0032a
2180

m0033a
1042

m0034a
1058

m0035a
2333

m0036a
2455

m0037a
3103

m0039a
1544

m0040
1051

m0038a
1087

m0041

2271

m0042a

1096

m0043
2584

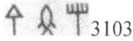

m0044a

3110

m0045a

1552

m0046a

3089

m0047a
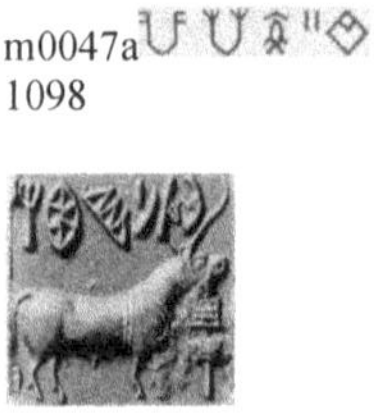
1098

m0048a

1186

m0049a

1047

m0050a

1557

m0051a

1555

m0052a

1540

m0053a

2128

m0054

2307

m0055a

2511

m0056

2406

m0057a

2340

m0058a
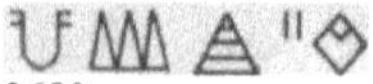
2680

m0059a

1029

m0060a

2124

m0061

m0062

3112

m0063

3068

m0064
2524

m0065
2440

m0066AC

1052

m0067
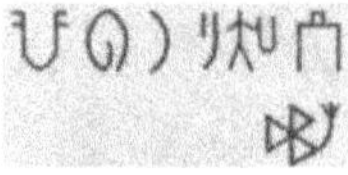
2264

m0068
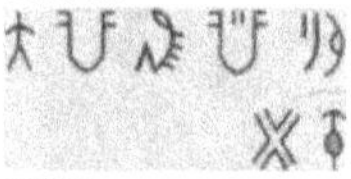
3108

m0069

1095

m0070

1048

m0071a

3083 [The second sign from left is an orthographic representation of the thigh of a bovid, perhaps a bull].

m0072a

2085

m0073

1046

m0074

2353

m0075

1019

m0076

m0077

3111

m0078

3118

m0079a

2083

m0080

2635

m0081a

1180

m0082

2451

m0083a

2267

m0084a

1108

m0085a

2365

m0086

2208

m0087

2148

m0088

1075

m0089

3116

m0090

3039

m0091

2429

m0092

2407

m0093

2305

m0094

2594

m0095

2657

m0096

2698

m0097

2549

m0098

2012

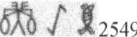

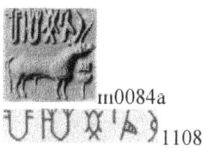

m0099
2475

m0100
1115

m0101
1537

m0102
1129

m0103
1076

m0104
2574

m0105
2337

m0106
2459

m0107
2593

m0108
1110

m0109
1151

m0110
2031

m0111
2029

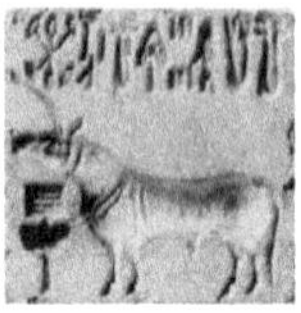
m0112
2099

m0113
2115

m0114
2166

m0115
3087

m0116
2481

m0117
1105

m0118
1104

m0119a
2018

m0120a
1099

m0121a
1188

m0122a
2015

m0123a

2702

m0124

1120

m0125

m0126

2311

m0127

1119

m0128a

2284

m0129

2193

m0130a

2285

m0131

2263

m0132

2082

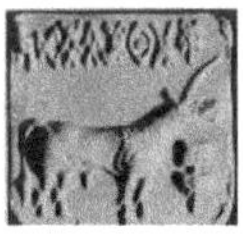

m0133a

2052

m0134

2187

m0135

1168

m0136

2233

m0137

2261

m0138

2381

m0139

2185

m0140

2563

m0141

2543

m0142

2630

m0143

2002

m0144

2048

m0145

1118

m0146

1100

m0147

3097

m0148

1245

m0149

1233

m0150

1236

m0151

2323

m0152
2102

m0153

2361

m0154
2373

m0155

1187

m0156

m0157
2022

m0158
2198

m0159
2355

m0160
2286

m0161
2088

m0162
2486

m0163
1543

m0164
2403

m0165
2687

m0166
1080

m0167.
1297

m0168a [The second sign may be an orthographic variant for a thigh of a bovid?]
2442

m0169
1113

m0170
2237

m0171
1149

m0172
1071

m0173
1161

m0174
1114

m0175

1291

m0176

1193

m0177

m0178

2354

m0179

m0180

2014

m0181
2490

m0182

2154

m0183
3113

m0184

2634

m0185

m0186

2161

m0187

2382

m0188

1287

m0189

1195

m0190
1205

m0191

1288

m0192

1206

m0193

2113

m0194
2254

m0195
2415

m0196
2474

m0197

2371

m0198

2363

m0199
2647

m0200
1148

m0201

2678

m0202
2625

m0203
1556

m0204
2623

m0205

1221

m0206

m0207 2458

m0208

2047

m0209
2375

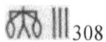

m0210

2656

m0211
1214

m0212

2577

m0213
1150

m0214

2571

m0215

3081

m0216

3036

m0217
2087

m0218

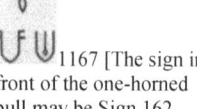

2175

m0219

2433

m0220a
3093

m0221a
3164

m0222
1194

m0223

1167 [The sign in front of the one-horned bull may be Sign 162

]

m0224

2215

m0225

2199

m0226
2152

m0227 2226

m0228 2502

m0229

3075

m0230.
1295

m0231
2444

m0232
2234 'Unicorn' with two horns! "Bull with two long horns (otherwise resembling the 'unicorn')", generally facing the standard. That it is the typical 'one-horned bull' is surmised from two ligatures: the pannier on the shoulder and the ring on the neck.

m0233

m0234.
1321

m0235

2689

m0236
2123

m0237

m0238AC
2534

m0239
2238

m0240.
1324

m0241
1536

m0242
2216

m0243
2390

m0244
2399

m0245
2290

m0246.
1317

m0247
2298

m0248.
1310

m0249
2378

m0250.

1308

m0251
2370

m0252
2423

m0253
2701

m0254
2090

m0255

2409 [The second sign is diamond-shaped?]

m0256
1332

m0257

2314

m0258a.
1340

m0259
2132

m0260
2567

m0261
2535

m0262 Zebu
2249

m0263

1336

m0264

2607

m0265

2155

m0266.
1306

m0267 Water-buffalo
2257

m0268 Water-buffalo
2445

m0269

2663

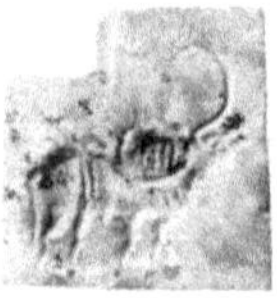

m0270

m0271 Goat-antelope with horns turned backwards and a short tail

m0272 Goat-antelope with horns bending backwards and neck turned backwards
2554

m0273 2673

m0274

1342

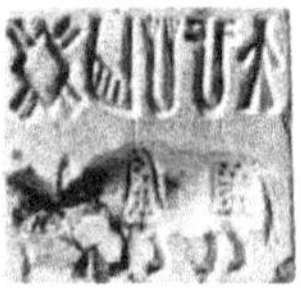

m0275
2131

m0276AC

3122

m0277

2309

m0278

2648

m0279

3060

m0280

1373

m0281

3115

m0282

2304

m0283

2127

m0284

2195

m0285

1367

m0286

2517

m0287

m0288

2518

m0289

3121

m0290

2527

m0291 Tiger
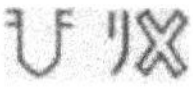
3069

m0292 Gharial

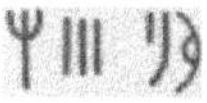
1361

m0293 Gharial

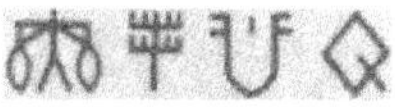
1360

m0294 One-horned bull?; elephant

1376

m0295 Pict-61: Composite motif of three tigers joined together.

1386

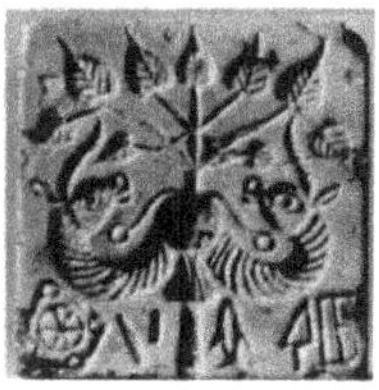

m0296 Two heads of one-horned bulls with neck-rings, joined end to end (to a standard device with two rings coming out of the top part?), under a stylized pipal tree with nine leaves.

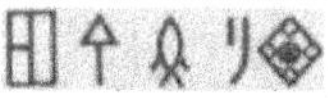
1387

m0297a Head of a one-horned bull attached to an undentified five-point symbol (octopus-like?)

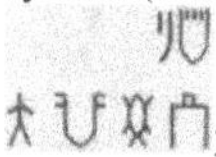
2641

m0298

m0299 Composite animal with the body of a ram, horns of a bull, trunk of an elephant, hindlegs of a tiger and an upraise serpent-like tail.

1381

m0300
Pict51: Composite animal: human face, zebu's horns, elephant tusks and trunk, ram's forepart, unicorn's trunk and feet, tiger's hindpart and serpent-like tail.

2521

m0301 Composite motif: human face, body or forepart of a ram, body and front legs of a unicorn, horns of a zebul, trunk of an elephant, hindlegs of a tiger and an upraised serpent-like tail.

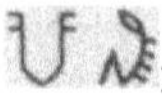2258

m0302 Composite animal with the body of a ram, horns of a bull, trunk of an elephant, hindlegs of a tiger and an upraise serpent-like tail.

1380

m0303 Composite animal.

2411

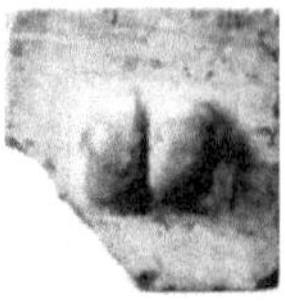

m0304B

m0304AC Pict-81: Person (with three visible faces) wearing bangles and armlets seated on a platform (with an antelope looking backwards) and surrounded by five animals: rhinoceros, buffalo, antelope, tiger and elephant.

2420

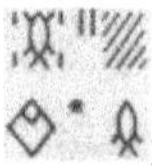

m0305AC 2235
Pict-80: Three-faced, horned person (with a three-leaved pipal branch on the crown with two stars on either side), wearing bangles and armlets.

m0306 Person grappling with two tigers standing on either side of him and rearing on their hindlegs.

2086

m0307 Person grappling with two tigers standing on either side of him and rearing on their hindlegs.

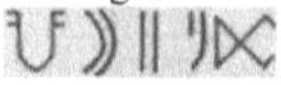2122

m0308AC Pict-105: Person grappling with two tigers standing on either side of him and rearing on their hindlegs.

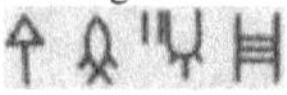2075

[The third sign from left may be a stylized 'standard device'?]

m0309 Pict-109: Person with hair-bun seated on a tree branch; a tiger looks at the person with its head turned backwards.

2522

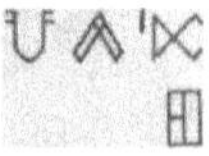

m0310AC
1355

m0311 Pict-52: Composite motif: body of a tiger, a human body with bangles on arms, antelope horns, tree-branch and long pigtail.

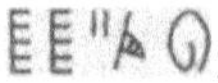2347

m0312 Persons vaulting over a water-buffalo.

m0313

2637

m0314

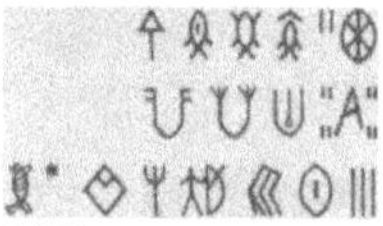
1400

m0315

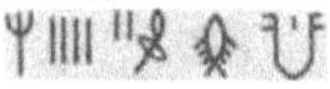1395

m0316 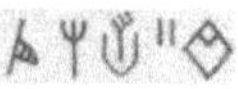
2408

m0317silver

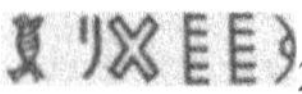2016

Mohenjodaro FEM, Pl. LXXXVIII, 316

2316

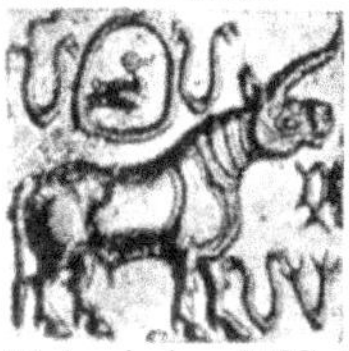

Mohenjodaro MIC, Pl. CVI,93

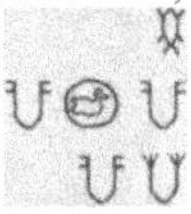1093

Mohenjo-daro. Copper seal. National Museum, New Delhi. [Source: Page 18, Fig. 8A in: Deo Prakash Sharma, 2000, *Harappan seals, sealings and copper tablets*, Delhi, National Museum].

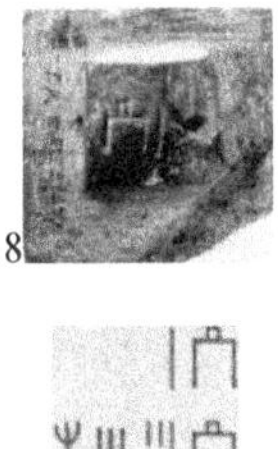

m0318

m0318B 2626

m0319

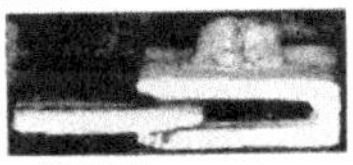

m0319C

2260

m0320

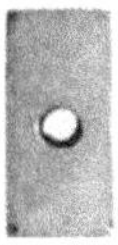

m0320D

2449

m0321

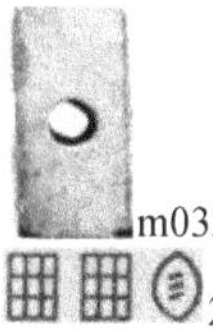

m0321D

2173

m0322

m0322D

1192

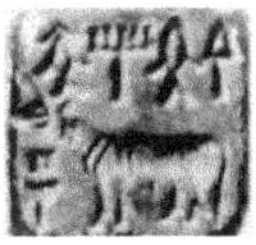

m0323

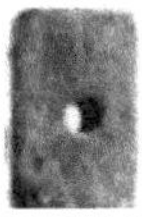

m0323D

1277

m0324A

m0324B

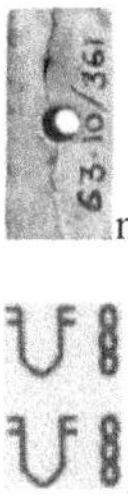

m0324D

1252

m0325A

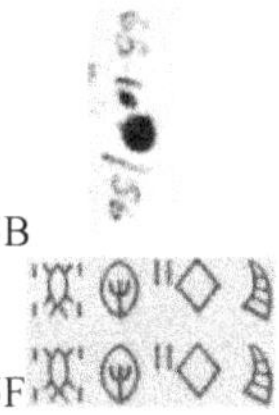

m0325B

m0325F

3106

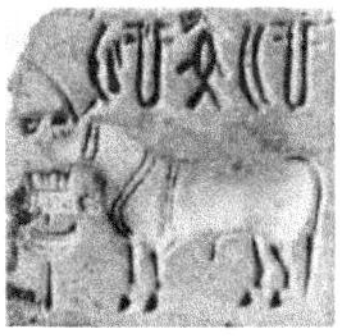

m0326A

m0326B

m0326C

m0326D

m326E

m0326F

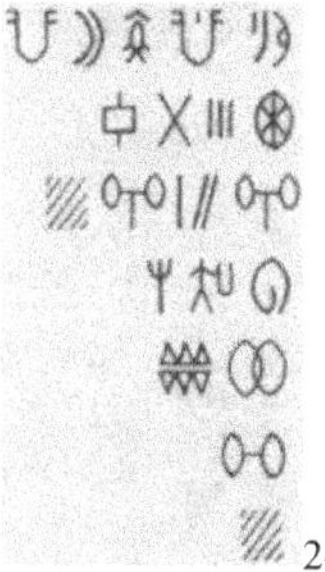

2405

m0327

2631

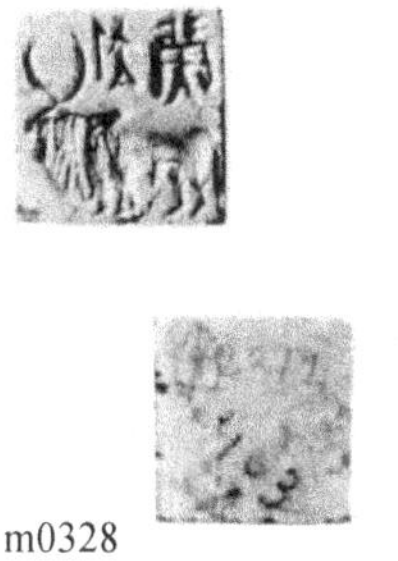

m0328

m0328B 2108

m0329

1477

m0330A

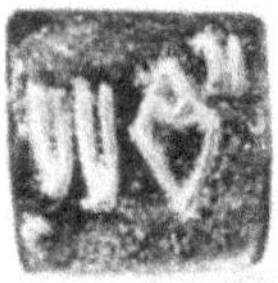

m0330B Perforated through the narrow edge of a two-sided seal

1475

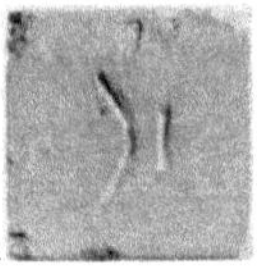

m0331A

m0331B

m0331D

m0331F Cube seal

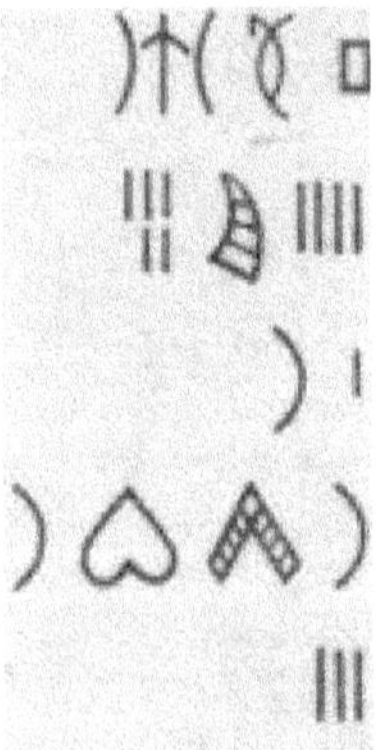

1471

m0332AC

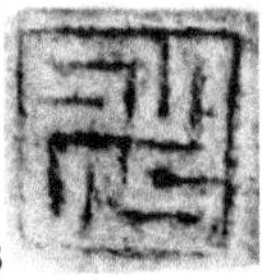

m0333

m0334

m0335

m0336

m0337

m0338

m0339

m0340

m0341

m0342

m0343

m0344

m0345

m0346

m0347

m0348

m0349

m0350

m0351

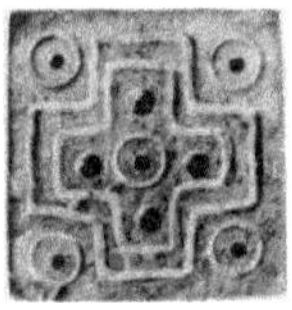

m0352A

m0352C

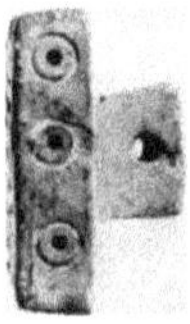

m0352D

m0352E
m0352F

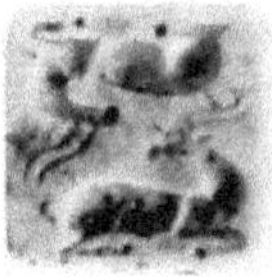

m0353

m0354

1403

m0356

1406

m0357

1401

m0358

2297

m0360
3102

m0361

2101

m0362

1466

m0363
1469

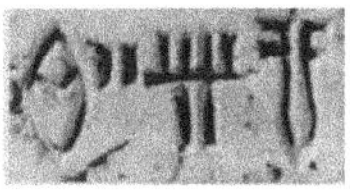
m0364
1465

m0365
2273

m0366
2077

m0367

2044

m0368
2336

m0370
2138

m0371
2461

m0372
1438

m0373
2043

m0374
2097

m0375

m375AC

m0376
1426

m0377
3120

m0378
1402

m0379
2159

m0380
2470

m0381

2162

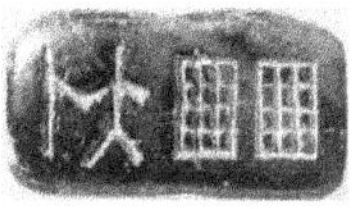
m382AC
1437

m0383
2240

m0384
2302

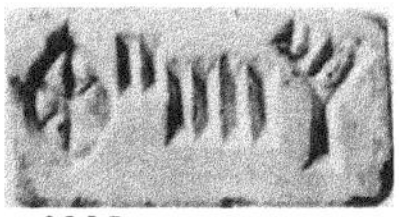
m0385

2387

m0386

1449

m0387
2041

m0388

2200

m0389
2397

m0390

1444

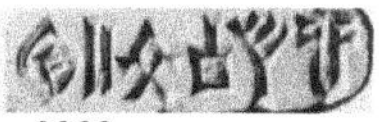
m0392
2046

m393AC
2120

m0394

2213

m0395
2183

m0396

1421

m0397

1415

m0398

2308

m399AC

1414

m0400

3088

m0401

2346

m0402

2395

m0403

1410

m404AC

1422

m0405

2221

m0406

1399

m0407

2643

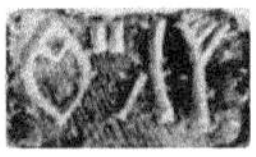

m0408

2100

m0409

2699

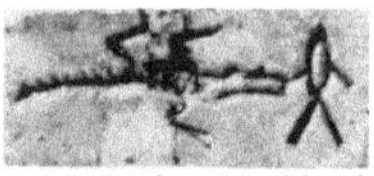

m0410 Pict-64: Gharial snatching, with its snout, the fin of a fish

2133

m0411

1431

m0412

1450

m0413

2319

m0414A

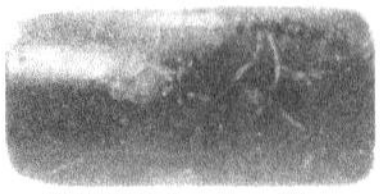

m0414B Seal with incision on obverse

2004

m0415a Bison

2500

m0416 Bison .

1309

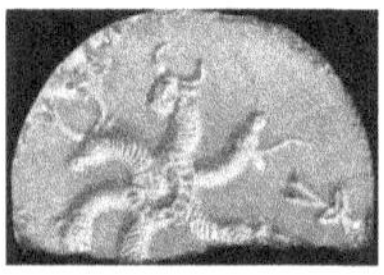

m417AC Pict-62: Composition: six heads of animals: of unicorn, of short-horned bull (bison), of antelope, of tiger,
and of two other uncertain animals) radiating outward from a hatched ring (or 'heart' design).

1383

m0418acyl

m0419acyl

m0419dcyl

m0419fcyl

m0420A1si

m0420A2si

3236

m0421A1si

m0421A2si

3237

m0422A1si

m0422A2si

m0423A1si

m0423A2si

3221

m0424A1si

m0424A2si

m0425A1si

m0425A2si

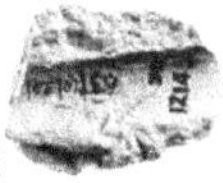
m0426Asi

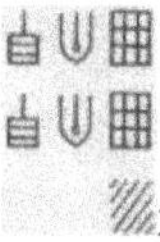
m0426Bsi 2809

m0427t

1630

m0428At

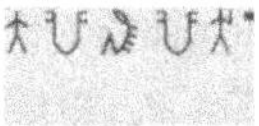
m0428Bt
1607 Pict- 132:
Radiating solar symbol.

m0429 Text 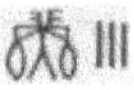 2862

m0430At

m0430Bt

2862

m0431At

m0431Bt
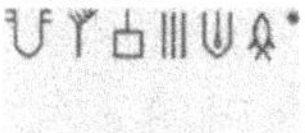
3239

m0432At

m0432Bt
1624

m0433At

m0433Bt
3233

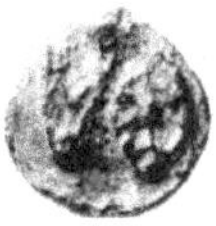
m0434At

m0434Bt

3248

m0435t

m0436At

m0436Bt

2804

m0437t

2867

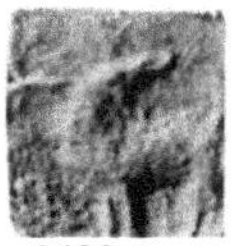
m0438atcopper

m0439t

m440AC

m0441At

m0441Bt

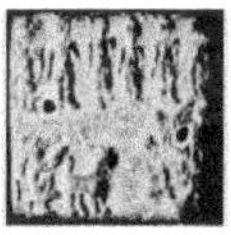

m0442At

m0442Bt

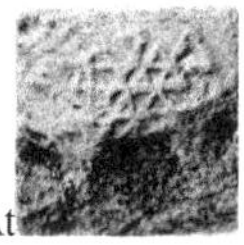
m443At

m443Bt

m444At 
3223

m445Bt 

m445AC

2821

m446At 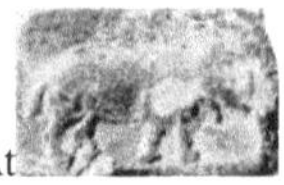

m446Bt

2854

m447At 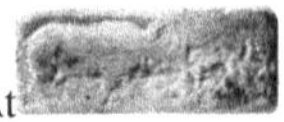

m447Bt

m448t

m449Bt

m449AC

2836

m450At

m450Bt

2864

m0451At

m0451Bt

3235

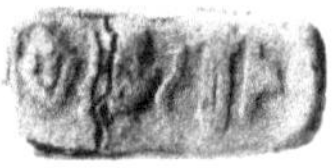
m0452At

m0452Bt

2855

m0453At

m453BC

1629 Pict-82 Person seated on a pedestal flanked on either side by a kneeling adorant and a hooded serpent rearing up.

m0455At

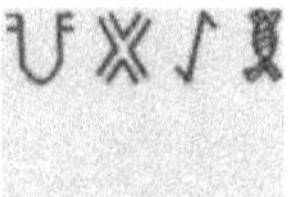
1619

m0456At
3219

m0457At 

m0457Bt

m0457Et

m0458At 

m0458Bt

3227

m0459At 

m0459Bt

3225

m0460At 

m0460Bt

3228

m0461At 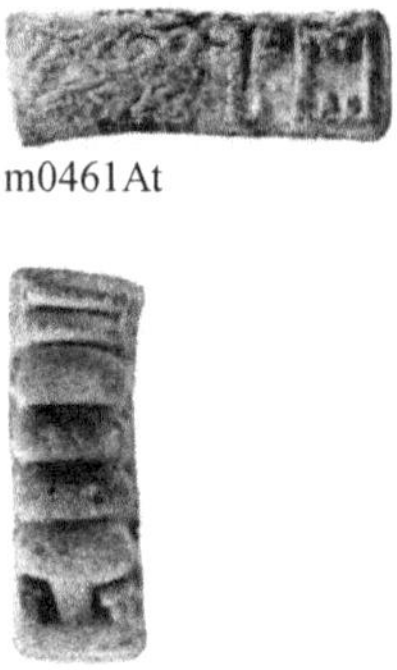
m0461Bt

2806 Pict-73: Alternative 1. Serpent (?) entwined around a pillar with capital (?); motif carvd in high-relief. Alternative 2. Ring-stones around a pillar with coping stones in a building-structure as at Dholavira?

m0462At

m0462Bt

3215

m0463At

m0463Bt

2813

m0464At

m0464Bt

3216

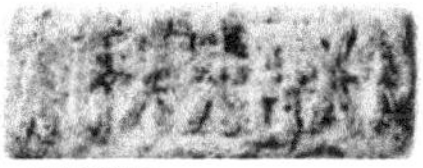

m0465At

m0465Bt

3220

m0466At

m0466Bt

m0467At

m0467Bt

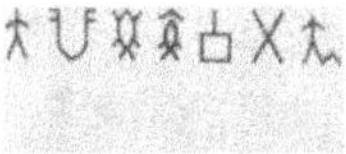

3209

m0468At

m0468Bt

3249

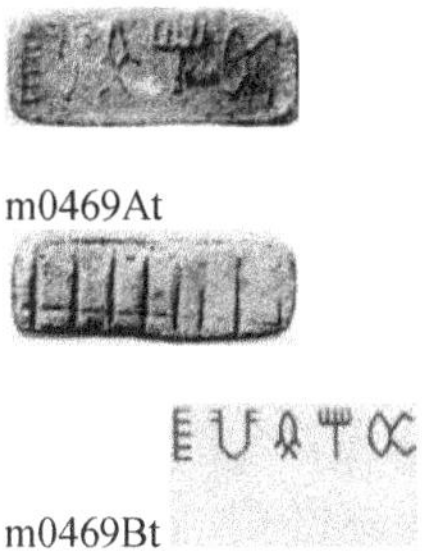

m0469At

m0469Bt

2830

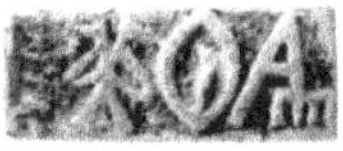

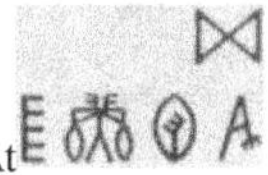

m0470At

2810

m0471At

m0471Bt

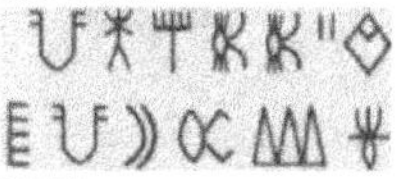

3232

m0472At

1615

m0473At

2848

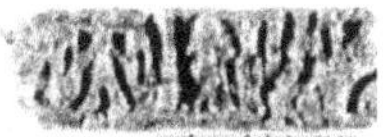

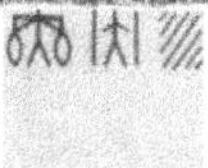

m0474At

3243

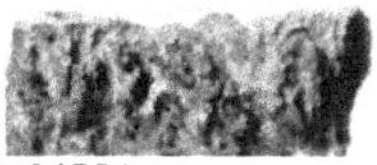

m0475Atcopper

3247

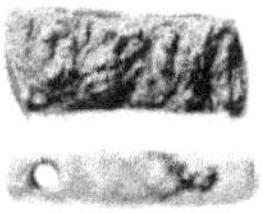

m0476At

m0476Ct

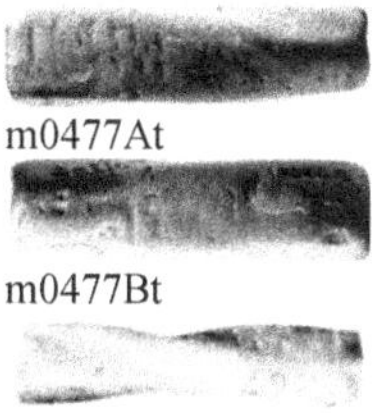

m0477At

m0477Bt

m0477Ct

2844

Two rhinoceroses, one at either end of the text (Pict-29).

m0478At

m0478Bt

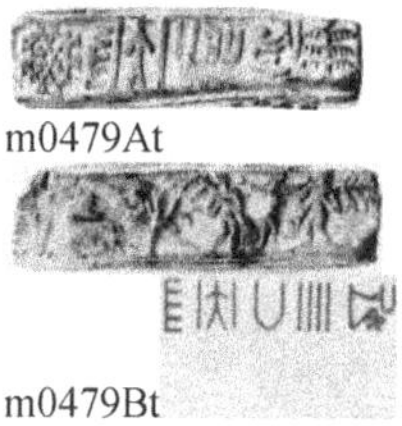

m0479At

m0479Bt

3224

m0480At

m0480Bt Tablet in bas-relief. Side a: Tree Side b: Pict-111: From R.: A woman with outstretched arms flanked by two men holding uprooted trees in their hands; a person seated on a tree with a tiger below with its head turned backwards; a tall jar with a lid.

Is the pictorial of a tall jar the Sign 342  with a lid? Sign 45  seems to be a kneeling adorant offering a pot (Sign 328 )

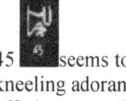

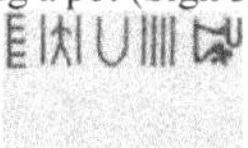

2815 Pict-77: Tree, generally within a railing or on a platform.

3230

m0481At
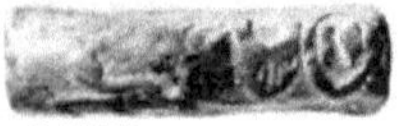
m0481Bt
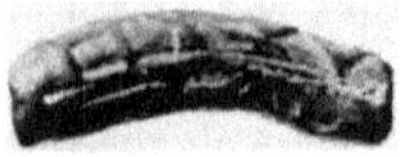
m0481Ct
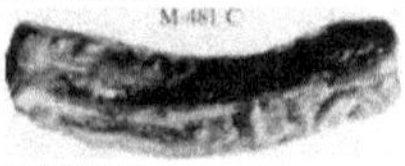
m0481Et

2846 Pict-41: Serpent, partly reclining on a low platform under a tree

m0482At

m0482Bt

1620 Pict-65: Gharial, sometimes with a fish held in its jaw and/or surrounded by a school of fish.

m0483At

m0483Bt

m0483Ct

m0483Et

2866

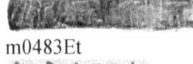

Pict-145: Geometrical pattern.
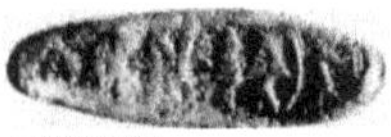
m0484At

m0484Bt

2861

m0486at

m0486bt

m0486ct
1625

m0487At
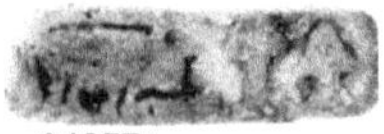
m0487Bt

m0487Ct
2852

m0488At

m0488Bt

m0488Ct

2802 Prism: Tablet in bas-relief. Side b: Text +One-horned bull + standard. Side a: From R.: a composite animal; a person seated on a tree with a tiger below looking up at the person; a svastika within a square border; an elephant (Composite animal has the body of a ram, horns of a zebu, trunk of an elephant, hindlegs of a tiger and an upraised serpent-like tail). Side c: From R.: a horned person standing between two branches of a pipal tree; a ram; a horned person kneeling in adoration; a low pedestal with some offerings.

m0489At
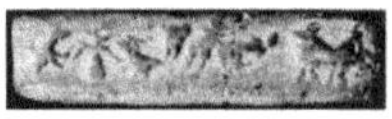
m0489Bt
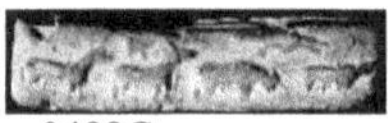
m0489Ct
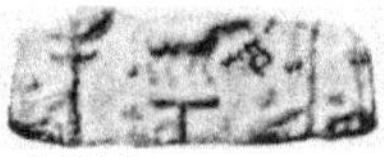
m0490At

m0490BCt

1605

m0491At

m0491BCt

1608 Pict-94: Four persons in a procession, each carrying a standard, one of which has the figure of a one-horned bull on top.

m0492At
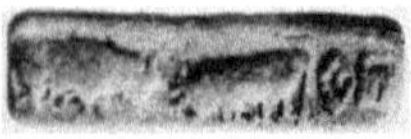
m0492Bt Pict-14: Two bisons standing face to face.
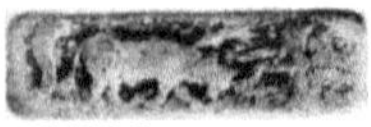
m0492Ct

2835
Pict-99: Person throwing a spear at a bison and placing one foot on the head of the bison; a hooded serpent at left.

m0493At

m0493Bt Pict-93: Three dancing figures in a row.
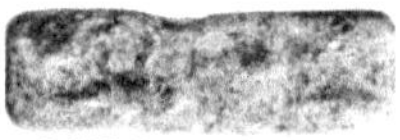
m0493Ct

2843

m0494At

m0494BGt Prism Tablet in bas-relief.

1623

m0495At

m0495Bt

m0495gt

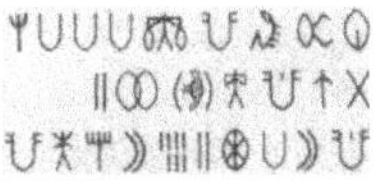
2847b

m0496At

m0496Bt

m0496Dt

m0497At

m0497Bt

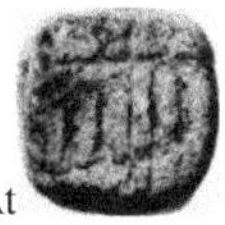
m0498At

m0498Bt

m0498Dt

m0499At

m0500at

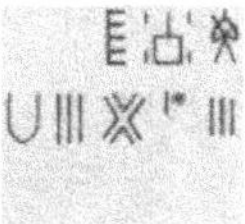
m0500bt
2604 Pict-76: Tree, generally within a railing or on a platform.

m0501At

m0501Bt

1412

m0502At
m0502Bt
3345

m0503 Text

3346

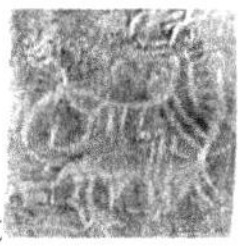
m0504At

m0504Bt
3323

m0505At
m0505Bt
1702

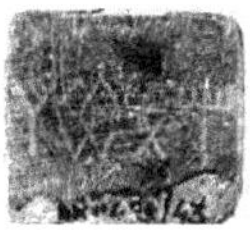

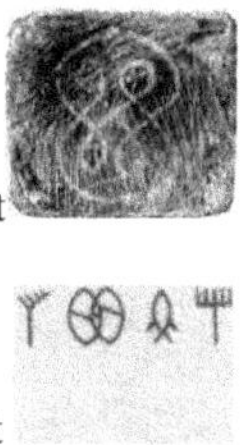
m0507At
m0507Bt
3350

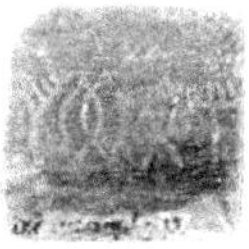

m0508At
m0508Bt
3352

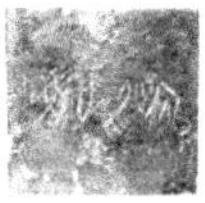
m0509At

m0509Bt

3320

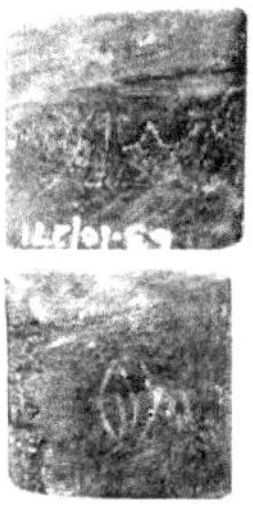
m0510At

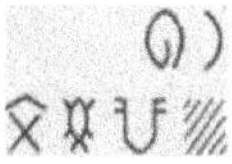
m0510Bt
3319

m0511At
m0511Bt
2905

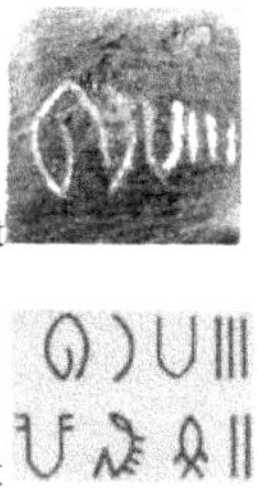
m0512At
m0512Bt
2906

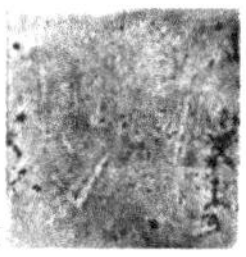
m0513At

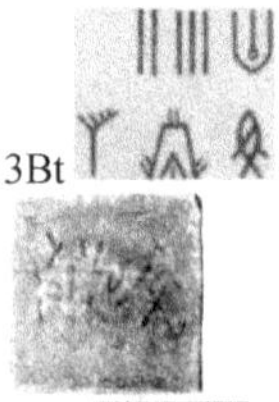

m0513Bt

3364

m0514At

m0514Bt
3302

m0515 Text

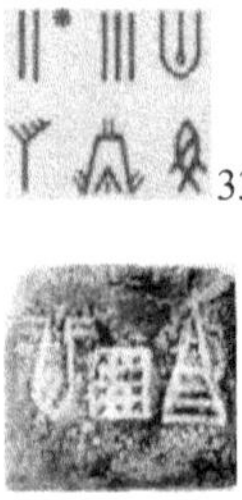

3335

m0516At

m0516Bt

3398

m0517At

m0517Bt 3334

m0519At

m0519Bt

1710

m0520
At, Bt

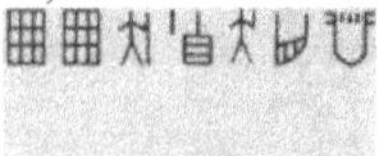

2916 m0521

3407

m0522At

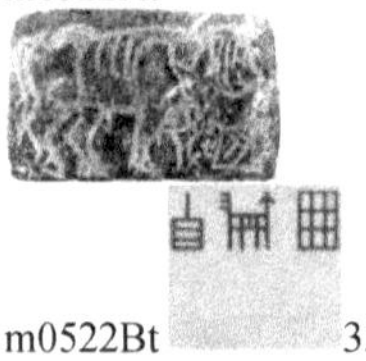

m0522Bt 3378

m0523At

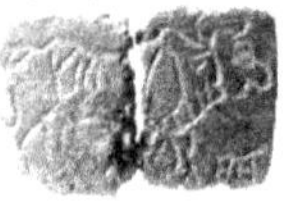

m0523Bt

1714

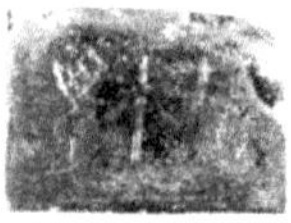

m0524At

m0524Bt

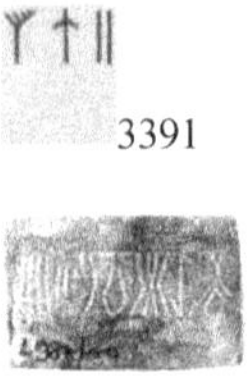

3391

m0525At

m0525Bt

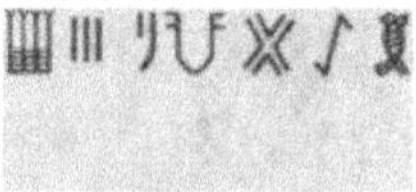

1713 Buffalo

m0526At

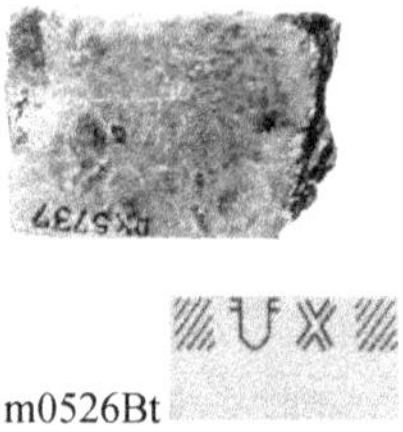

m0526Bt
3329 Buffalo

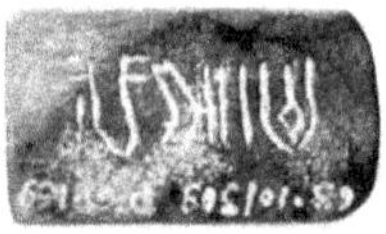

m0527At

m0527Bt
3336

m0528At

m0528Bt
3368

m0529At

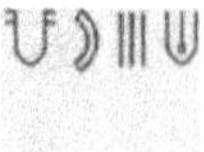

m0529Bt
3392

m0530At

m0530Bt
3356

m0531At

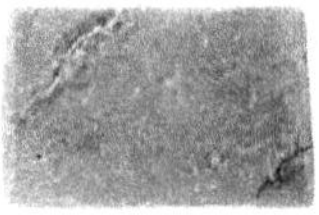

m0531Bt

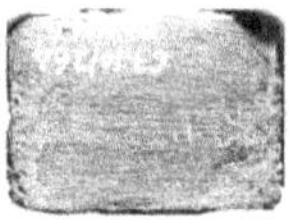

m0532At

m0532Bt
3349

m0534At

m0534Bt

3304

m0535At
m0535Bt

3355

m0536At
m0536Bt

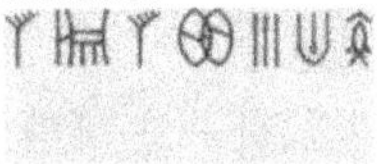
3312

m0537At
m0537Bt

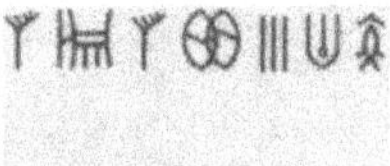
1705

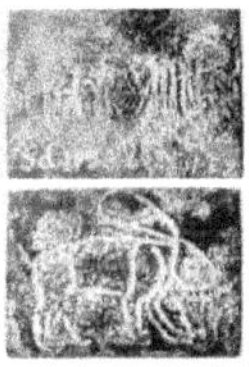
m0538At
m0538Bt

3384

m0539At
m0539Bt

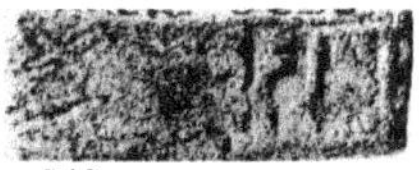
m540t

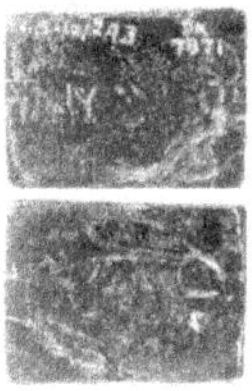
m0541At
m0541Bt

3331

m0542At
m0542Bt

3326 Hare?

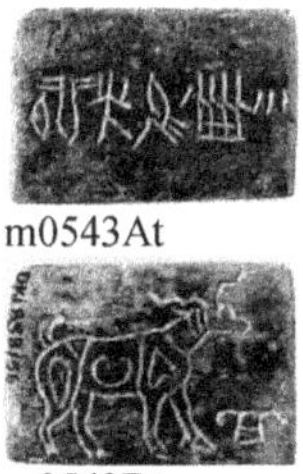
m0543At
m0543Bt

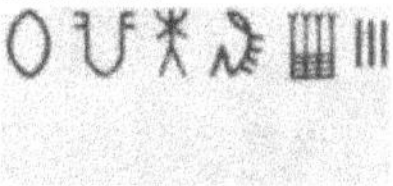
3363 [Note the 'heart' orthograph on the body of the antelope. This is comparable to Sign 323

]

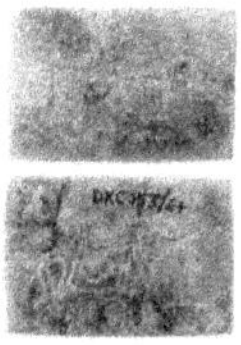
m0544At
m0544Bt

3357

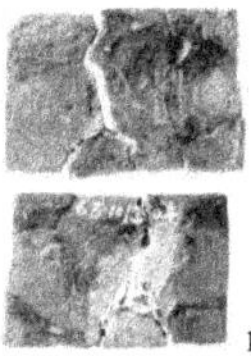
m0545At
m0545Bt

3301

m0546At
m0546Bt

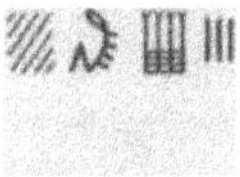
3383

m0547At
m0547Bt

3303

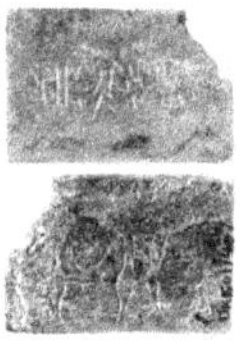
m0548At
m0548Bt

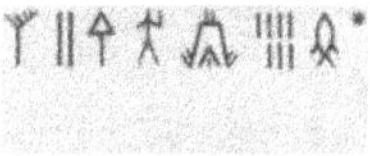
3305

m0549At
m0549Bt

3373

m0550At

m0550Bt
3351

m0551At
m0551Bt

1708 Ox-antelope with long tail.

m0552At

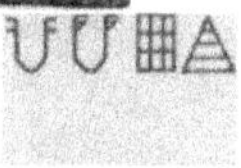
m0552Bt
3306

m0553At

m0553Bt
3353

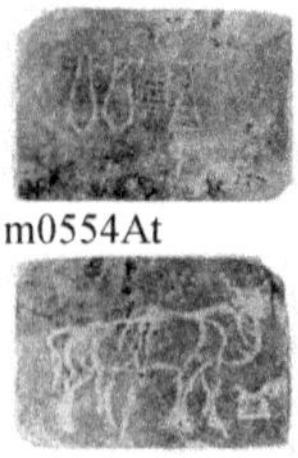
m0554At
m0554Bt

1712

m0555At
m0555Bt
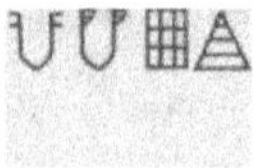
3314

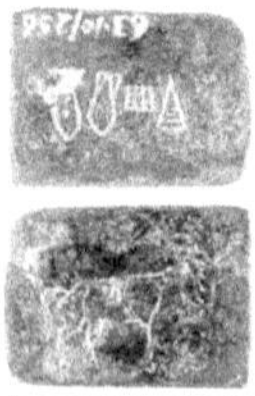
m0556At
m0556Bt
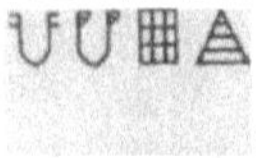
3404

m0557At
m0557Bt

3341

m0558At
m0558Bt

3342

m0559At
m0559Bt
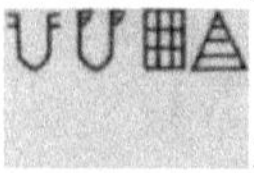
2909

m0560At
m0560Bt

3386

m0561At
m0561Bt

3339

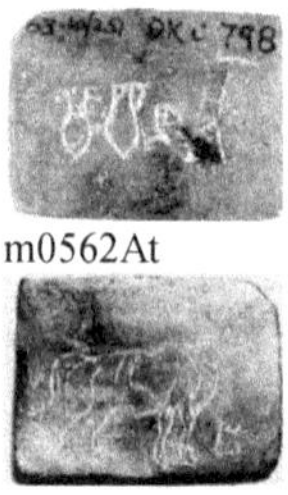
m0562At
m0562Bt

3361

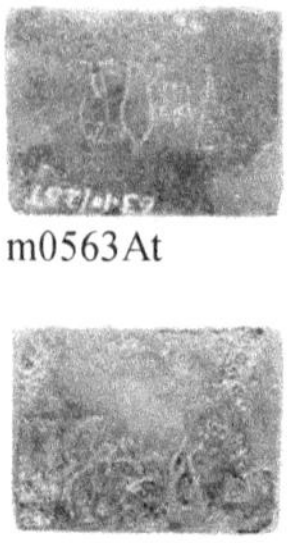
m0563At
m0563Bt
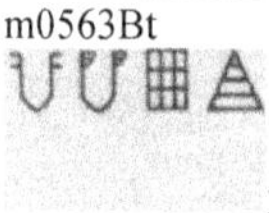
3379

m0564At
m0564Bt

3371

m0565At
m0565Bt
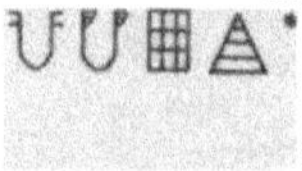
3403

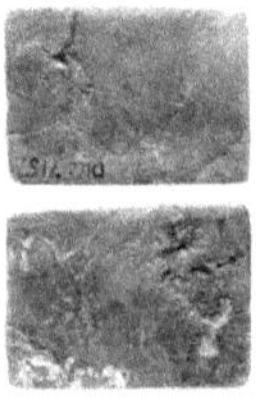
m0566At
m0566Bt
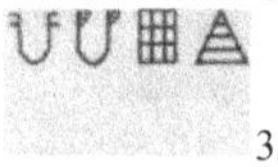
3359

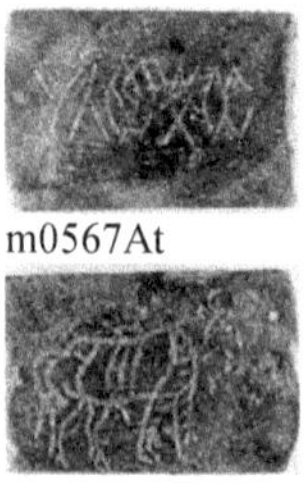
m0567At
m0567Bt
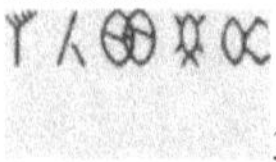
3322
Bison.

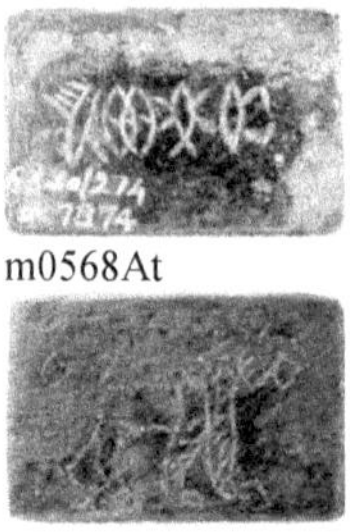
m0568At
m0568Bt

3332

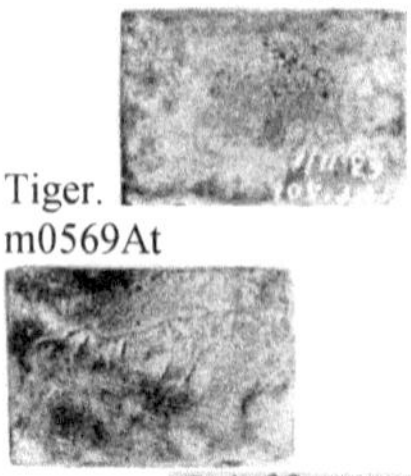
Tiger.
m0569At
m0569Bt

3372

m0571At
m0571Bt

2913 Horned elephant. Almost similar to the composition: Body of a ram (with inlaid 'heart' sign), horns of a bull, trunk of an elephant, hindlegs of a tiger and an upraised serpent-like tail

m0572At
m0572Bt

3317

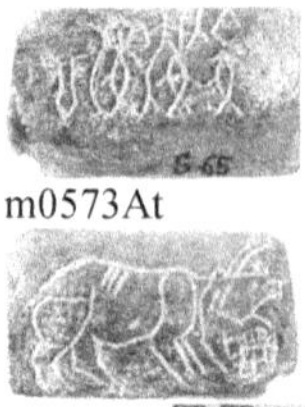
m0573At
m0573Bt
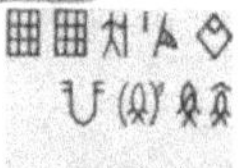
3415

m0574At

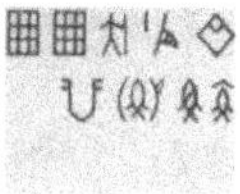
m0574Bt
3318

m0575At
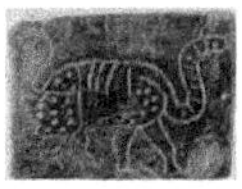
m0575Bt

3316

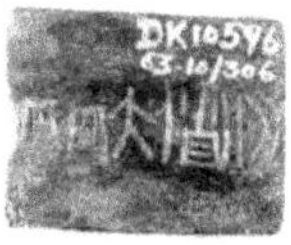
m0576At

m0576Bt

3344

m0577At

m0577Bt
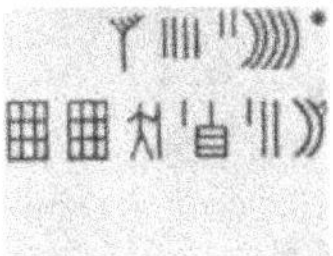
3347

m0578At

m0578Bt
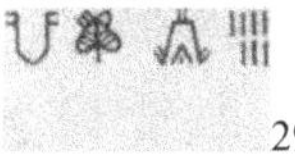
2908

m0580At

m0580Bt

3321

m0581At
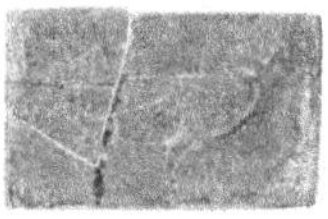

m0581Bt
3340

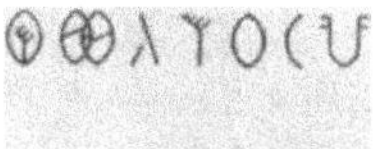

2914

Pict-89: Standing person with horns and bovine features, holding a bow in one hand and an arrow or an uncertain object in the other.

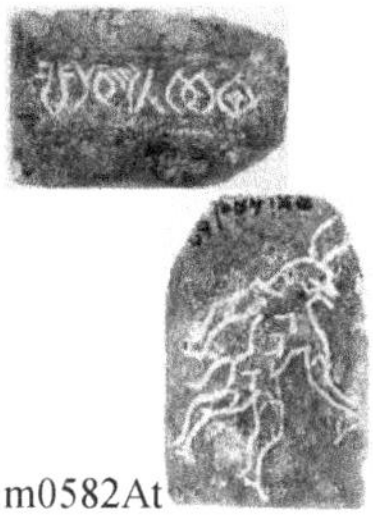
m0582At
m0582Bt
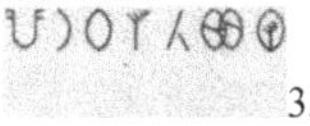
3358

m0583At
m0583Bt
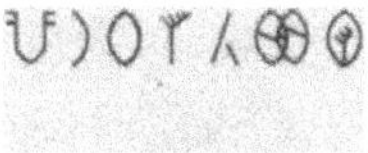
3387

m0584At
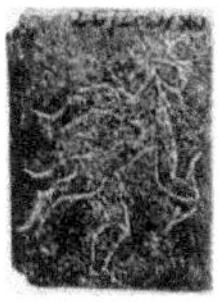
m0584Bt

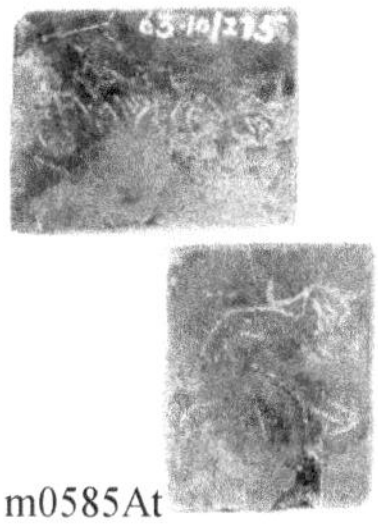
m0585At
m0585Bt
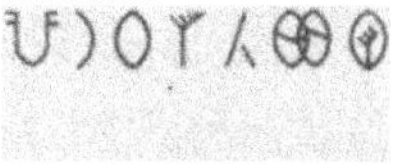
3369

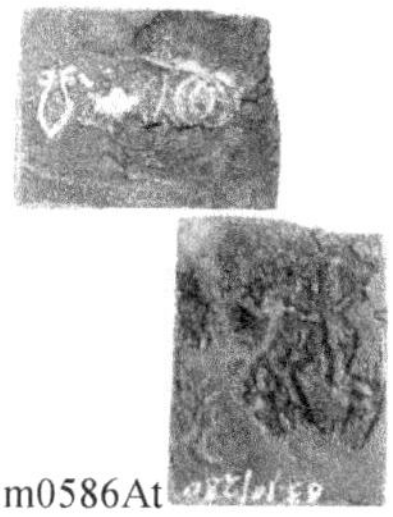
m0586At
m0586Bt

3406

m0587At
m0587Bt

3365 Horned Archer?

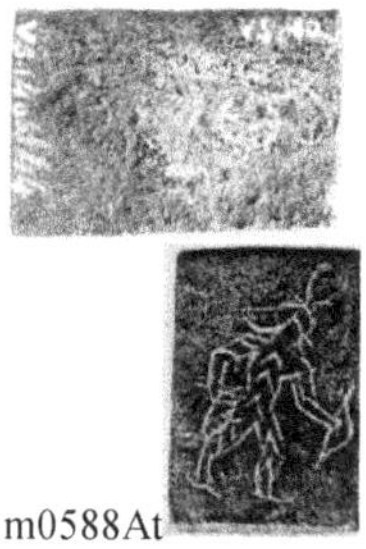
m0588At

m0588Bt Horned archer.

m0592At
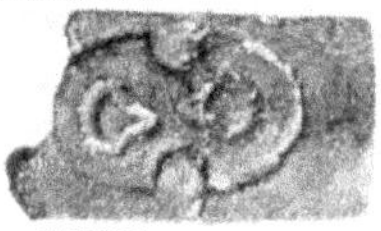
m0592Bt

3413 Pict-133: Double-axe (?) without shaft. [The sign is comparable to the sign which appears on the text of a Chanhudaro seal: Text 6402, Chanhudaro Seal 23].

m0593At
m0593Bt
3337

m0594At
m0594Bt

m0595A
m0595B
1010

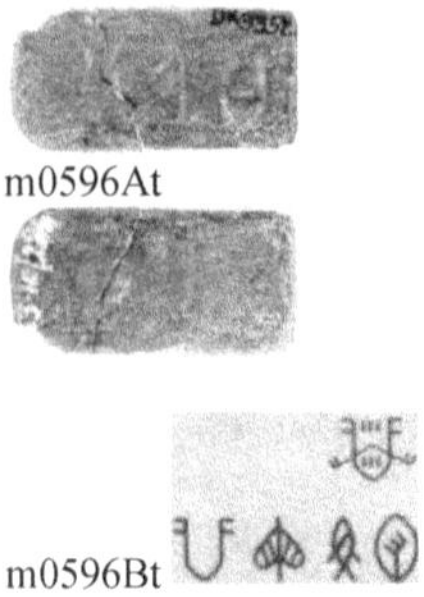
m0596At
m0596Bt
3313

m0598 Text
3410

m0599At

m0599Bt

3360

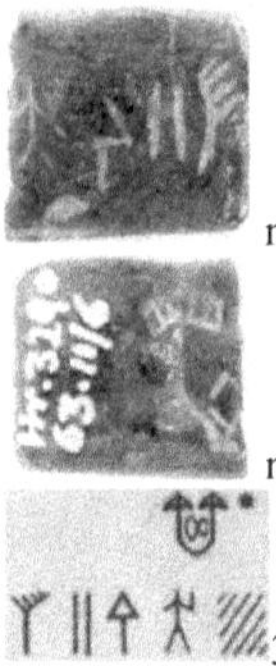
m0600At
m0600Bt
3375

m0601At
m0601Bt

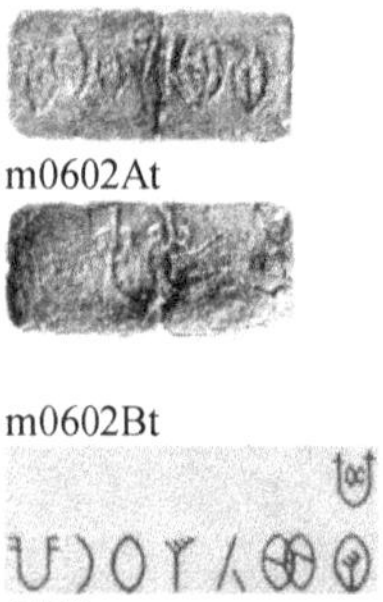
m0602At
m0602Bt
3414

m0604At
m0604Bt
3315

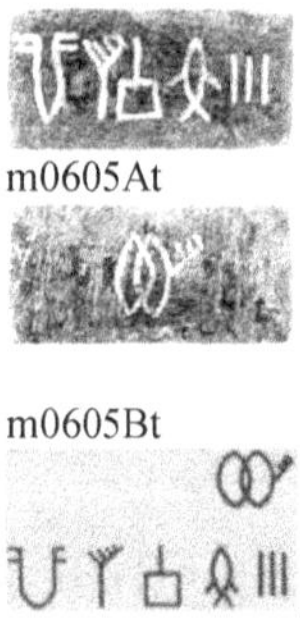
m0605At
m0605Bt
2902

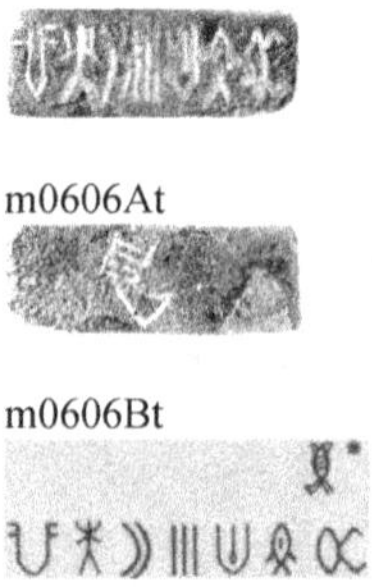
m0606At
m0606Bt
2918

m0608At
m0608Bt

m0614
1904

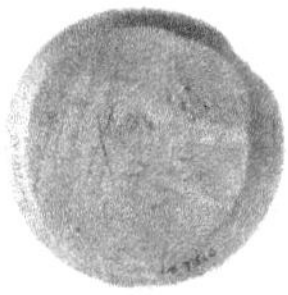
m0615

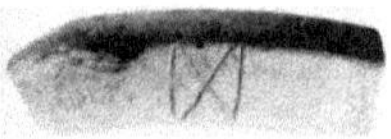
m0618

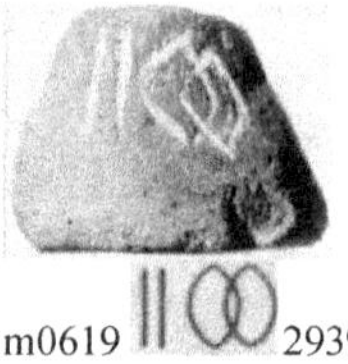
m0619
2939

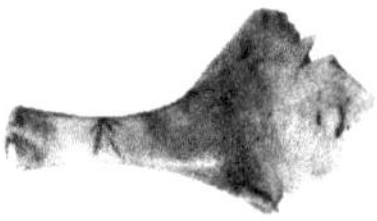
m0620

m0621
2367

m0622
m0623

m0624
1015

m0625
1027

m0626
1012

m0627
1004

m0628
1033

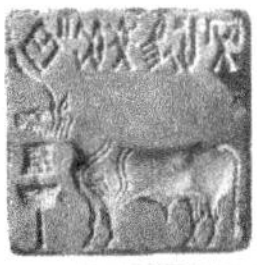
m0629

m0630A

m0631
1008

m0632
1017

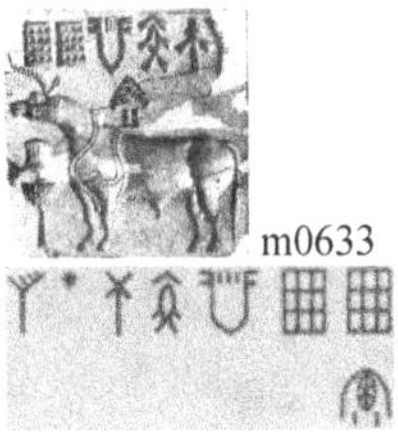
m0633
1016

m0634
2069

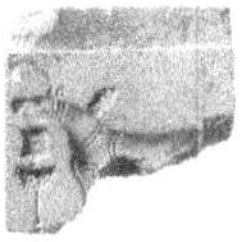
m0635a

m0636
2019

m0637
1034

m0638 One-horned bull
1404

m0639

m0640

m0641

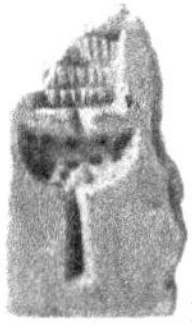
m0642

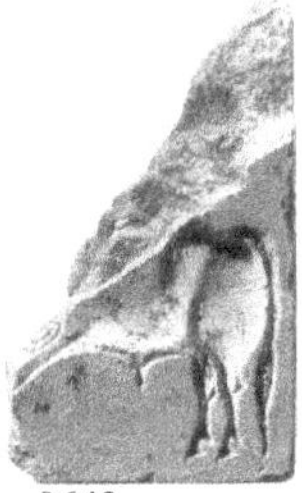
m0643

m0644
1553

m0645

m0646A1

m0646a12

m0646A2
2653

m0647
1024

m0648
3104

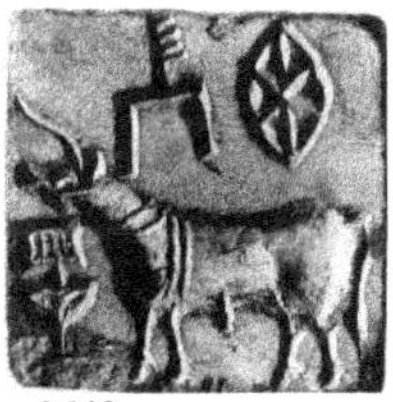
m0649

2530

m0650
1032

m0651
2578

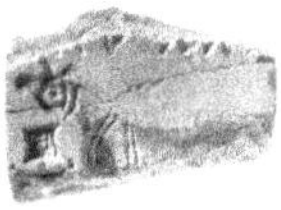
m0652

m0653

1057
m0654 2561

m0655
2098

m0656

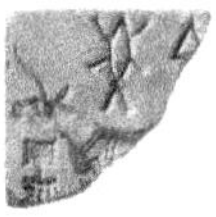
m0657

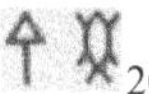
2026

m0658
1039

m0659

m0661
2207

m0662
1061

m0663
2597

m0664
2628

m0665
1139

m0666
2243

m0667
1111

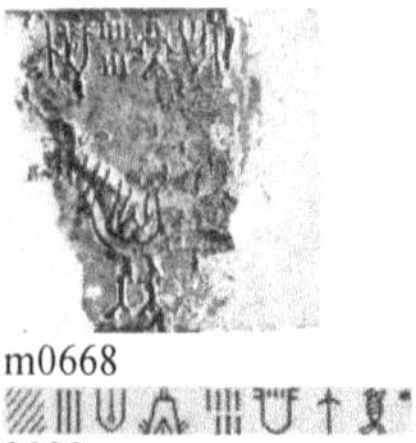
m0668
2032

m0669
2686

m0670
1030

m0671
1021

m0672
1040

m0673 1025

m0674
1068

m0675
2197

m0676

m0677

m0678
1066

m0679

m0680A1

m0681 2182

m0682

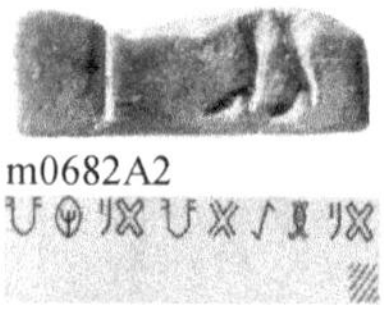
m0682A2
2690

m0683a

m0683A1

m0683A2
2174

m0684

m0685

1276

m0686
2324

m0687
m0688

m0689

m0690

m0691

m0692
1031

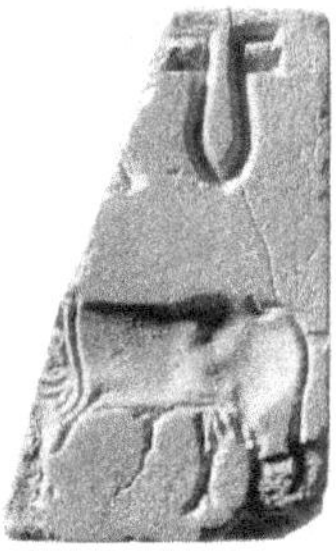

m0693

m0694

m0695

m0696
m0697

m0698

m0699
1050

m0700

m0701
1059

m0702
2206

m0703
2438

m0704
2351

m0705
2272

m0706
1097

m0707

m0708
2666

m0709
2071

m0710
3159

m0711
1166

m0712
1091 Note Sign391 ligatured on the animal's neck; this may be a logonym (i.e. two heiroglyphs – rings and spoked circle -- representing the same lexeme) for the rings on the neck?

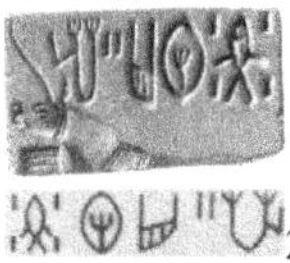

m0713
2432

m0715
2681

m0716

2076 [Are there signs following these two signs?]

m0717
1078

m0714

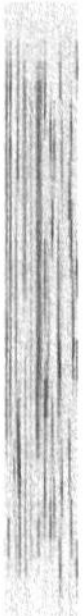

2446
m0718

2209

m0719
2137

m0720
1082

m0721
1165

m0722
1014

m0723
2054

m0724

m0725

m0726

m0727a

m0727A1

m0727A2

2168

m0728
2691

m0729

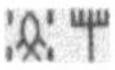
1177

m0730

m0732
2674

m0733
2519

m0734
1539

m0735
1060

m0736
2562

m0737
1112

m0738

2644

m0739

m0740

1090
m0741

2421

m0742
2595

m0743

m0744

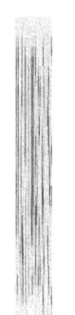

m0745
1175

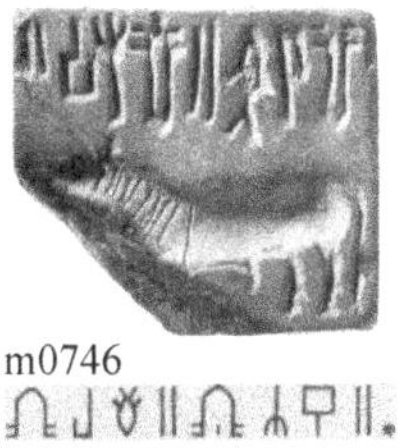
m0746
1081

m0747

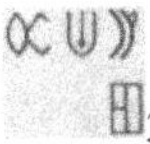
2471

m0748

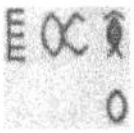
1135

m0749

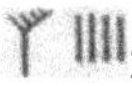
2008

m0750
2065

m0751

1102

m0752a

m0753a

m0753A1

m0753A2
2589

m0754
1145

m0755

m0756a
1028

m0757
2507

m0758a
2184

m0759 One-horned bull.

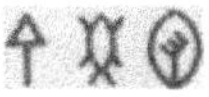
2384

m0760

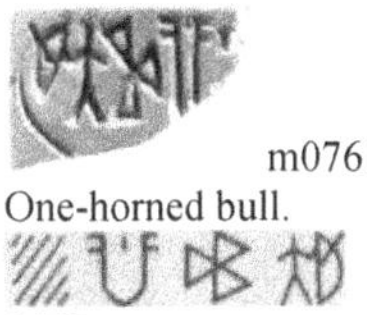
m0761
One-horned bull.
1417

m0762a

2645

m0763

m0764

m0765

m0766
m0767

m0768
1176

m0769
2034

m0770a
1138

m0771
2676

m0772
2453

m0773
m0774

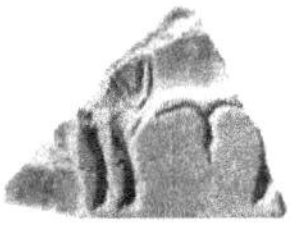
m0775

m0776
1146

m0777
2536

m0778
2425

m0779
2622

m0780
1178

m0781
2251

m0782
1122

m0783
1127

m0784
1128

m0785
1181

m0786
1107

m0787
2503

m0788

m0789
1185

m0790

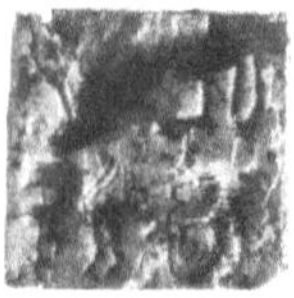
m0791

m0792
2013

m0793

m0794
2067

m0795
1228

m0796
2105

m0797

m0798
1084

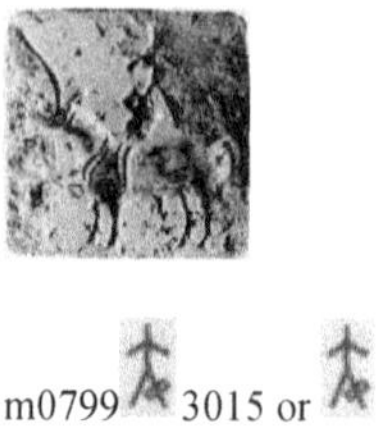
m0799 3015 or
3147

m0800

m0801
2104

m0802
1182

m0803
1131

m0804
2570

m0805
3041

m0806

m0807
2669

m0808
2146

m0809
2548

m0810
2364

m0811

2211

m0812

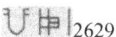

2629

m0813

m0814
2426

m0815
2555

m0816
2424

m0817
2435

m0818

1089

m0819
2081

m0820

m0821

1238

m0822

1249

m0823
1086

m0824

1164

m0825

1239

m0826

m0827
2513

m0828
2114

m0829

m0830

2274

m0831

2546

m0832

m0833

2281

m0834

2569b
m0835

2179

m0836

m0837
3085

m0838

2368

m0839
2476

m0840
2617

m0841

m0842

2704

m0843

m0844

1290

m0845
2202

m0846

1005

m0847
1156

m0848

2241

m0849
1121

m0850
2533

m0851
2660

99933333m0852a
2413

m0853
2255

m0854
2501

m0855
2473

m0856
1211

m0857
2091

m0858a

2189

m0859
2063

m0860

m0861
1123

m0862
2253

m0863
2621 Is the 'stubble' ligatured glyph a variant of Sign 162  ?]

m0864
1240

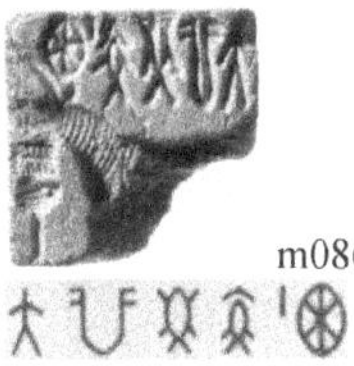

m0865
1109

m0866
2646

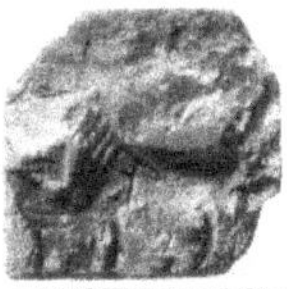

m0867

m0868
3160

m0869

m0870
1160

m0871

m0872

m0873

1170
m0874

3092
m0875
1189

m0876
m0877

m0878
1092

m0879

2121

m0880

m0881
1242

m0882
2312

m0883

m0884

3158

m0885

m0886
3072

m0887a

1169

m0888
1155

m0889
1126

m0890
2117

m0891

1073

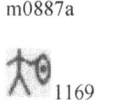

m0892

1247

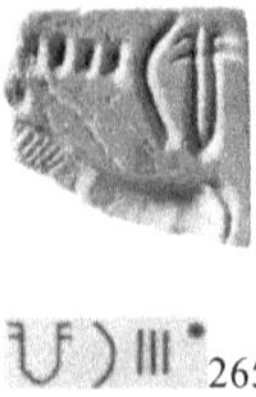

m0893

2659 One-horned bull.

m0894

2393

m0895

2262

m0896

2134

m0897

2545

m0898 2167

m0899

2242

m0900

2335

m0901

2276

m0902a

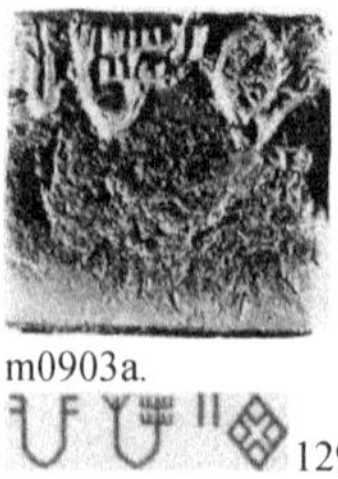

m0903a.

1294

m0904

m0905

m0906

m0907

2192

m0908

m0909

3028

m0910

m0911

m0914

2143

m0915

1218

m0916

1204

m0917

1224

m0918

m0919

2343

m0920

1219

m0921

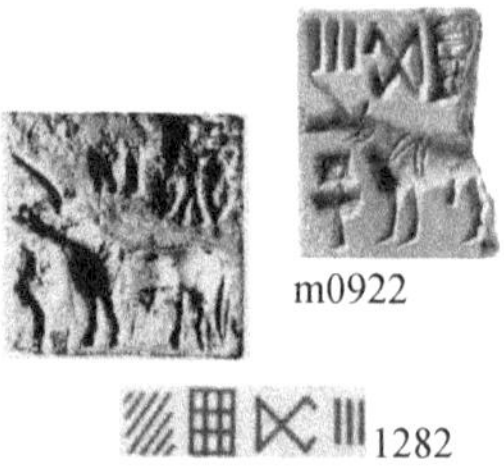

m0922

1282

m0923

m0924

2591

m0925

1292

m0926

2219

m0927

1171

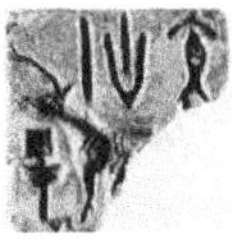
m0928

1202

m0929a

1144

m0930

3020
m0931
3091

m0932

3022

m0933

2160

m0934

1158

m0935

2144

m0936

1197

m0937

2066

m0938

2158

m0939a

2652

m0940a

2060

m0941

2256

m0942

1296

m0943

2282

m0944

2419

m0945

1208

m0946

2358

m0947

2404

m0948

2250

m0949A

m0949C

1271 Also, Sign 141

m0950a

1013

m0951

1263

m0952

2265

m0953

2582

m0954

1262

m0955
2547

m0956
1251

m0957
1026

m0958
2348

m0959
1147

m0960
1388

m0961
1163

m0962
3074

m0963
1232

m0964
2010

m0965
1222

m0966
2070

m0967
2460

m0968
2300

m0969
2239

m0970a
2116

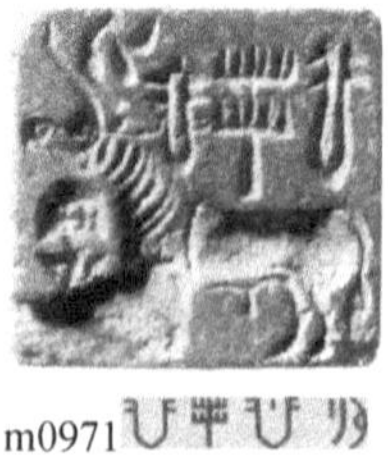
m0971
1234

m0972a
2557

m0973a
2585

m0974a
2650

m0975
2295

m0976
1203

m0977
3152

m0978

m0979
2564

m0980
2317

m0981

m0982a
2021

m0983

m0984
1143

m0985

m0986a 2341

m0987a

1007

m0988

m0989

m0990

2472 One-horned bull.

m0991

2203

m0992

2464

m0993a

1267

m0994a

2165

m0995

m0996

2299 One-horned bull.

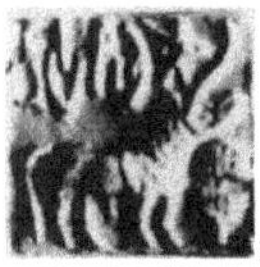

m0997a

3105

m0998

2176

m0999

2452

m1000a

1487 One-horned bull.

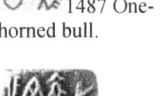

m1001a

1283

m1002

m1003

1275

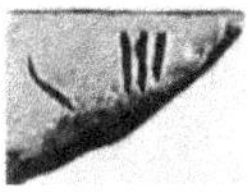

m1004

m1005 1001

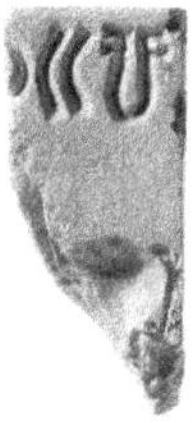

m1006

1499 Bovid.

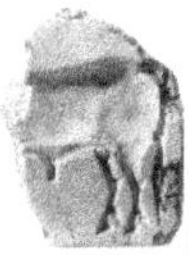

m1007

m1008

m1009

2627

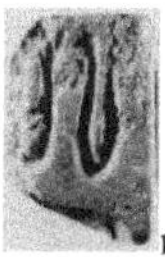

m1010

2672 Bovid.

m1011

m1012

m1013

m1014

One-horned bull?

1397 m1015

m1016.

1348

m1017.

1300

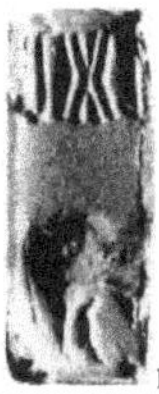

m1018a

2483 Bovid.

m1019.
1298

m1020
2496

m1021a.
1299

m1022

m1023

m1024

m1025a

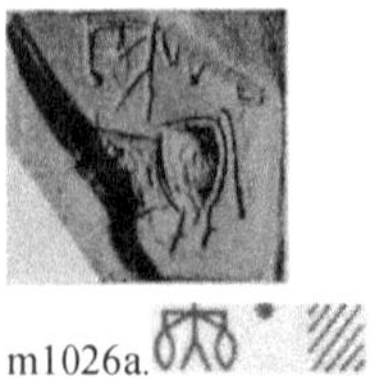
m1026a.
1307

m1027

m1028
2671 Bovid.

m1029
1265

m1030
3145

m1031
2053

m1032
2217

m1033

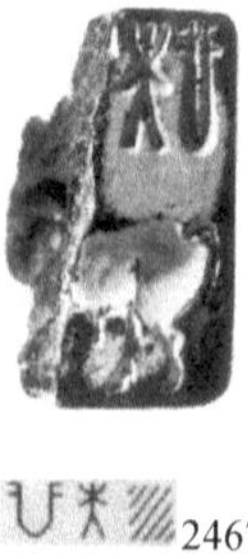
m1034
2467

m1036

m1037

m1038

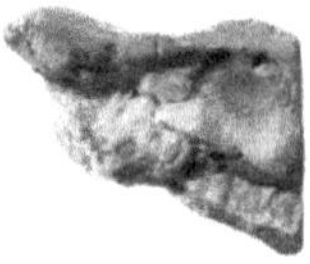
m1039

m1040

m1041

m1042

m1043

m1044a
1551 Bovid.

m1045
2447 Bovid.

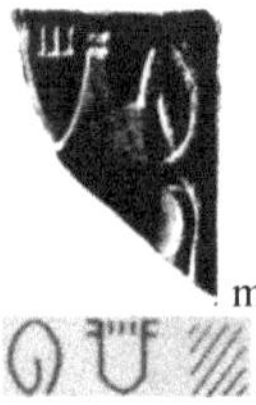
m1046
3058

m1047
1281

m1048

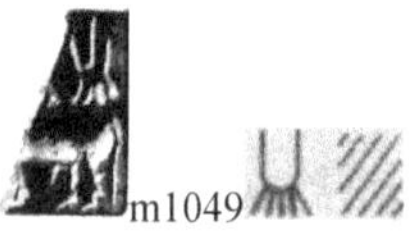
m1049
3032

m1050
1196

m1051

m1052

3100

m1053
2163

m1054

2448

m1055
2529

m1057
2566

m1058a

1392

m1059

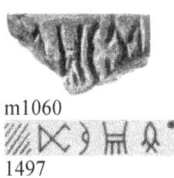

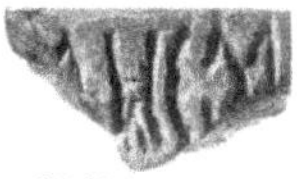
m1060
1497

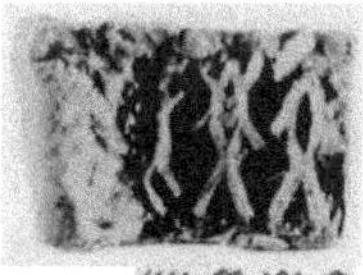
m1061a
1379

m1062
2089

m1063

2357

m1064
1492

m1065

2151

m1066
1547

m1067a
1496

m1068

m1069
1390

m1070
2040

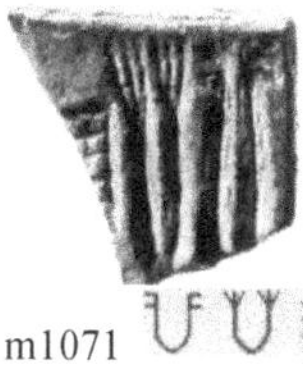
m1071
1488

m1072a
1443

m1073
1489

m1074

m1075a
1479

m1076

m1077a

2359

m1078

m1079
2655

m1080
1542

m1081a
2129

m1082.
1349

m1083

m1084
1316 Bison.

m1085.
1322

m1086a

3070

m1087a.
1319

m1088

2268

m1089a.
1315

m1090
2675

m1091

m1092

1312

m1093

m1094

m1095
2495
Bison

m1096
2410

m1097
2313

m1098

1301

m1099
1313

m1100
2201 Bison

m1101
2431
Zebu.

m1102

m1103.
1337

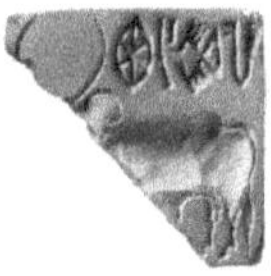
m1104
1335

m1105

m1106

2331 Zebu

m1107a
2306

m1108
1339

m1109
1327 Zebu

m1110
1334

m1111.
1333

m1112

2366 Zebu.

m1113
2441

m1114.
1331

m1115
1328 Zebu

m1116.
1329

m1117a
2615

m1118
3157

m1119
2463

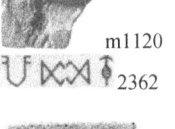

m1120
2362

m1122
2610

m1126
2332

m1127
2696

m1128a
3163

m1129a .
1302 Markhor.

m1130

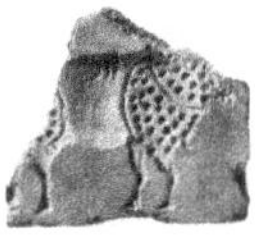
m1131

m1132
1545
Rhinoceros.

m1133
1343

m1134
2651

m1135
2140
Pict-50 Composite animal: features of an ox and a rhinoceros facing the standard device.

m1136

m1137
2531
Rhinoceros.

m1138.
1344

m1139.
1341

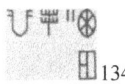

m1140a
2188
Rhinoceros.

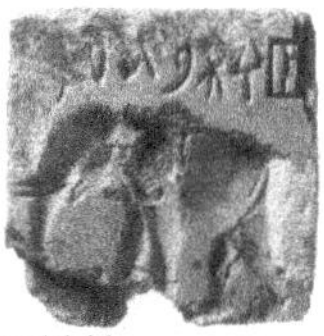
m1141
2169

m1142

m1143

m1144

m1145

m146a
1374 Elephant

m1147

m1148
2590

m1149
1368 Elephant.

m1150 
1534

m1151 
1535

m1152

1369

m1154

1362 Elephant.

m1155

2573

m1156

1370

m1157a

2110

m158

m1159
2171

m1160
2057

m1161

2504

m1162

2058

m1163
2640 Tiger.

m1164
2665
Tiger.

m1165a
2064

m1166.
1351

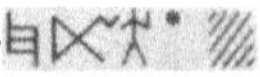

m1167
2484 Tiger.

m1168

2360

Seal showing a horned tiger. Mohenjodaro. (After Scala/Art Resource).

Tiger with long (zebu's) horns?

1385

Pict-

49 Uncertain animal with dotted circles on its body.

1626

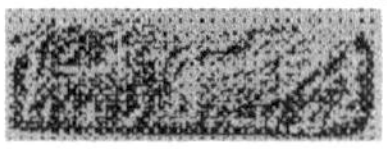

Pict-47 Row of uncertain animals in file.

m1169a

2024

Pict-58: Composite motif: body of an ox and three heads: of a one-horned bull (looking forward), of antelope (looking backward), and of short-horned bull (bison) (looking downward).

m1170a
1382 Composite animal

m1171
Composite animal

m1172

m1173
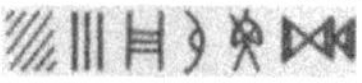
1191

m1175a

2493
Composite animal: human face, zebu's horns, elephant tusks and trunk, ram's forepart, unicorn's trunk and feet, tiger's hindpart and serpent-like tail.

m1176

m1177

2450
Composite animal:

human face, zebu's horns, elephant tusks and trunk, ram's forepart, unicorn's trunk and feet, tiger's hindpart and serpent-like tail.

m1178

2559

m1179

2606 Human-faced markhor with long wavy horns, with neck-bands and a short tail.

m1180a .
1303 Human-faced markhor

m1181A
2222 Pict-80: Three-faced, horned person (with a three-leaved pipal branch on the crown), wearing bangles and armlets and seated on a hoofed platform

Padri . Head painted on storage jar from Padri, Gujarat (c. 2800 BCE). Details of body with multiple hands (?) Similar horned-heads painted on jars are found at Kot Diji, Burzhom and Kunal (c. 3rd millennium BCE). [Source: Page 21, Figs. 10A and B in: Deo Prakash Sharma, 2000, *Harappan seals, sealings and copper tablets*, Delhi, National Museum].

m1182a

m1183a

m1184

m1185

Pict-103 Horned (female with breasts hanging down?) person with a tail and bovine legs standing near a tree fisting a horned tiger rearing on its hindlegs.

1357

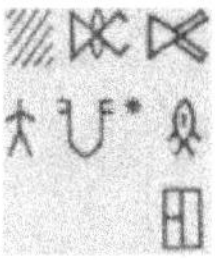

m1186A
2430 Composition: horned person with a pigtail standing between the branches
of a pipal tree; a low pedestal with offerings (? or human head?); a horned person kneeling in adoration; a ram with short tail and curling horns; a row of seven robed figures, with twigs on their pigtails.

m1187

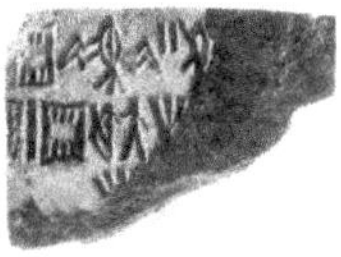

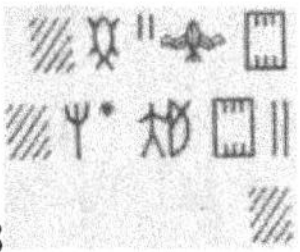

m1188
2228

m1189

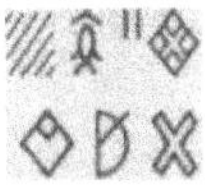

1396

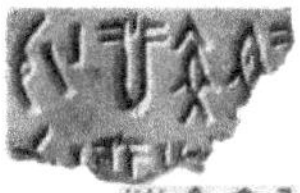

m1190
2558

m1191

1389

m1192

1495

m1193a

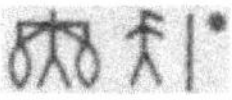

2401

m1194a

3066

m1195

2181

m1196

m1197

m1198

1482

Silver

m1199A
2520

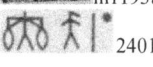

m1200A

m1200C

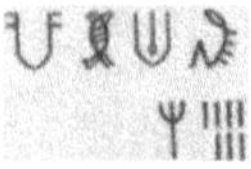

3078

m1201

m1202A

m1202C.

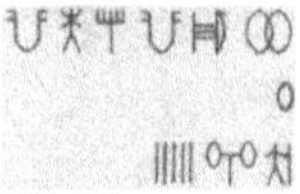

1325 Space on the side of the seal was used to inscribe a third line

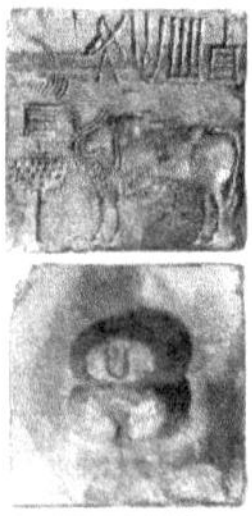

m1203A

m1203B

1018

m1204

2095

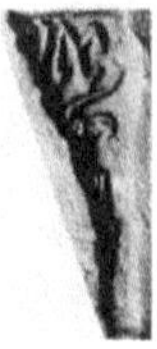

m1205a

m1205c

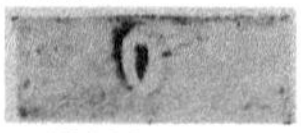

m1205f

1293 + Two signs on the sides of the seal.

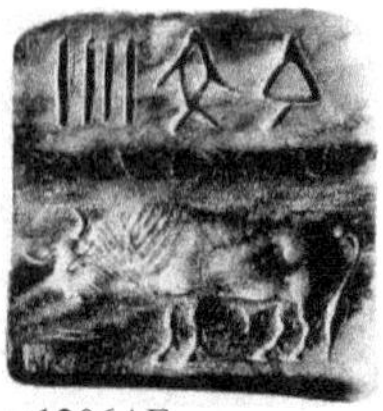

m1206AE

m1206e1

m1206F

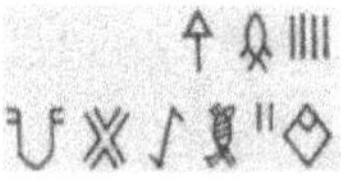

2229 Seal with a projecting knob containing the top three signs; m1206e is inscribed on the top edge of the lower indented frame which depicts the bison.

m1208

m1221

m1222

1268

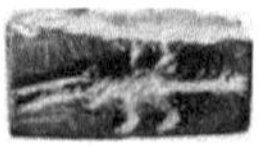

m1223

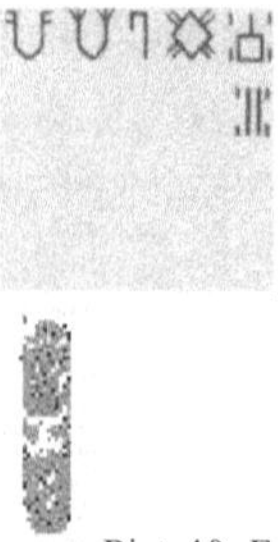

2045

Pict-40: Frog.

2565

Pict-37 Goat-antelope with a short tail

m1224A

m1224B

1224

m1224e

Pict-88

1227

Standing person with horns and bovine features (hoofed legs and/or tail).

m1225A

m1225B.

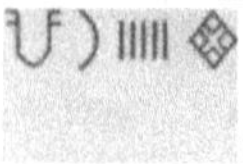

1311 Cube seal with perforation through the breadth of the seal Pict-118: svastika_, generally within a square or rectangular border.

m1226A.

1326 Unfinished seal.

m1227

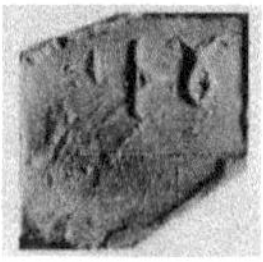

m1228a

1394

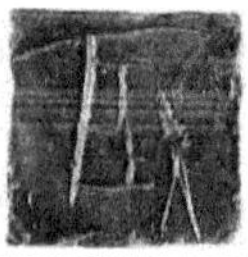

m1230a

1358

m1231

2321

Unfinished seal?

m1232a

2497 Unfinished seal

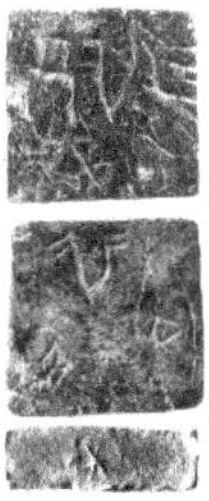
m1233A
m1233B

m1233cd
2352

m1234a
m1234b

m1234d
m1234e

m1235a
m1235bc

2394 Unfinished seal
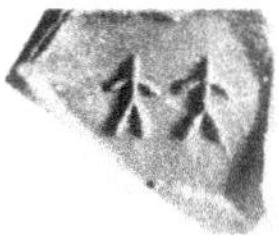
m1236

1483 Unfinished seal?

m1239
m1240
m1241
m1242

m1243
m1244
m1245
m1246

m1247
m1248
m1249
m1250
m1251

m1252
m1253

m1254
m1255
m1256
m1257

m1258
m1259
m1260
m1261

m1262

2301

m1263
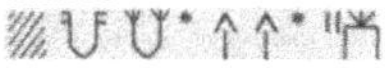
1391

m1264a

1405

m1265

2227

m1266

1470

m1267

1494

m1268

2288

m1269

m1270

1464

m1271

2603

m1272

m1273

2679

m1274

2106
m1275

3161

m1276
2428

m1277

m1278

2028

m1280a

1462

m1281
2266

m1282

m1283

m1284a
2477

m1285a
2204

m1286
1455

m1287
1454

m1288

3086

m1289
1452

m1290

1463

m1291a
2688

m1292

1461

m1293a
2388

m1294

2291

m1295
1458

m1296a

3144

m1297

1445

m1298

3037

m1299a
1456

m1300
2350

m1301

m1302a
1432

m1303a
1398

m1304
1423

m1305

2289

m1306
1430

m1307

m1308
2697

m1309

2579

m1310
1418

m1311
2485

m1312

2318

m1313

2093

m1314a
1439

m1315

2345

m1316a

m1317

3095

m1318
1416

m1319

m1320
1447

m1321
1446

m1322a

3079

m1323

2006

m1324

2682

m1325

2118

m1326

3143

m1327
1408

m1328
2392

m1329A 

m1329C
2439

m1330
1409

m1331a
2303

m1332

m1333
1434

m1334a
2170

m1335a
2072

m1336a
2515

m1337
2055

m1338a
2020

m1339 2025

m1340
2369

m1341
2092

m1342a
1393

m1343
1433

m1344

2315

m1346a

m1349B

m1349A

m1350
2599

m1351

2142

m1353

1459

m1354a

1498

m1355a

2568

m1356

m1357

2356

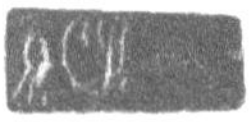

m1358

m1359

2575

m1360

1442

m1361a

1474

m1362A

m1362C

2230

m1363

2372

m1364A

m1364C

2542

m1365A

m1365B

2658 Cricket, spider or prawn?

m1366

2094

m1367a 2661 Two bisons standing face-to-face

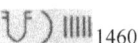

m1368

1460

m1369

1478

m1370a

2509 Cylinder seal; tree branch

m1371A1

m1371A2

m1372A1

m1372A2

m1373A1

m1373A2

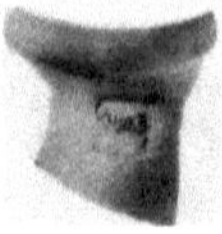

m1374A1

m1374A2

m1375A1

m1375A2

1560 Seal impression on pot

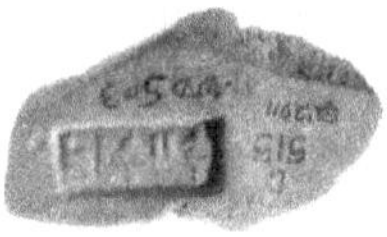

m1376A1

m1376A2

m1378A1

m1378A2

m1379A2

m1380A2

m1381A1

m1381A2 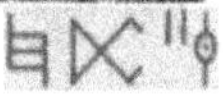

1559 Seal Impression on a pot

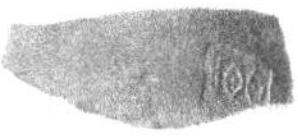

m1382A1

m1382A2 Seal impression on a potsherd

 3244

m1383

m1384si

m1385A14

m1385A2

m1385A3

m1386si

m1387t

m1388t

 2856

m1389t

m1390At

m1390Bt

2868 Pict-74: Bird in flight.

m1391t

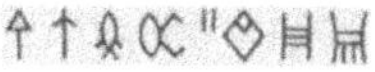

2826

m1392t 

2837

m1393t

m1394t

m1395At

m1395Bt

m1396t

m1397At

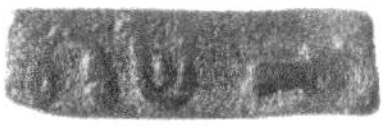

m1397Bt

m1398t

2807

m1400At

m1400B

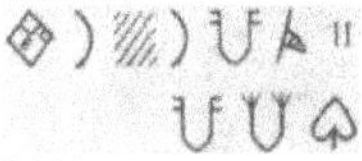

2851

m1401t

2822

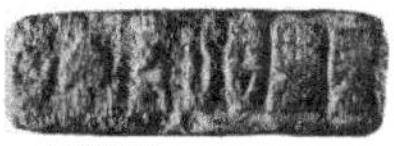

m1402At

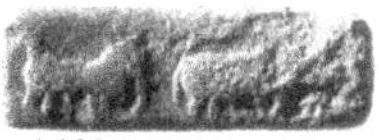

m1402Bt

m1403At

m1403Bt

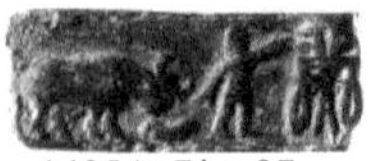
m1405At Pict-97: Person standing at the center pointing with his right hand at a bison facing a trough, and with his left hand pointing to

the sign Obverse: A tiger and a rhinoceros in file.

m1405Bt Pict-48 A tiger and a rhinoceros in file

2841

m1406At

m1406B 2827 Pict-102: Drummer and people vaulting over? An adorant?

m1407At
m1407Bt

m1408At

m1409At
m1409Bt Serpent (?) entwined around a pillar with capital (?) or ring-stones stacked on a pillar?; the motif is carved in high relief on the reverse side of the inscribed object.

m1410At m1410Bt

m1411At m1411Bt

m1412At

m1412Bt

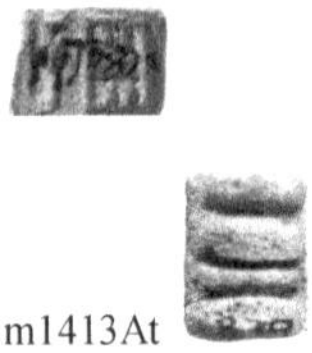
m1413At

m1413Bt

m1414At

m1414Bt

m1415At
m1415Bt

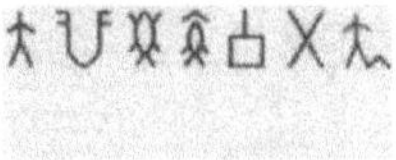
2825

m1416At

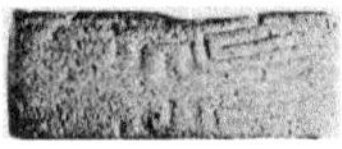
m1416Bt

2818

m1417t
3242

m1418At
m1418Bt

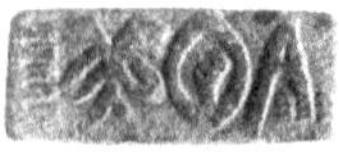
m1419At

m1419Bt
2812

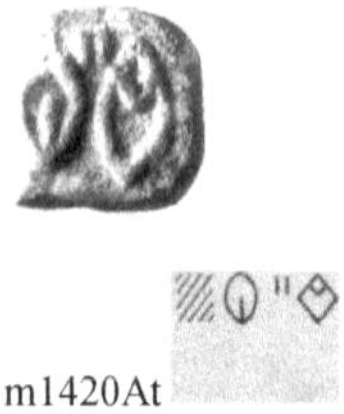
m1420At
2865

m1421At

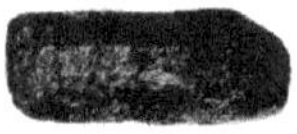
m1421Bt

m1422At

2845

m1423At

m1423Bt Elephant shown on both sides of the tablet.

m1424Atc

m1424Btc

3234

m1425At

m1425Bt

m1427At

m1427Bt

2860

m1428At

m1428Bt

m1428Ct

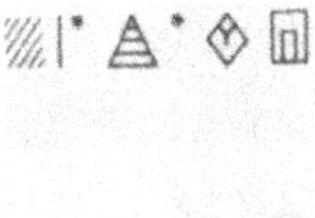

2842

m1426

1621 

m1429At

m1429Bt Pict-125: Boat.

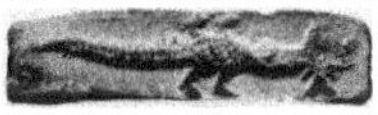

m1429Ct

3246 Gharial holding a fish in its jaws.

Pict-100 Person throwing a spear at a buffalo and placing one foot on the head of the buffalo.

2279

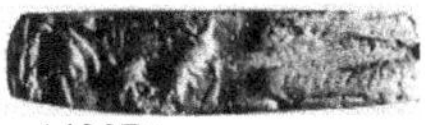

m1430Bt

m1430C

m1430At Pict-101: Person throwing a spear at a buffalo and placing one foot on its head; three persons standing near a tree at the center.

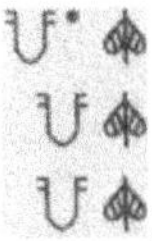

2819 Pict-60: Composite animal with the body of an ox and three heads [one each of one-horned bull (looking forward), antelope (looking backward) and bison (looking downwards)] at right; a goat standing on its hindlegs and browsing from a tree at the center.

m1431A

m1431B

m1431C

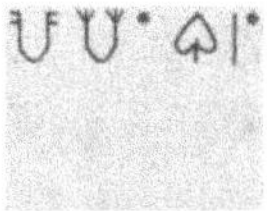

m1431E 2805 Row of animals in file (a one-horned bull, an elephant and a rhinoceros from right); a gharial with a fish held in its jaw above the animals; a bird (?) at right. Pict-116: From R.—a person holding a vessel; a woman with a platter (?); a kneeling person with a staff in his hands facing the woman; a goat with its forelegs on a platform under a tree. [Or, two antelopes flanking a tree on a platform, with one antelope looking backwards?]

m1432At

m1432Bt

m1432Ct

m1433At

m1433Bt

m1433Ct

m1436it

m1438it

m1439it

 3132

m1440 it 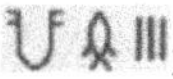 2374

m1441it

m1442it

m1443it

 3213

m1444Ait

m1444Bit 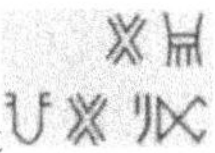
2339

m1445Ait

m1445Bit

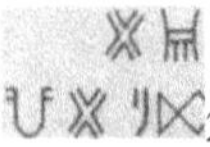
2505

m1447Ait

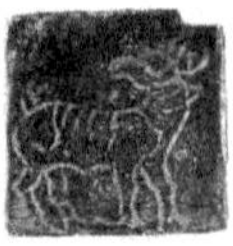
m1448Act

m1448Bct

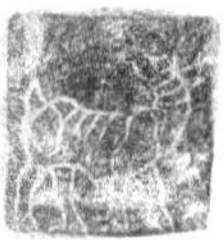
m1449Act

m1449Bct (obverse of inscription) Incised copper tablet (two sides) Markhor with head turned backwards

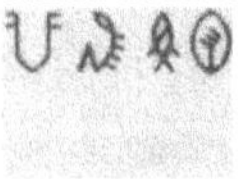
1801

m1450Act

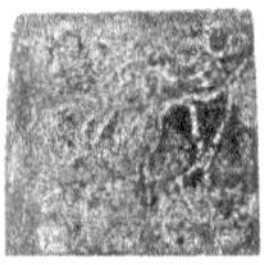

m1450Bct

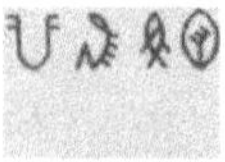

1701

m1451Act

m1451Bct

m1452Act

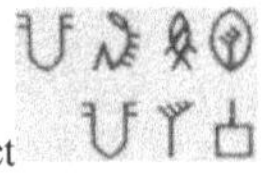
m1452Bct
2912

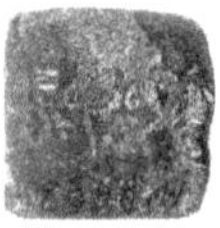

m1453Act

m1453Bct

m1456Act
1805

m1457Act

m1457Bct
2904 Pict-124: Endless knot motif.

m1458Act

m1461Act

m1462Act

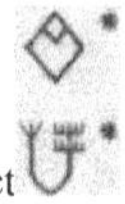
m1463ABct 2919

m1465Act 2921

m1470Act

m1472Bct

m1474Act

m1474Bct

m1475Act

m1475Bct

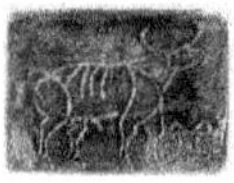

m1476Bct

m1477Act

m1477Bct

m1482Act

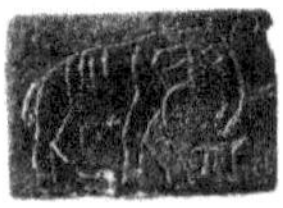

m1482Bct

m1483Act

m1483Bct

m1484Act

m1484Bct

m1485Bct

m1486Act

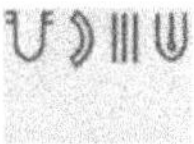

m1486Bct
1711 Incised copper tablets.Elephant

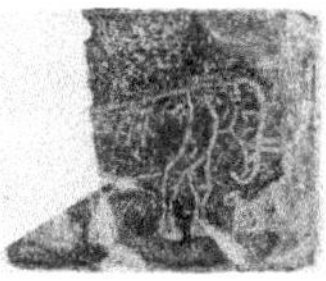

m1488Bct

m1491Act

m1491Bct

m1492Act

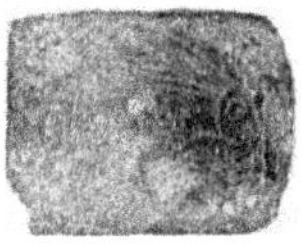

m1492Bct

m1493Bct

m1494

1706 Hare

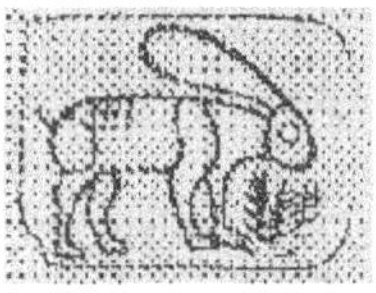

Pict-42

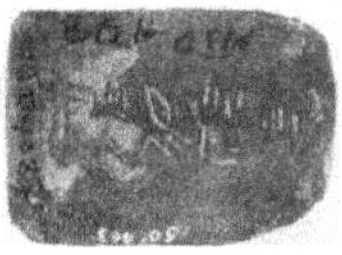

m1497Act

m1498Act

m1498Bct

2917

1803

Pict-30

1804

Pict-39 Ox-antelope with a long tail; a trough in front.

m1501Bct

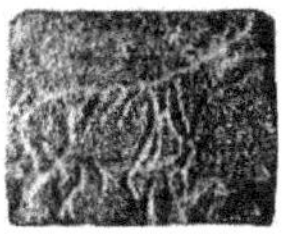

m1502Bct

m1503Act

m1503Bct

m1505Act

m1505Bct

m1506Act

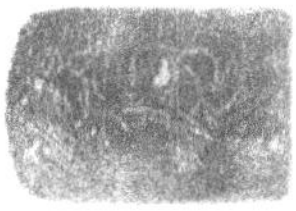

m1506Bct

m1508Act

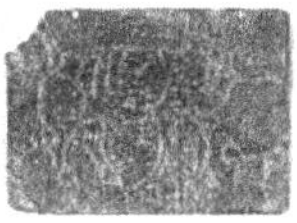

m1508Bct

1708

m1511Act

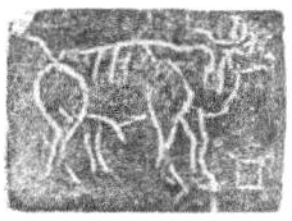

m1511Bct

m1512Act

m1512Bct

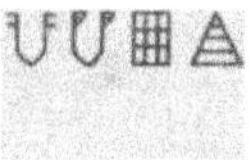

m1513
1712

m1514

1715

m1515Act

m1515Bct

2910

m1516Act

m1516Bct

m1517Act

m1517Bct

m1518
1709

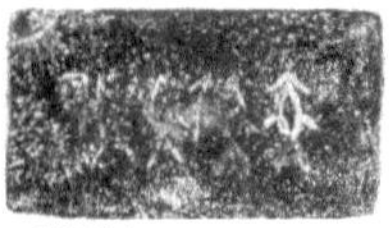

m1520Act

m1520Bct

2907

m1521Act

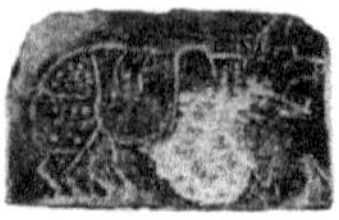

m1521Bct

m1522Act

m1522Bct

m1523Act

m1523Bct

m1524

3396

m1528Act

m1529Act

2920

m1529Bct

m1532Act

m1532Bct

m1534Act

m1534Bct

1703

Composition:

Two horned heads one at either end of the body. Note the dottings on the thighs which is a unique artistic feature of depicting a rhinoceros (the legs are like those of a rhinoceros?). The body apparently is a combination of two rhinoceroses with heads of two bulls attached on either end of the composite body.

m1535Act

m1535Bct

m1540Act

m1540

m1547Act

1547Bct

m1548A

m1548Bct

m1549Act

m1549Bct

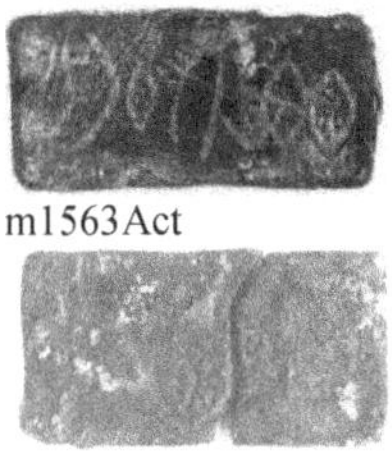

m1563Act

m1563Bct

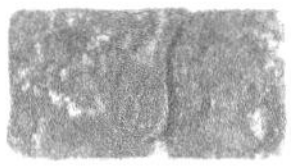

m1566Bct

m1568Act

m1568Bct

m1569 3333

m1575

m1576

m1578 3251

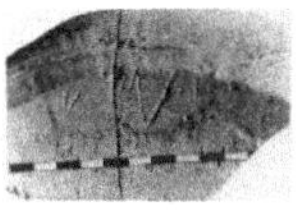

m1591

1592

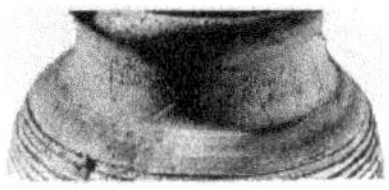

m1597

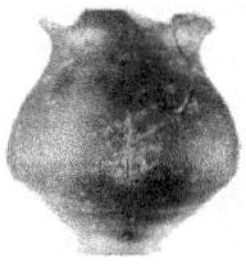

m1598

m1601

3252

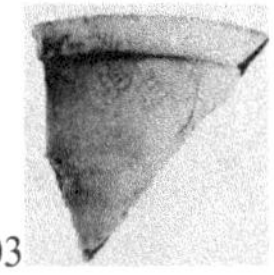

m1603

m1609

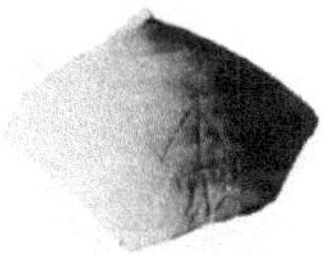

m1611

m1626 3245

m1629bangle

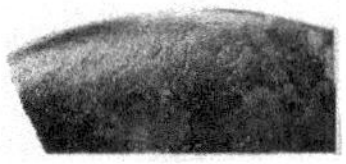

m1630bangle

m1631bangle

m1632bangle

m1633bangle

m1634bangle

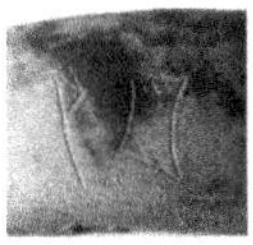

m1635bangle

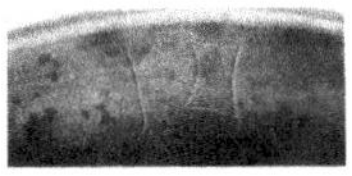

m1636bangle

m1637bangle

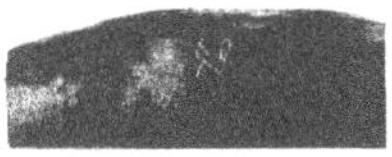

m1638bangle

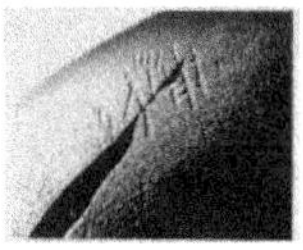

m1639bangle

m1640bangle

m1641bangle

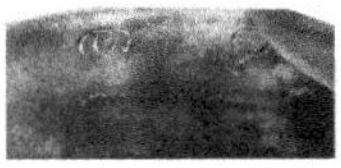

m1643bangle

m1645bangle

m1646bangle

m1647bangle

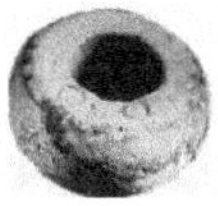

m1648shell

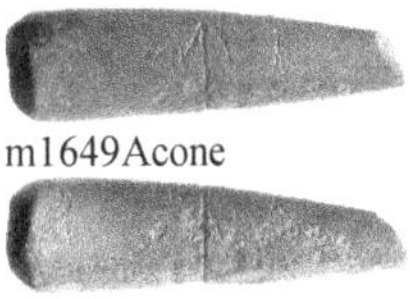

m1649Acone

m1649Bcone

3253

m1650ivory stick

3505

Pict-144: Geometrical pattern.

Pict-141: Geometrical pattern. 2942

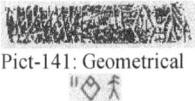

Pict-142: Geometrical

pattern.
2943 Ivory or bone rod

Pict-143: Geometrical pattern.Ivory stick

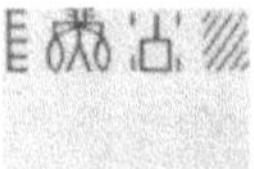

2948

Ivory rod, ivory plaque with dotted circles. Mohenjodaro. [Musee National De Arts Asiatiques Guimet, 1988-1989, *Les cites oubliees de l'Indus Archeologie du Pakistan*.]

m1652A ivory stick

m1653 ivory plaque

1905

m1654A ivory cube

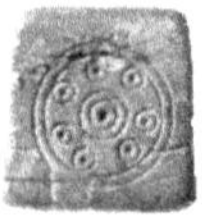

m1654B ivory cube

m1654D ivory cube

m1655faience ornament

m1656 steatite ornament

m1657A steatite

m1657B steatite

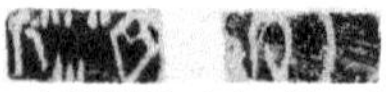

m1658AB etched bead

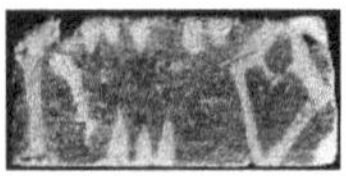

m1658
2952 Etched Bead

m1659 bangle

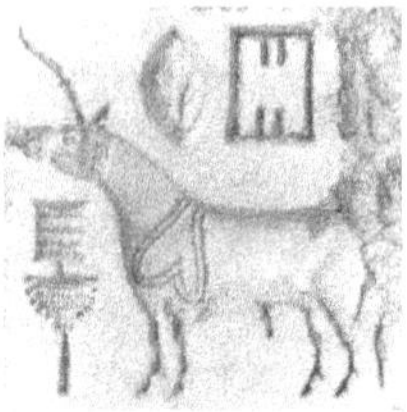

m1660

m1661a

m1662

m1663a

m1664a

m1665a

m1666a

m1667

m1668a

m1669a

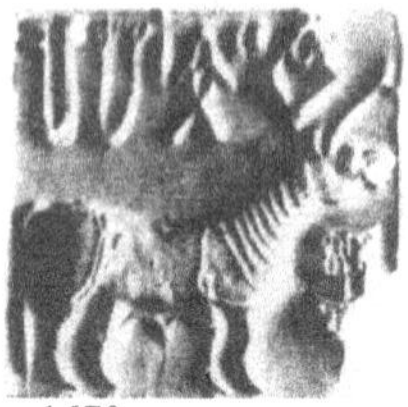

m1670a

m1671a

m1672

m1673a

m1674a

m1675a

m1676a

m1677a

m1678a

m1679a

m1680a

m1681a

m1682a

m1683a

m1684a

m1685a

m1686a

m1687a

m1688a

m1689a

m1690a

m1691a

m1692a

m1693a

m1694a

m1695a

m1696a

m1697

m1698

m1699a

m1700a

m1701a

m1702a

m1703a

m17054

m1705a

m1706a

m1707a

m1708a

m1709a

m1710a

m1711a

m1712a

m1713a

m1714a

m1715a

m1716a

m1717a

m1718

m1719a

m1720

m1721

m1722a

m1723a

m1724a

m1725a

m1726a

m1727

m1728a

m1729a

m1730a

m1731a

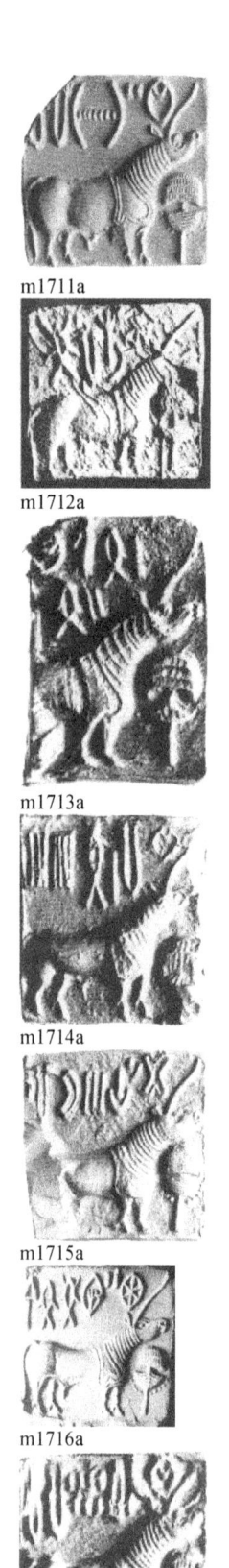

m1732

m1733a

m1734

m1735

m1736a

m1737a

m1738

m1739a

m1740

m1741a

m1742

m1743

m17441

m1745a

m1746

m1747a

m1748

m1749a

m1750a

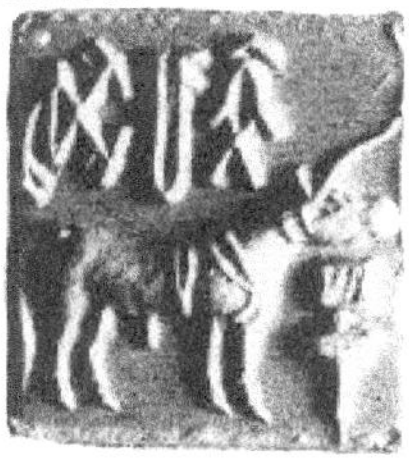
m1751a

m1752a

m1753a

m1754a

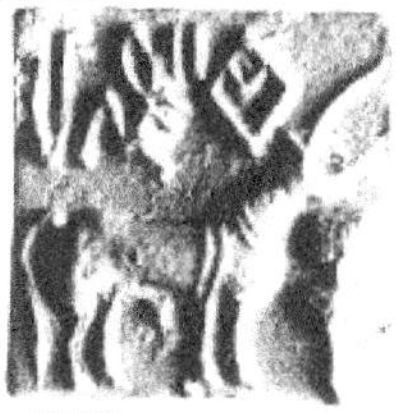
m1755a

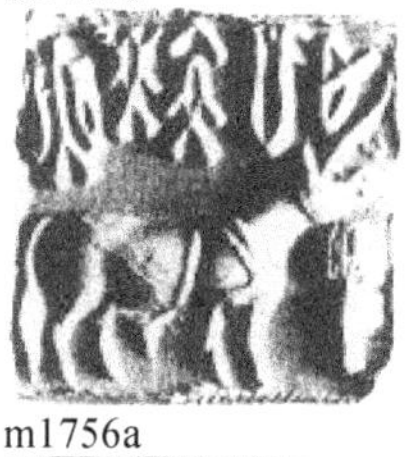
m1756a

m1757a

m1758a

m1759a

m1760

m1761a

m1762a

m1763a

m1764a

m1765

m1766a

m1767

m1768a

m1769a

m1770a

m1771a

m1772a

m1773

m1774a

m1775a

m1776

m1777a

m1778

m1779a

m1780a

m1781a

m1782

m1783a

m1784a

m1785a

m1786a

m1787a

m1788a

m1789a

m1790a

m1791a

m1792a

m1793a

m1794

m1795a

m1796a

m1797

m1798a

m1799

m1800

m1801a

m1802

m1803a

m1804a

m1805a

m1806a

m1807a

m1808a

m1809a

m1810a

m1811a

m1812a

m1813a

m1814

m1815a

m1816a

m1817a

m1818

m1819

m1820

m1821a

m1822a

m1823

m1824a

m1825a

m1826

m1827

m1828a

m1829a

m1830

m1831a

m1832a

m1833a

m1834a

m1835a

m1836

m1837a

m1838a

m1839a

m1840

m1841a

m1842a

m1843a

m1844

m1845a

m1846a

m1847a

m1848

m1849

m1850a

m1851

m1852

m1853a

m1854a

m1855a

m1856a
m1857

m1858

m1860a

m1863a

m1864
m1865a

m1866
m1868a

m1869
m1872a

m1876a

m1877

m1878a

m1879a

m1880a

m1881

m1882

m1883

m1884a

m1885a

m1886a

m1887

m1888a

m1889

m1890

m1891a

m1892a

m1893

m1894

m1895a

m1896a

m1897

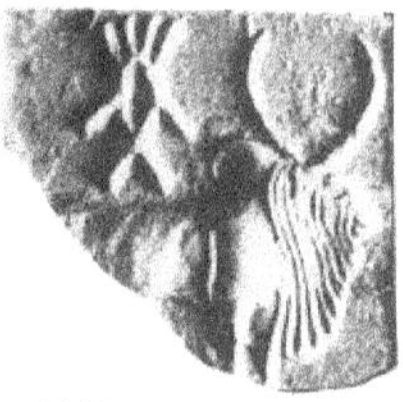

m1898a

m1899a

m1900a

m1901

m1092a

m1903a

m1904a

m1905a

m1906

m1907a

m1909

m1910

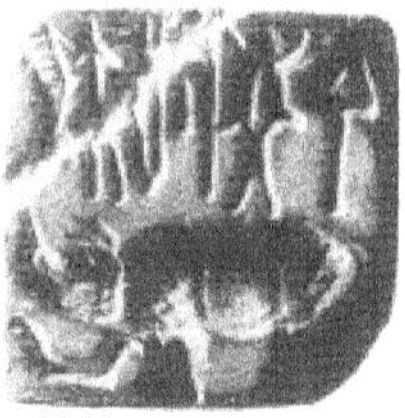

m1911a

m1912

m1912

m1913

m1914

m1915a

m1916a

m1917

m1918a

m1919

m1920a

m1921a

m1922a

m1923a

m1923c

m1923d

m1923e

m1927a

m1927b

m1928a

m1928b

m1930A

m1930B

m1931

m1932

m1933

m1934a

m1935

m1936

1937

m1938
m1939a

m1940

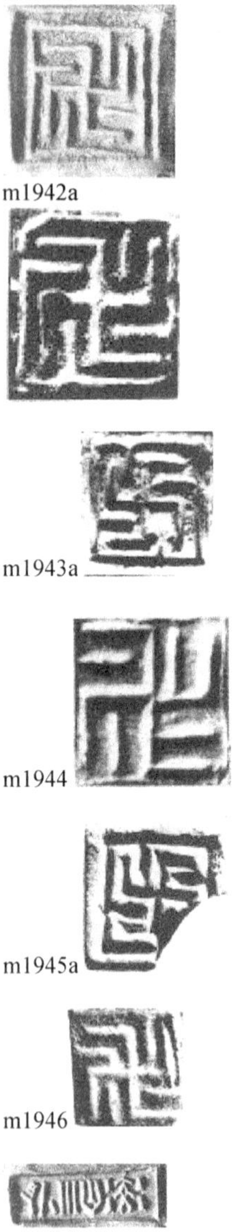

m1941a
m1942a
m1943a
m1944
m1945a
m1946

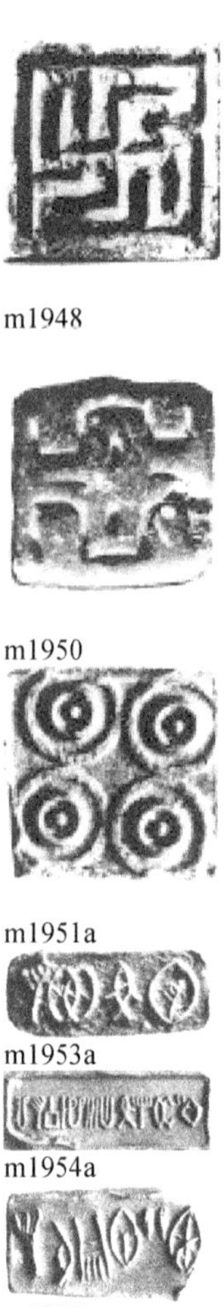

m1947a
m1948
m1950
m1951a
m1953a
m1954a
m1955a

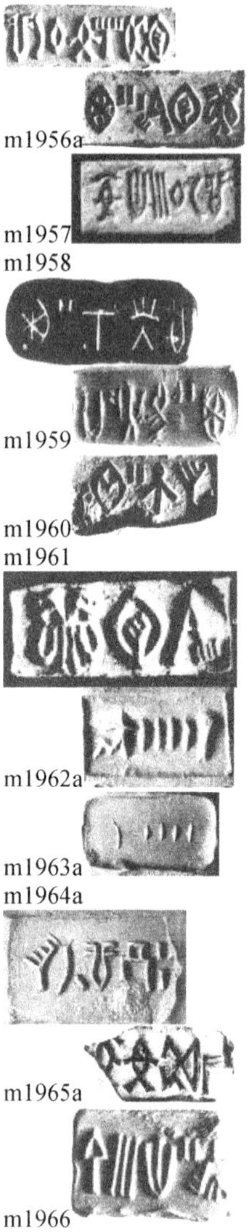

m1956a
m1957
m1958
m1959
m1960
m1961
m1962a
m1963a
m1964a
m1965a
m1966

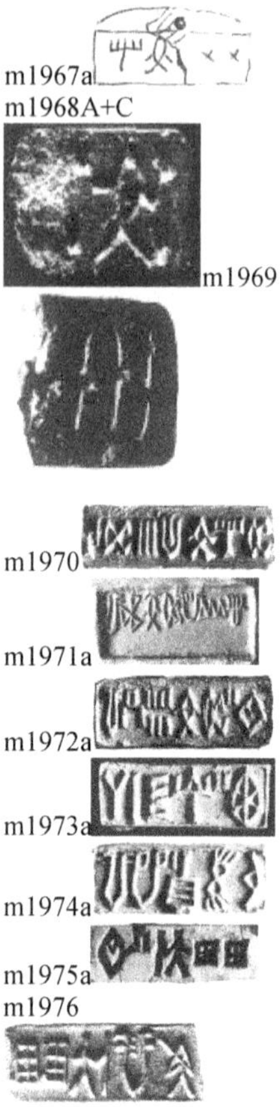

m1967a
m1968A+C
m1969
m1970
m1971a
m1972a
m1973a
m1974a
m1975a
m1976
m1977a

m1978a
m1979a
m1980
m1981a
m1982a

m1983a
m1984a
m1985a

m1986a
m1987a
m1988a
m1989a
m1989b

m1990a
m1990b
m1991a

m1992a

m1993a

m1994A

m1995A

m1996A

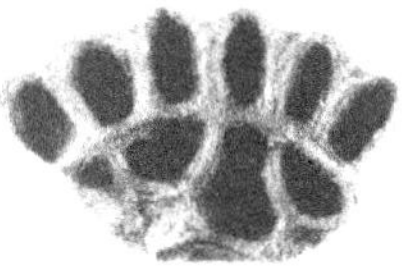

m1997A

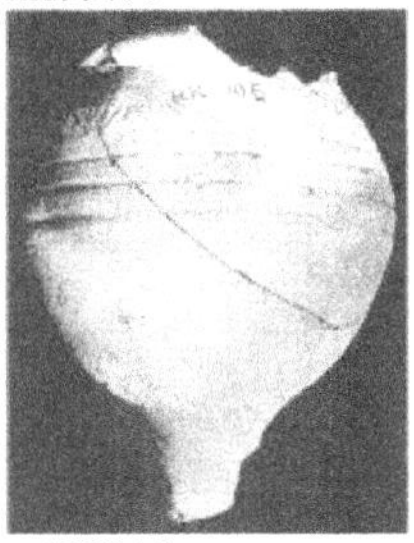

m1998A1

m1998A2

m1999A1

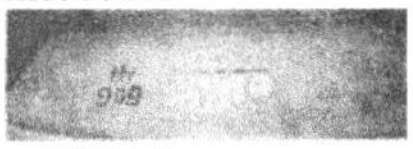

m1999A2

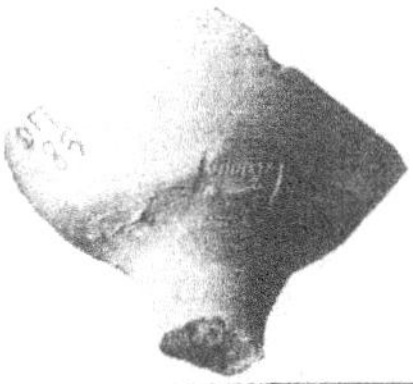

m2000A1

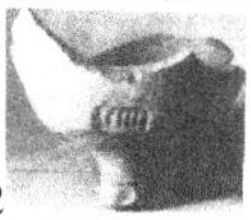

m2000A2

m2001A1

m2001A2

m2002

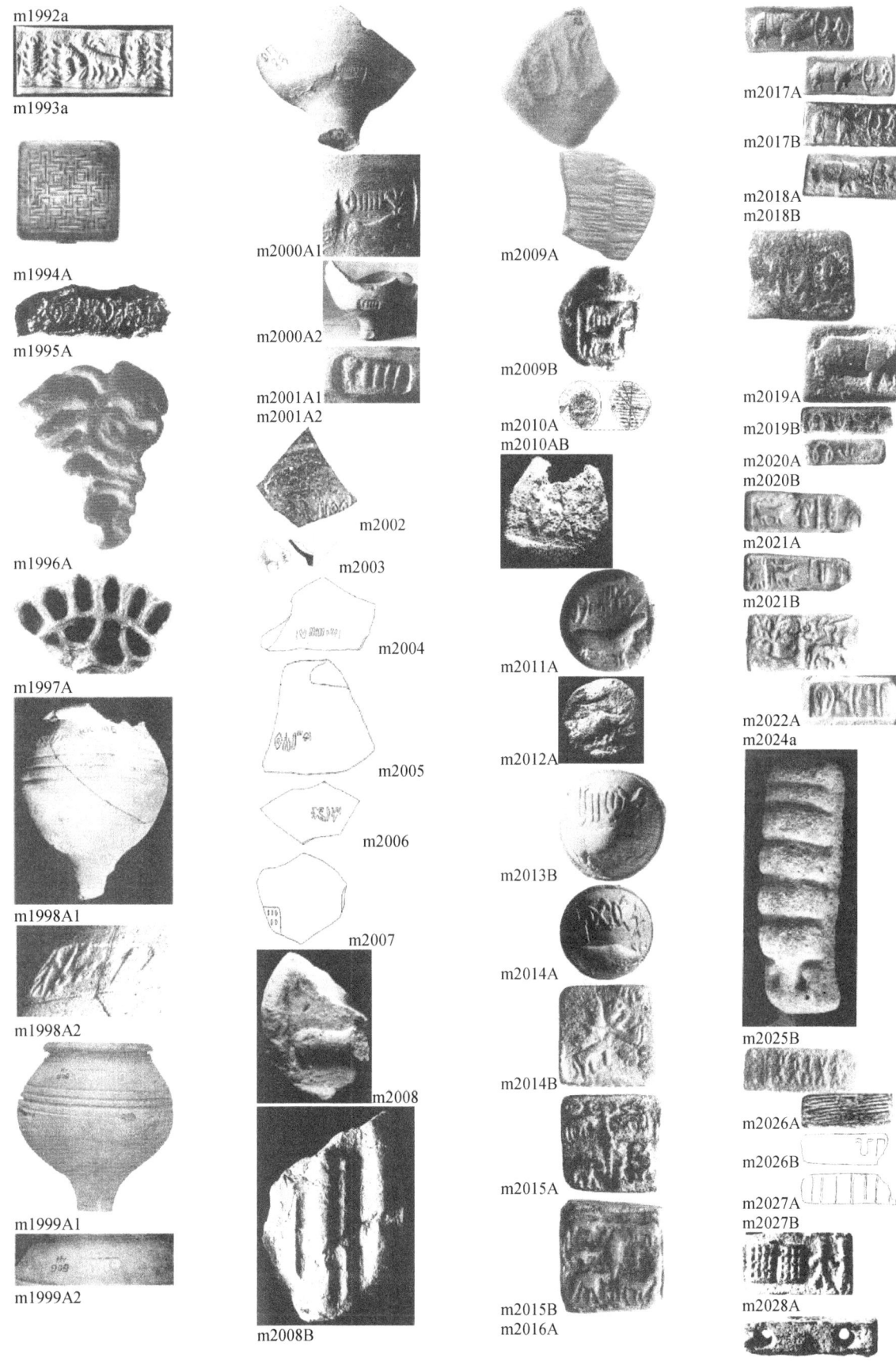

m2003

m2004

m2005

m2006

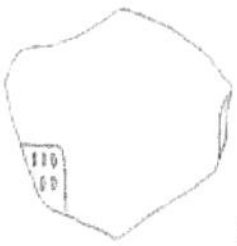

m2007

m2008

m2008B

m2009A

m2009B

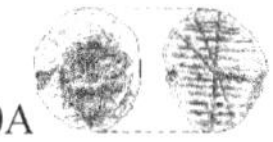

m2010A

m2010AB

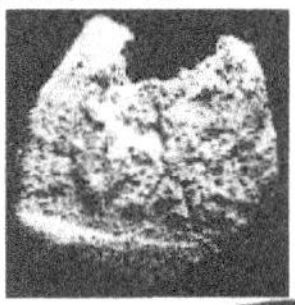

m2011A

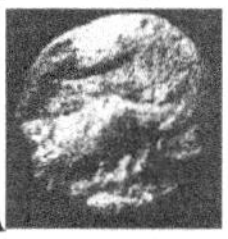

m2012A

m2013B

m2014A

m2014B

m2015A

m2015B

m2016A

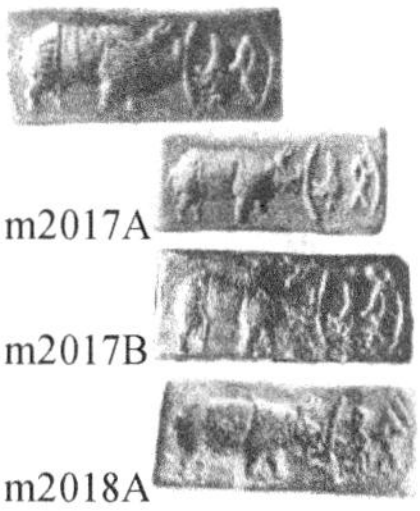

m2017A

m2017B

m2018A

m2018B

m2019A

m2019B

m2020A

m2020B

m2021A

m2021B

m2022A

m2024a

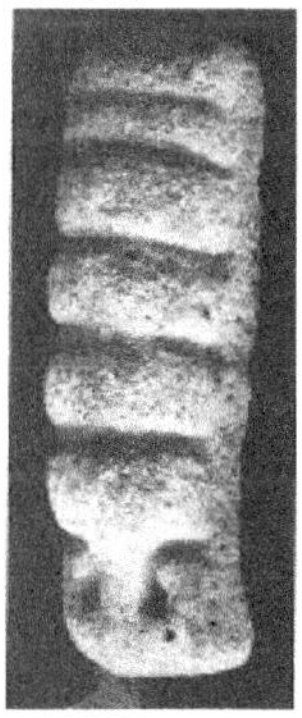

m2025B

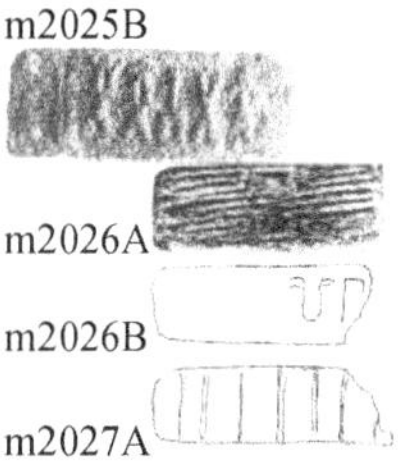

m2026A

m2026B

m2027A

m2027B

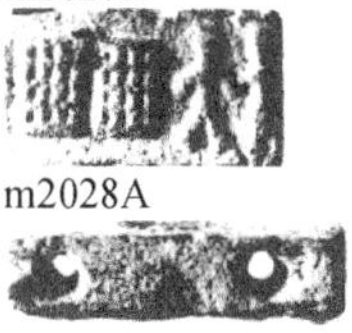

m2028A

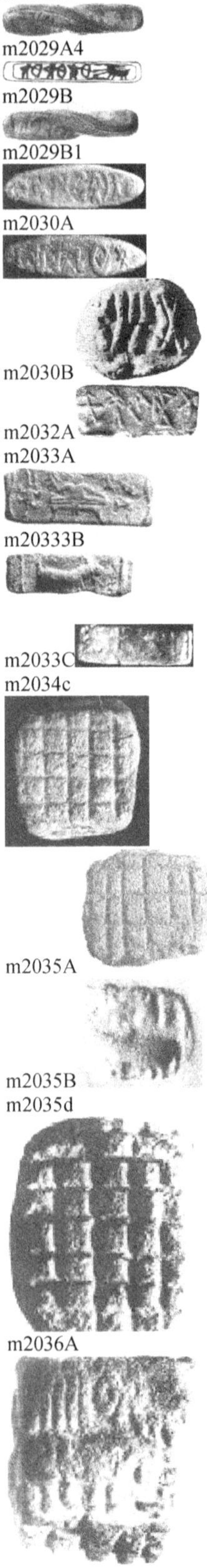

m2028C
m2029A4
m2029B
m2029B1
m2030A
m2030B
m2032A
m2033A
m20333B
m2033C
m2034c
m2035A
m2035B
m2035d
m2036A

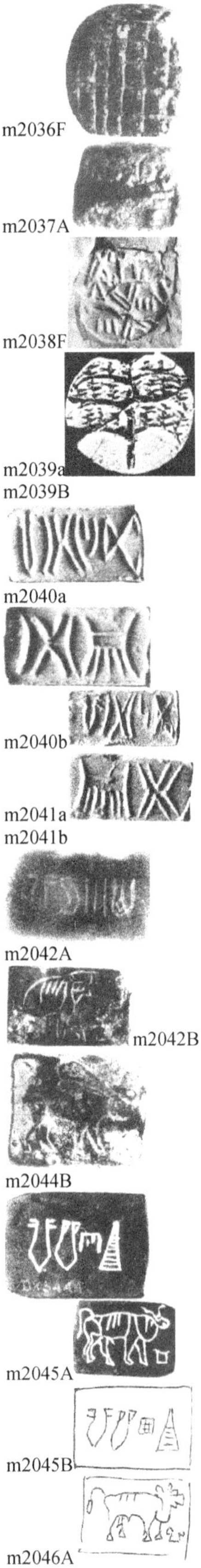

m2036F
m2037A
m2038F
m2039a
m2039B
m2040a
m2040b
m2041a
m2041b
m2042A
m2042B
m2044B
m2045A
m2045B
m2046A
m2046B

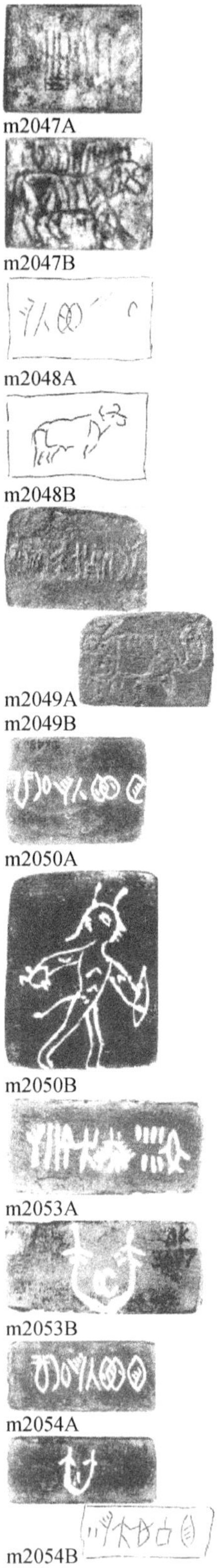

m2047A
m2047B
m2048A
m2048B
m2049A
m2049B
m2050A
m2050B
m2053A
m2053B
m2054A
m2054B
m2055A

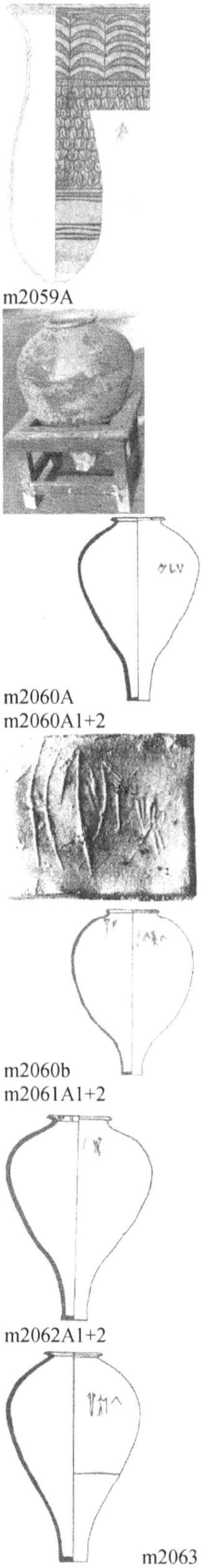

m2059A
m2060A
m2060A1+2
m2060b
m2061A1+2
m2062A1+2
m2063

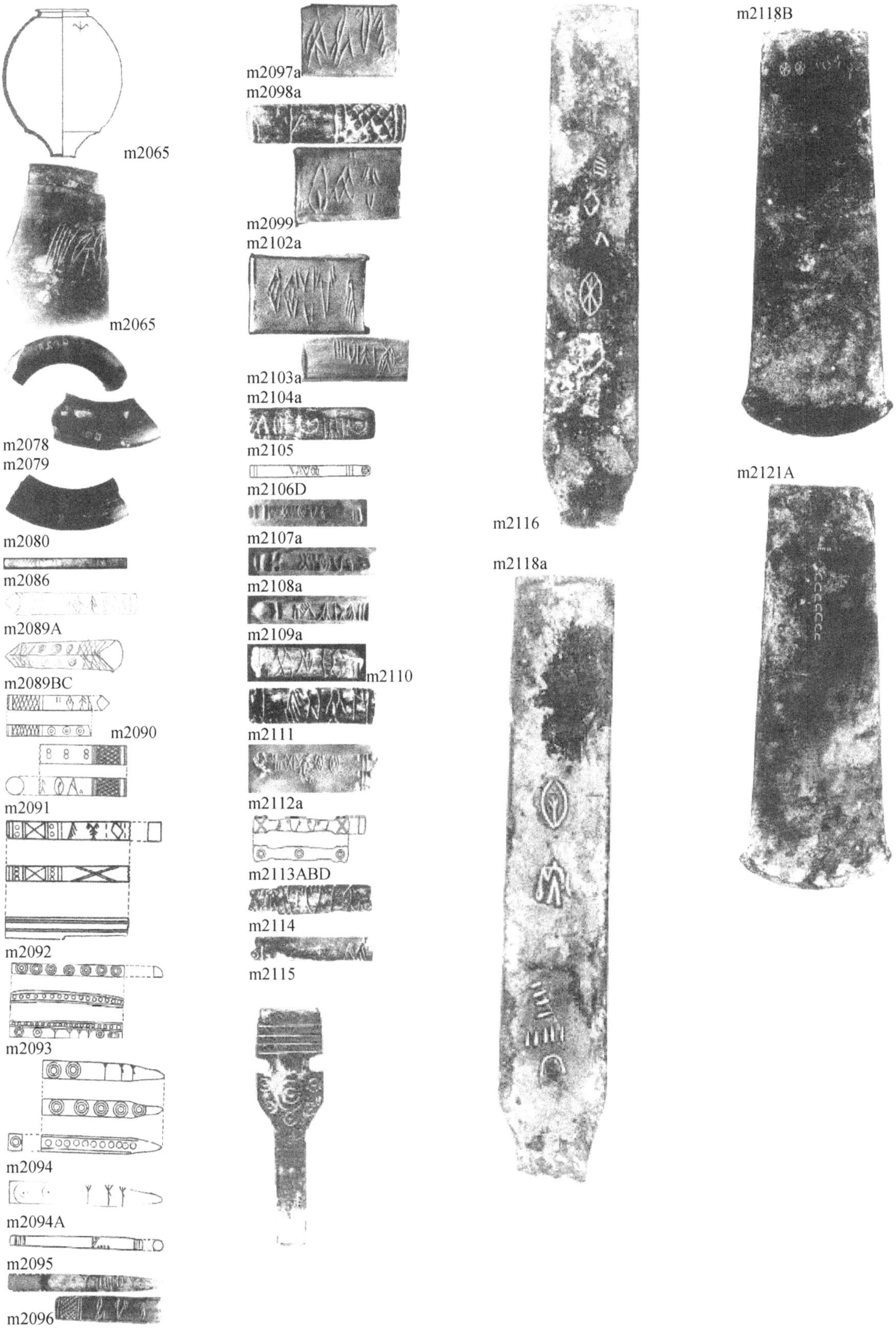
m2065
m2065
m2078
m2079
m2080
m2086
m2089A
m2089BC
m2090
m2091
m2092
m2093
m2094
m2094A
m2095
m2096
m2097a
m2098a
m2099
m2102a
m2103a
m2104a
m2105
m2106D
m2107a
m2108a
m2109a
m2110
m2111
m2112a
m2113ABD
m2114
m2115
m2116
m2118a
m2118B
m2121A

m2121B

m2123

m2124

m2125

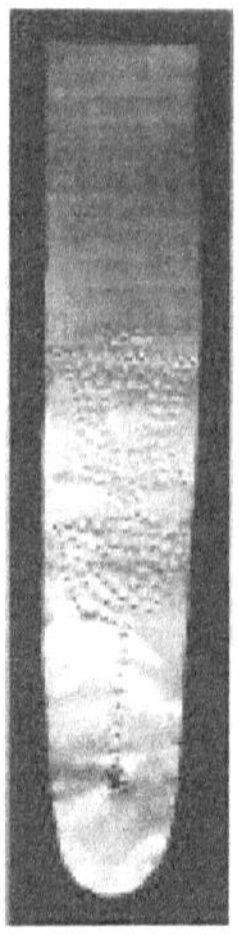

m2125A1

m2128A1

m2129A1

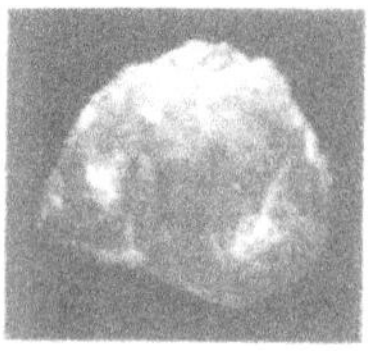

M-2131A

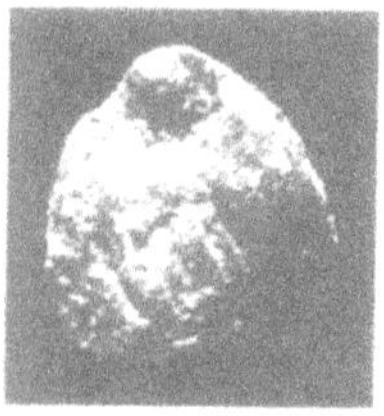

M-2131B

Photograph from ASI: Sindh series Photo archive of ASI, Janpath, New Delhi. Si. 5:6639, 5:6640. Rattle? Bulla?

Mohenjodaro Texts either not illustrated or not linked with inscribed objects:

1002

1003

1020

1036

1041

1043 1044

1045

1049

1053

1055

1065

1070

1072

1074

1077

1079

1083

1088

1094

1101

1103

1106

1116

1117

1125

1130

1429

m1651Aivory stick

m1651D

m1651F

2947

1132

1133

1134

1136

1137

1141

1142

1154

1157

1159

1162

1172

1173

1174

1179

1183

1190

1198

1199

1200

1201

1207

1209

1212

1213

1215 1217

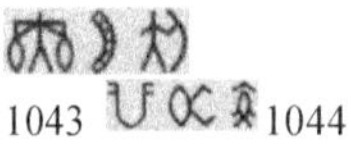

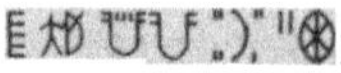
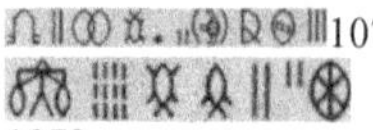
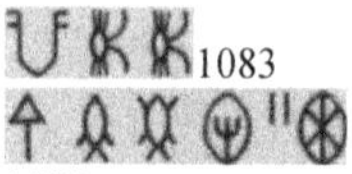

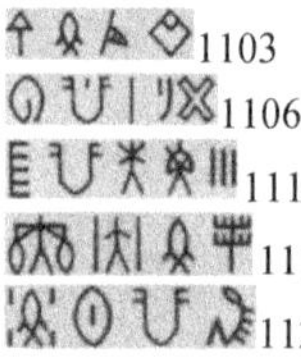
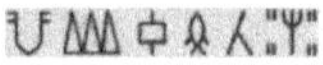

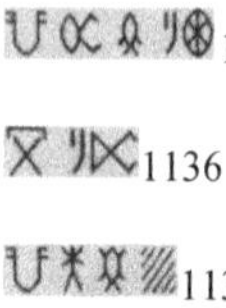

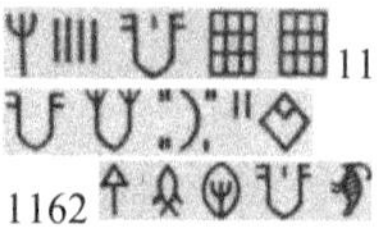

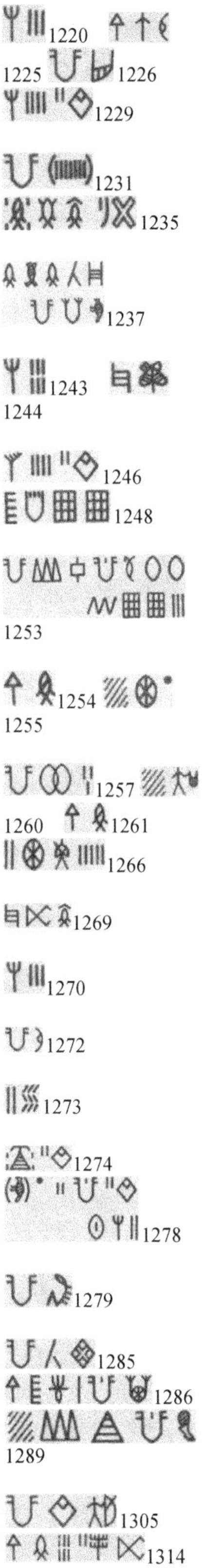

1220 1225 1226 1229 1231 1235 1237 1243 1244 1246 1248 1253 1254 1255 1257 1260 1261 1266 1269 1270 1272 1273 1274 1278 1279 1285 1286 1289 1305 1314

1318 1320 1323 1330 zebu bull 1338 1345 1346 1347 1350 1365 1366 1372 1407 1411 1419 1420 1424 1425 1427 1435 1436 1441 1448 1451 1453 1457 1467 1468 1480 1484 1486 1490 1491

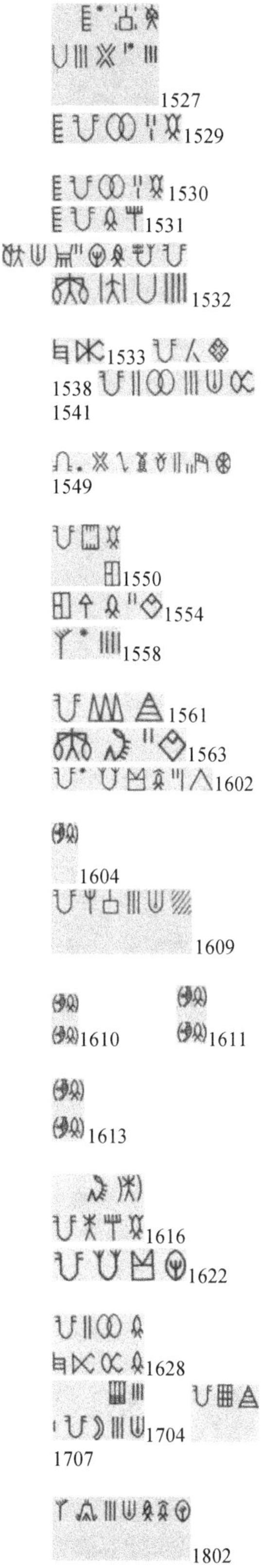

1527 1529 1530 1531 1532 1533 1538 1541 1549 1550 1554 1558 1561 1563 1602 1604 1609 1610 1611 1613 1616 1622 1628 1704 1707 1802

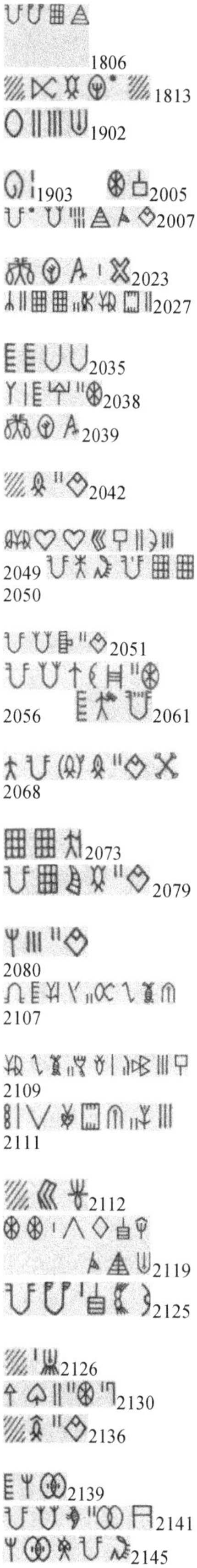

1806 1813 1902 1903 2005 2007 2023 2027 2035 2038 2039 2042 2049 2050 2051 2056 2061 2068 2073 2079 2080 2107 2109 2111 2112 2119 2125 2126 2130 2136 2139 2141 2145

2147
2153
2154
2157
2164
2186
2190
2191
2196
2205
2214
2220
2224
2231
2232
2236
2244
2246
2252
2269
2270
2275
2277
2278
2280
2283
2292
2293
2294
2296
2310
2322
2326

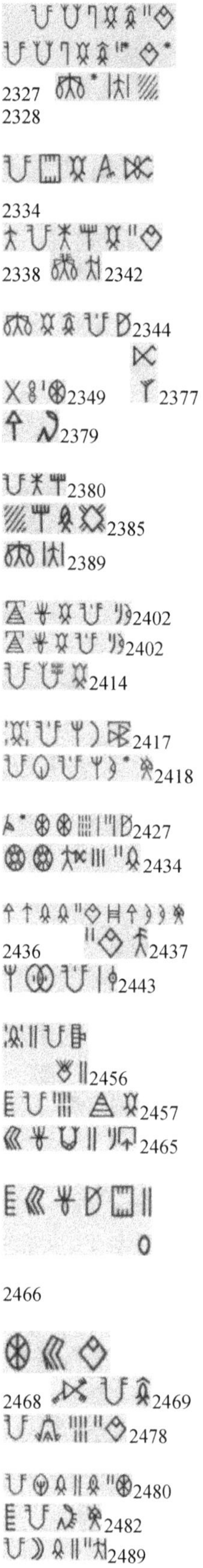
2327
2328
2334
2338
2342
2344
2349
2377
2379
2380
2385
2389
2402
2402
2414
2417
2418
2427
2434
2436
2437
2443
2456
2457
2465
2466
2468
2469
2478
2480
2482
2489

2491
2492
2498
2499
2506
2508
2512
2514
2516
2523
2525
2526
2528
2532
2538
2539
2540
2541
2551
2552
2556
2560
2572
2580
2581
2583
2587
2588
2592
2596

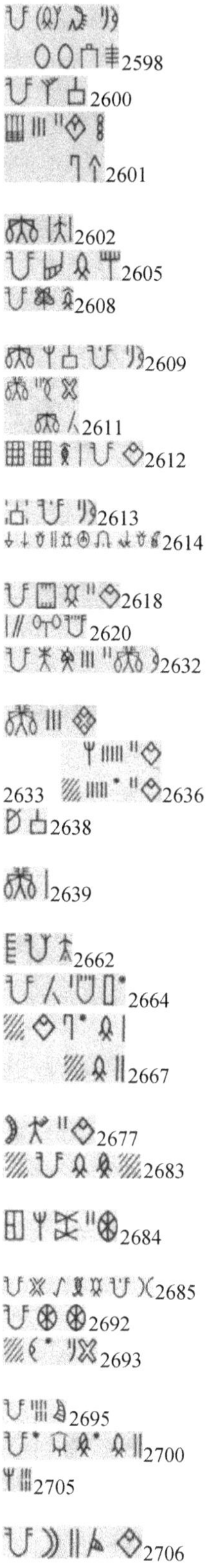
2598
2600
2601
2602
2605
2608
2609
2611
2612
2613
2614
2618
2620
2632
2633
2636
2638
2639
2662
2664
2667
2677
2683
2684
2685
2692
2693
2695
2700
2705
2706

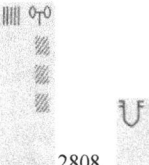

2808
2814

2820

2824

2831
2839

2849

2857

2858
2901
Incised copper tablet

2903 Incised copper tablet

2911 Incised copper tablets. Markhor.

2915

2923
Inscribed bronze implement (MIC Plate CXXVI-2)

2924
Inscribed bronze implement (MIC Plate CXXVI-3)

2925
Inscribed bronze implement (MIC Plate CXXVI-5)

2926
Inscribed bronze implement (MIC Plate CXXVII-1)

2928
Inscribed bronze implement (MIC Plate CXXXIII-1)

2929 Incised on pottery

2930 Graffiti on pottery

2931 Graffiti on pottery

2934 Graffiti on pottery

2935 Graffiti on pottery

2936 Graffiti on pottery

2937 Seal impression on pot

2938Mohenjodaro, Pottery graffiti. Boat.

2940 Ivory or bone rod

2941 Ivory or bone rod Geometrical patterns followed by inscription.

2944 Ivory or bone rod

2945 Ivory or bone rod

2947

2949 Dotted circles

2950

2951

3001

3002

3010

3016

3019

3021

3023
3024

3035

3038

3042
3044

3051
3052

3056
3063
3064

3067
3069

3080
3090
3094

3096
3098
3099
3114
3123
3151
3153
3154

3155
3156
3162
3165
3202

3203
3206

3207
3217
3218
3222

3226
3238
3307

3309

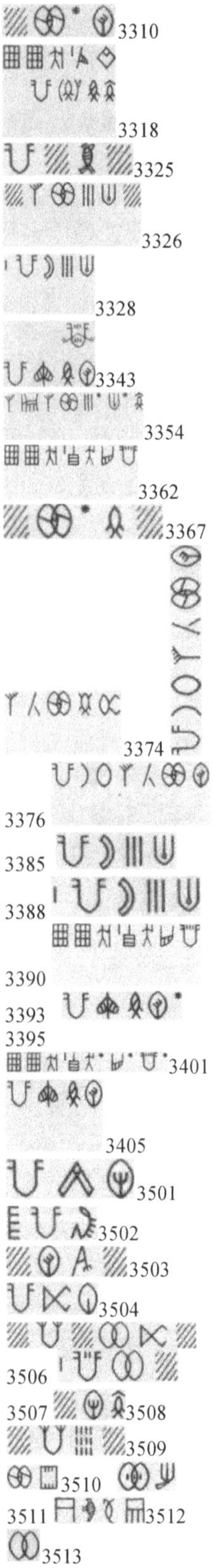

3310
3318
3325
3326
3328
3343
3354
3362
3367
3374
3376
3385
3388
3390
3393
3395
3401
3405
3501
3502
3503
3504
3506
3507
3508
3509
3510
3511
3512
3513

Nindowari-damb01 
Squirrel sign

Nindowari-damb02

Nindowari-damb03

Nausharo01

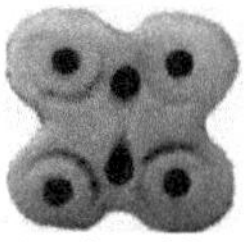

Nausharo02

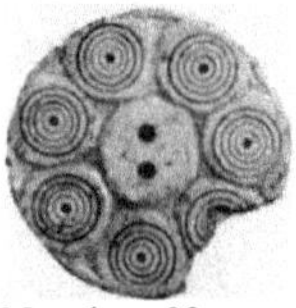

Nausharo03

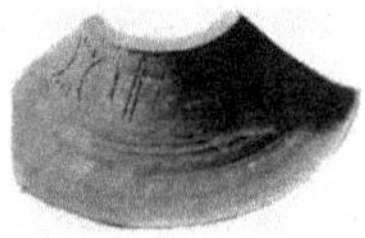

Nausharo04

Nausharo05

Nausharo06

Nausharo07
Nausharo08

Nausharo09

Nausharo10

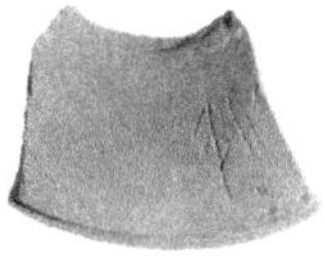

Naro-Waro-dharo01

Naro-Waro-dharo02

Naro-Waro-dharo03

Pabumath

Prabhas Patan (Somnath) 1A

Prabhas Patan (Somnath)1B

Pirak1

Pirak12

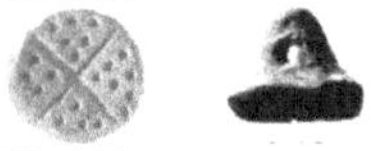

Pirak13

Pirak15

Pirak16

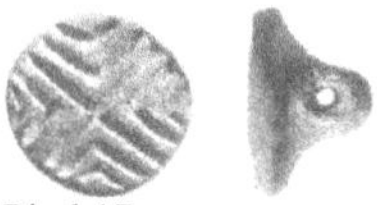

Pirak17

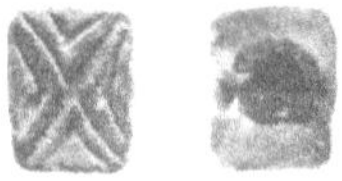

Pirak18

Pirak18A

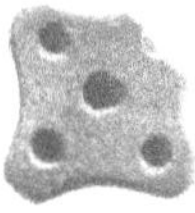

Pirak19

Pirak2

Pirak20

Pirak24

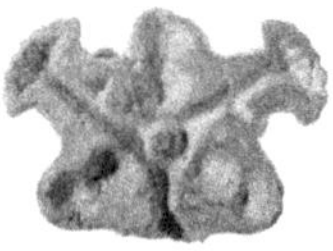

Pirak26Ac
Pirak27

Pirak28

Pirak35

Pirak38

Pirak3 post-harappan

Pirak40

Pirak4

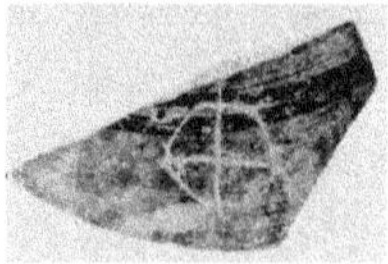

Rangpur

Rakhigarhi1

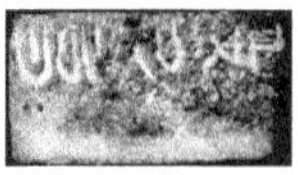

Rakhigarhi 2

9111

Rakhigarhi 65

Rahman-dheri01A

Rahman-dheri01B

Rahman-dheri120

Rahman-dheri126

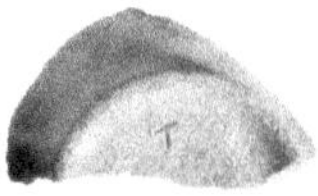

Rahman-dheri127

Rahman-dheri150

Rahman-dheri153

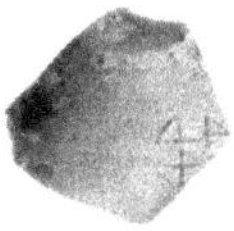

Rahman-dheri156

Rahman-dheri158

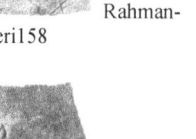

Rahman-dheri216

Rahman-dheri241

Rahman-dheri242

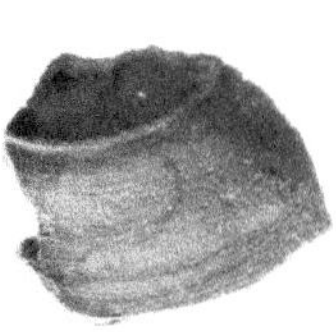

Rahman-dheri243

Rahman-dheri254

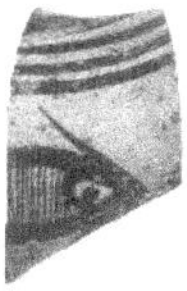

Rahman-dheri255

Rahman-dheri257

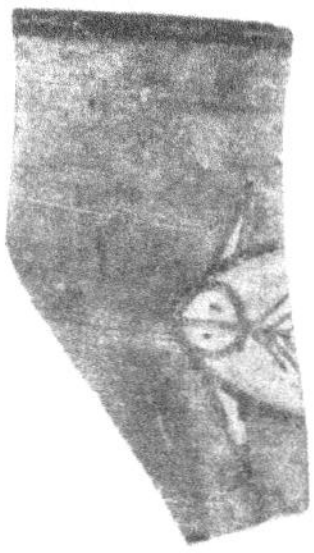

Rahman-dheri258

Rahman-dheri259

Rahman-dheri260

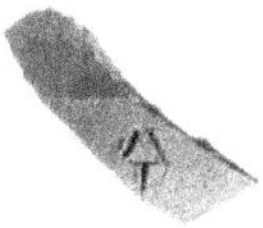

Rahman-dheri90

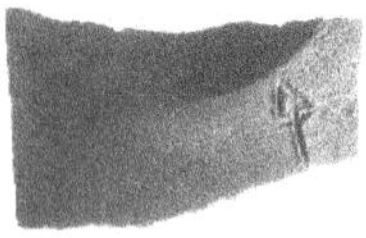

Rahman-dheri92

Rohira1

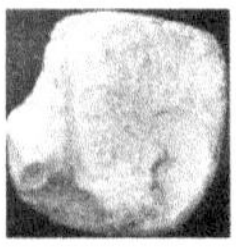

Rohira2

Rojdi

9041

9042

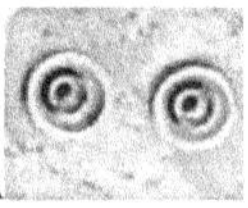

Rupar1A

Rupar1B

9021

9022

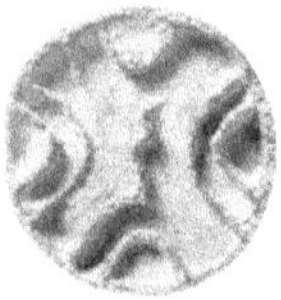

Shahi-tump

Sibri-damb01A

Sibri-damb01B

Sibri-damb02a

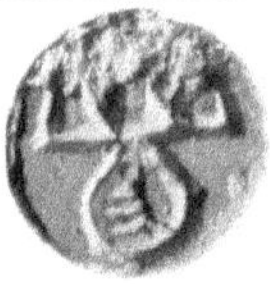

Sibri-damb02E

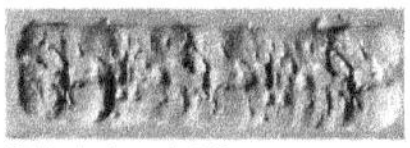

Sibri-damb03a

sibri cylinder seal zebu

Surkotada1

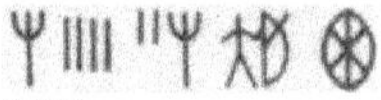

9091

Surkotada

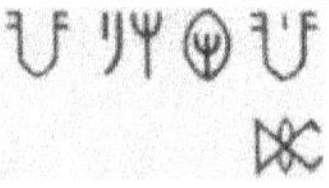

2

9092

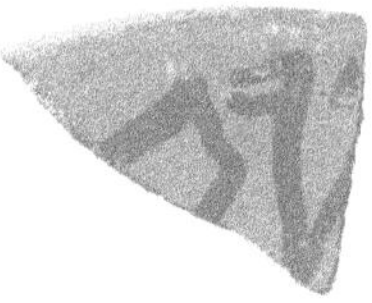

Surkotada3c

9093

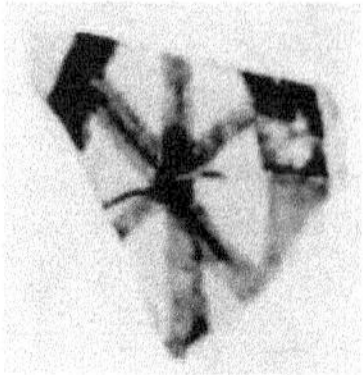

Surkotada 4  9094

Surkotada 6 

9095

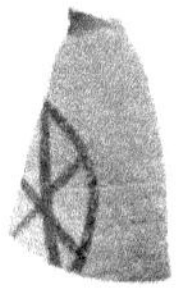

Surkotada 7

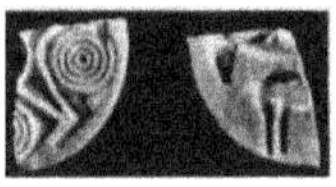

Tarkhanewala-dera1AB

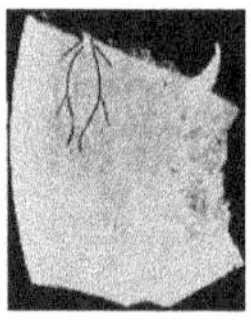

Tarkhanewala-dera 3

9031

Tarakai Qila01A

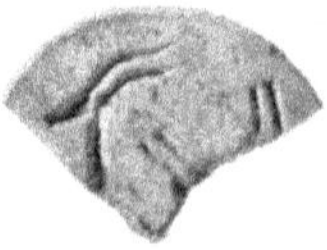

Tarakai Qila01B

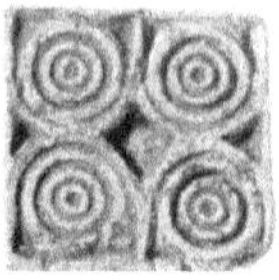

Tarakai Qila02

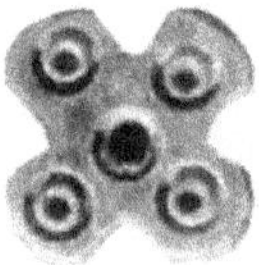

Tarakai Qila03

Tarakai Qila04

Tarakai Qila06

(provenance) unkn01

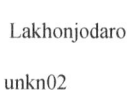

Lakhonjodaro

unkn02

*****

unkn03

unkn04 unkn05A unkn06

Seau l'nde. Musee des Arts Asiatique, Guimet, France

Mohenjo-daro. Copper tablet DK 11307 (SC 63.10/262).

Mohenjodaro; limestone; Mackay, 1938, p. 344, Pl. LXXXIX:376.

Mohenjodaro; Pale yellow enstatite; Mackay 1938, pp. 344-5; Pl. XCVI:488; Collon, 1987, Fig. 607.

Rakhigarhi: Cylinder Seal (ASI), Lizard or gharial?

Rojdi. Ax-head or knife of copper, 17.4 cm. long (After Possehl and Raval 1989: 162, fig. 77

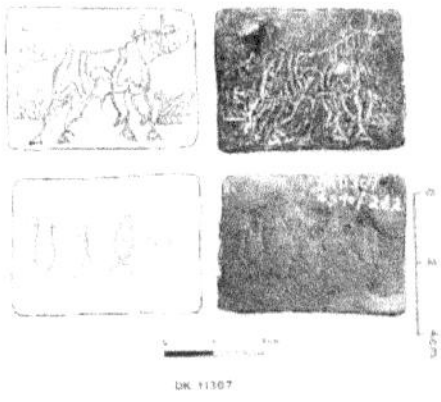

m03 552654

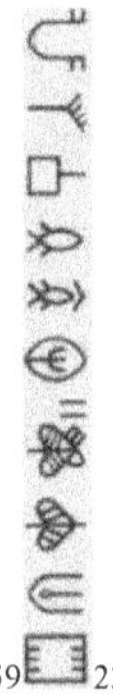

m0359 2325

m0369

2537

m0391

310

Identifying Harappans as Mleccha, Meluhha, Indo-European speakers

Decipherment of Harappa Script justifies a rewrite of the received wisdom of the region of Indo-European (IE) speakers who contributed to the Bronze Age metalwork revolution. The IE speakers included Meluhha-Mleccha speakers from Bharata *sprachbund*. This is consistent with what John Marshall calls 'Pre-Sanskrit civilization' in reference to the script inscriptions comparable to Egyptian hieroglyphs.

Ancient Bharata texts use terms such as म्लेच्छदेशः, कामरूपवङ्गादिः Meluhha, Mleccha, Kamarupa, Vanga etc. which signify areas coterminous with Sarasvati civilization region and many contact areas populated by Meluhha or Indo-European speakers who created over 8000 Harappa Script inscriptions data archiving metalwork catalogues. It is clear from the decipherment that many vocables and expressions used in the catalogues are from Bharata *sprachbund* (which included Indo-European, Munda (Austro-Asiatic) and Dravidian streams). Now, based on the plain texts derived from cipher texts of inscriptions, we can firmly identify Harappans as Mleccha, Meluhha, Indo-European speakers.

Shu-ilishu (Interpreter of Meluhha) cylinder seal ca. 3rd millennium BCE shows Meluhha merchant holding an antelope followed by a person carrying a liquid measure. *Melh* 'goat' rebus; *milakkhu* 'copper', *meluhha*; *ranku* 'liquid measure' rebus: *ranku* 'tin'.

Prof. Shivaji Singh in his Maulana Azad Memorial Lecture of November 11, 2016 cites references from Rigveda to attest Anus, Druhyus, Pūrus, Yadus, and Turvasas as the five peoples, Panca-janah, etc. mentioned in the ancient text. I suggest that based on the details provided in the Great Epic and other ancient texts of Bharata, the term 'five peoples' should not be restricted to identifying'five groups' but should be interpreted as an idiomatic expression signifying, as a collective noun, a community of people in general.

The way the Great Epic Mahabharata describes groups of peole as mleccha, it appears as though the entire Bharatavarsha consisted of mlecchas, simply identified with reference to their propensity to use phonetic forms of speech at variance from Samskrtam. Thus, Tocharians (Tushara) who speak a version of Indo-European, become mleccha-speakers, so do others who mispronounce in the vernacular. Thus, Anus, Yadavas, Pauravas were also followers of Vedic culture and Indo-European mleccha-speakers. In later times, some mlecchas also get characterized as those who do not follow the traditions of Aryavarta. Mleccha region is distinguished from Arya, 'virtuous' region framed on deviations from the Vedic, शिष्ट śiṣṭa 'virtuous' conduct or s'iSTAcAra 'virtuous cultural practices' शिष्टा* चार [p= 1076,3] *m.* practice or conduct of the learned or virtuous , good manners , proper behaviour (VasiSTha)

— म्लेच्छदेशः स विज्ञेय आर्य्यावर्त्तस्ततः परमिति AND म्लेच्छति शिष्टाचारहीनो भवत्यत्रम्लेच्छः

Harappa Script decipherment has demonstrated that the artisans, metalworkers were followers of the Vedic culture and traditions, in particular reference to the conduct of Soma SamsthA yAgas after installing an octagonal yupa (See Binjor evidence of yajnakunda, octagonal yupa and Harappa Script seal, all from Sarasvati River Basin 4 MSR site). The following ancient texts affirm this by 1. equating godhuma on yupa (caSAla) as mlecchabhojanah and 2. Equating soma and mleccha:

चणकस्तु कलायः स्याद् गोधूमो म्लेच्छभोजनः ॥ (३३४.२) https://sa.wikisource.org/wiki/अष्टाङ्गनिघण्टु

पारिखा षा खरी शावा षिः शङ्खो वर्तुलः स्मृतः । षीः पुत्रः षुः खुरः पूरः षूः सोमो म्लेच्छ एव च।।
92 https://sa.wikisource.org/wiki/एकार्थनाममाला

For example, भारतवर्ष-स्यान्तः शिष्टाचाररहितः कामरूपवङ्गादिः is an expression attested in शब्दकल्पद्रुमः. This expression refers to regions such as Kamarupa and Vanga which are explained together with other regions such as Anga, Kalinga, Pundra, Suhma etc.:

> **कामरूपः**, अयं खलु वङ्गदेशस्य ईशानभागे आसामप्रदेशस्यपश्चिमभागे वर्त्तते ॥...
>
> अङ्गस्याङ्गो भवेद्देशो वङ्गो वङ्गस्य च स्मृतः ।कलिङ्गविषयश्चैव कलिङ्गस्य च स स्मृतः ॥पुण्ड्रस्य पुण्ड्रा प्रख्याताः सुह्मा सुह्मस्य च स्मृताः ।
> https://sa.wikisource.org/wiki/शब्दकल्पद्रुमः/

Another example is attested in Baudhayana Srauta Sutra, specifying east-west migrations of people. Translation by Kashikar is as follows: Ayu moved (from Kurukshetra) towards the east. Kuru-Pancala and Kasi-Videha were his regions. This is the realm of Ayu. Amavasu proceeded towards the west. The Gandharis, Spars'us and Arattas were his regions. This is the realm of Amavasu.

Map shows the 'seven rivers and Sarasvati'; various sites with Harappan artifacts far from Sapta Sindhu; also, the two movements eastward by Ayu and westward by Amavasu. http://www.newsgram.com/vedic-sanskrit-older-than-avesta-baudhayana-mentions-westward-migrations-from-india-dr-n-kazanas/

Pañcāla पाञ्चाल *a.* (-ली *f.*) Belonging to or ruling over the Pañchālas. -लः 1 The country of the Pañchālas. -2 A prince of the Pañchālas. -लाः *m.* (pl.) 1 The people of the Pañchālas. -2 An association of five guilds (*i e.* of a carpenter, weaver, barber, washer- man, and shoe-maker).

पाञ्चालक pāñcālakaThe reference to Panchala is cognate with PancakammALa, five categories of artisans: பஞ்சகம்மாளர் *pañca-kammāḷar* , *n.* < *pañcantaṭṭāṉ, kaṉṉāṉ, ciṟpaṉ, taccaṉ, kollaṉ*; தட்டான், கன்னான், சிற்பன், தச்சன் கொல்லன் என்ற ஐவகைப் பட்ட கம்மாளர். (சங். அக.), i.e. goldsmith, coppersmith, sculptor, carpenter, smelter.

"The *Tukhara* were among Indo-European tribes that conquered Central Asia during the 2nd century BCE, according to both Chinese and Greek sources. Ancient Chinese sources refer to these tribes collectively as the *Da Yuezhi* ("Greater Yuezhi"). In subsequent centuries the *Tukhara* and other tribes founded the Kushan Empire, which dominated South Asia. The account in Mahabharata (Mbh) 1:85 depicts the Tusharas as mlechchas ("barbarians") and descendants of Anu, one of the cursed sons of King Yayati. Yayati's eldest son Yadu, gave rise to the Yadavas and his youngest son Puru to the Pauravas that includes the Kurus and Panchalas. Only the fifth son of Puru's line was considered to be the successors of Yayati's throne, as he cursed the other four sons and denied them kingship. The Pauravas inherited the Yayati's original empire and stayed in the Gangetic plain who later created the Kuru and Panchala Kingdoms. They were followers of the Vedic culture. The Yadavas made central and western India their stronghold. The descendants of Anu, known as the Anavas, are said to have migrated to Iran." https://en.wikipedia.org/wiki/Tushara_Kingdom

The archaeological context of the 2000+ sites of Sarasvati Civilization on the banks of Vedic River Sarasvati can also be viewed in the context of the details provided in the Great Epic which is a continuum from Rigveda and other Veda texts.

The total categories of people specifically named number 850 in the Great Epic, *Mahābhārata.*

Some typical names of groups of people are: Yavanas, Yadus, Yadavas, Vrishnis, Videhas, Vidarbhas, Vasatis, Vangas, (Bahlikas) Valhikas, Vahikas, Uttara-Kurus, Usinaras, Ulukas, Uddras, Tusharas, Trigartas, Tanganas, Swaitya, Surasenas, Srinjayas, Somakas, Sivis, Sindhus, Sinhalas, Sindhu-Sauviras, Satwatas, Samgakas, Samsaptakas, Salwas, Sakas, Saindhavas, Rohitakas, Pulindas, Pukkasas, Prabhadrakas, Paurava, Parvata, Paradas, Pandavas, Pancha-Nodas, Panchalas, Palhavas, Nishadas, Naimishas, Mlecchas, Matsyas, Marichipas, Manasas, Malavas, Mahishakas, Magadhas, Madras, Madrakas, Madhu, Madhavas, Lokapalas, Kusikas, Kunti, Kulindas, Kumaras, Kurds, Kukkuras, Kshudrakas, Kosalas, Kiratas, Kichakas, Khasas, Khandavas, Kekayas, Kausikas, Keralas, Kauravas, Kasmiras, Kaunteyas, Karushas, Karushakas, Kapas, Kankas, Kanadas, Kamvojas, Kambojas, Kalingas, Kalikeyas, Kaikeyas, Kacchas, Jamvuvan, Ikshwaku, Hunas, Hansas, Haihayas, Govasanas, Gopas, Gautami, Gandharas, Ganas, Dravidas, Dasarnas, Dasarha, Darvabhiaras, Daradas, Dandakas, Chinas, Chedis, Cholas, Bhutas, Bhojas, Bhargas, Bharatas, Avantis, Artayani, Arattas, Aratta-Vahikas, Angas, Andhras, Anarttas, Amvashthas, Ailas, Agniswattas, Abhishahas, Abhiras

In all these names which indicative collective groups of people, the most dominant use occurs for Bharatas in the Great Epic, *Mahabharata* (Bharata 2261, Bharatas 418 and Bharata's 668). Mleccha and related expressions are seen in 70 occurrences.

Thus, I suggest that the people identified as Bharatam Janam by Rishi Visvamitra (RV 3.53.12) refers to all the people in general, the Panca jana, 'the collective of five peoples, i.e. people in general'.

In the context of frequent references to mleccha as people inhabiting Bharata Varsha in many islands and in many parts of Bharata, mleccha speakers (mlecchavAcas) constituted the speaker of the lingua france, Meluhha, who pronounced Samskritam words in the colloquial tongue with mispronunciations and variant spellings. Hence, the pronunciation variations recorded in the *Indian Lexicon* which is a compendium of over 8000 semantic clusters encompassing Indo-Aryan, Dravidian and Mundarica streams, thus establishing the reality of Bharata *sprachbund* of the Bronze Age. Mleccha as the lingua franca, is attested by Manu and recognized by Bharata. These tongues are later categorized as Apabhrams'a or Apas'abda or as Desi in Hemacandra's Desi NAmamAlA, a Prakritam lexicon.

In the context of life-activities of the people of the Bronze Age, Bharatam Janam refers to metalcasters: bharata 'alloy of copper, pewter, tin'.

Mleccha is a language category while Bharata is a professional category. Panca-panca-janah or panca-janah is a collective category to signify the samajam or community of people, comparable to *nālu,* 'four' in Tulu. One idiom, *nālvar,* 'four people' (Kannada) refers to the four pall-bearers who carry the body to the cremation ground, i.e. an idiom which signifies samajam or community in general.

Thus. 'Pre-Sanskrit civilization' mentioned by John Marshall can be explained as a reference to the vernacular of Sanskrit, Meluhha which was the *lingua franca* of the civilization.

Excerpts from Prof. Shivaji Singh's lecture:

The ‘Five Peoples’ called Pañcha-janāḥ (Rv, 1.89.10; 3.37.9; 5.9.8; etc.) are the most frequently mentioned group of ethnic units in the Ṛigveda. They are designated also as Pañchajātā (Rv, 6.61.12), Pañcha-mānushāḥ (Rv, 8.9.2), Pañcha-chārshaṇyaḥ (Rv, 5.86.2; 7.15.2; 9.101.9), Pañcha-kṛishṭayaḥ (Rv, 2.2.10; 3.53.16; 4.38.10; etc.), and Pañcha-kshitayaḥ (Rv, 1.7.9; 1.76.3; 5.35.2; etc.).

Last three of these nomenclatures seem to refer to their gradually evolving stages of social formation. Thus, while ‘chārshaṇyaḥ’, from root *char* (to move), may point to their predominantly nomadic pastoral condition, ‘kṛishṭayaḥ’ from root *kṛish* to cultivate), may indicate their settled agricultural situation. Similarly, ‘kshitayah’, from root *kshi* (to possess, to have power over), may express their still more developed status when these peoples had acquired territorial consciousness about the area they occupied (Nandi 1986-87: 156-57).
The actual names of the ethnic units constituting this group of five peoples is not explicitly stated in the Ṛigveda resulting in certain wild speculations by some ancient and medieval authorities (Cf. Aitareya Brāhmaṇa, 3.31; Nirukta, 3.8; Sāyaṇa on Rv, 1.7.9, etc.). However, on circumstantial evidence, modern scholars in general agree that the Anus, Druhyus, Pūrus, Yadus, and Turvasas are the Rigvedic ‘Five Peoples’. They are clearly mentioned together in one verse (Rv, 1.108.8) and substituting Yakshu for Yadu in another hymn too (Rv, 7.18). It is also clear that initially all these five peoples lived on the banks of river Sarasvati (Rv, 6.61.12) though later on in the Rigvedic period itself several of them moved on to other areas.” (Singh, Shivaji 1997-98, Sindhu and Sarasvati in the Ṛigveda and Their Archaeological Implications, *Purātattva* 28:26-38; loc.cit. Maulana Azad Memorial Lecture, 10 November 2018). https://www.academia.edu/30071331/With_Veda_in_One_Hand_and_Spade_in_the_Other_Writing_Early_History_of_India_Afresh_--_Lecture_by_Prof._Shivaji_Singh

Excerpts from ancient texts explaining some select terms of *lingua franca*

म्लिष्टं, क्ली, (म्लेच्छ् + क्तः + “क्षुब्धस्वान्तध्वान्तलग्न-म्लिष्टविरिब्धेत्यादि ।” ७ । २ । १८ । इतिनिपातितम् ।) अस्पष्टवाक्यम् । तत्पर्य्यायः ।अविस्पष्टम् २ । इत्यमरः । १ । ६ । २१ ॥**म्लिष्टः**, त्रि, (म्लेच्छ + क्तः ।) अव्यक्तवाक् । म्लानः ।इति मेदिनी । टे, २५ ॥**म्लेच्छ**, कि देश्योक्तौ । इति कविकल्पद्रुमः ॥ (चुरा०-वा भ्वा०-पर०-अक०-सक० च-सेट् ।) देश्याग्राम्या उक्तिर्देश्योक्तिरसंस्कृतकथनमित्यर्थः ।कि, म्लेच्छयति म्लेच्छति मूढः । अन्तर्विद्यामसौविद्वान्न म्लेच्छति धृतव्रत इति हलायुधः ॥अनेकार्थत्वादव्यक्तशब्देऽपि । तथा चामरः ।अथ म्लिष्टमविस्पष्टमिति । म्लेच्छ व्यक्तायां वाचिइति प्राञ्चः । तत्र रमानाथस्तु । म्लेच्छति वटु-र्व्यक्तं वदतीत्यर्थः । अव्यक्तायामिति पाठे कुत्-सितायां वाचीत्यर्थः ।‘तत्सादृश्यमभावश्च तदन्यत्वं तदल्पता ।अप्राशस्त्यं विरोधश्च नञर्थाः षट् प्रकीर्त्तिताः ॥’इति भाष्यवचनेन नञोऽप्राशस्त्यार्थत्वात् इतिव्याख्यानाय हलायुधोक्तमुदाहृतवान् । इतिदुर्गादासः ॥

म्लेच्छं, क्ली, (म्लेच्छस्तद्देशः उत्पत्तिस्थानत्वेना-स्त्यस्य । अर्शआद्यच् ।) हिङ्गुलम् । इतिराजनिर्घण्टः ॥ (तथास्य पर्य्यायः ।“हिङ्गुलन्दरदं म्लेच्छमिङ्गुलञ्चूर्णपारदम् ॥”इति भावप्रकाशस्य पूर्ब्बखण्डे प्रथमे भागे ॥)

म्लेच्छः, पुं, (म्लेच्छयति वा म्लेच्छति असंस्कृतंवदतीति । म्लेच्छ् + अच् ।) किरातशवरपुलि-न्दादिजातिः । इत्यमरः ॥ पामरमेदः । पाप-रक्तः । अपभाषणम् । इति मेदिनी । छे, ६ ॥म्लेच्छादीनां सर्व्वधर्म्मराहित्यमुक्तं यथा, हरि-वंशे । १४ । १५ -- १९ ।“सगरः स्वां प्रतिज्ञाञ्च गुरोर्व्वाक्यं निशम्य च ।धर्म्मं जघान तेषां वै वेशान्यत्वं चकार ह ॥अर्द्धं शकानां शिरसो मुण्डयित्वा व्यसर्जयत् ।जवनानां शिरः सर्व्वं काम्बोजानान्तथैव च ॥पारदा मुक्तकेशाश्च पह्लवाः श्मश्रुधारिणः ।निःस्वाध्यायवषट्काराः कृतास्तेन महात्मना ॥शका जवनकाम्बोजाः पारदाः पह्लवास्तथा ।कोलसप्याः समहिषा दार्व्वाश्चोलाः सकेरलाः ।सर्व्वे ते क्षत्त्रियास्तात धर्म्मस्तेषां निराकृतः ॥वशिष्ठवचनाद्राजन् सगरेण महात्मना ॥”शकानां शकदेशोद्भवानां क्षत्त्रियाणाम् । एवंजवनादीनामिति । अत्र जवनशब्दस्तद्देशोद्भव-वाची चवर्गतृतीयादिः । जवनो देशवेगिनो-रिति त्रिकाण्डशेषाभिधानदर्शनात् ॥ * ॥ तेषांम्लेच्छत्वमप्युक्तं विष्णुपुराणे । तथाकृतान् जवना-दीनुपक्रम्य ते चात्मधर्म्मपरित्यागात् म्लेच्छत्वंययुरिति । बौधायनः ।“गोमांसखादको यश्च विरुद्धं बहु भाषते ।सर्व्वाचारविहीनश्च म्लेच्छ इत्यभिधीयते ॥”इति प्रायश्चित्ततत्त्वम् ॥ * ॥अपिच । देवयान्यां ययातेर्द्वौ पुत्त्रौ यदुः तुर्वसुश्च ।शर्म्मिष्ठायां त्रयः

पुत्राः द्रुह्युः अनुः पुरुश्च ।तत्र यदुप्रभृतयश्चत्वारः पितुराज्ञाहेलनं कृत-वन्तः पित्रा शप्ताः । ज्येष्ठपुत्रं यदुं शशाप तववंशे राजा चक्रवर्त्ती मा भूदिति । तुर्व्वसु-द्रुह्य्वनून् शशाप युष्माकं वंश्या वेदवाह्या म्लेच्छाभविष्यन्ति । इति श्रीभागवतमतम् ॥ * ॥("असृजत् पह्लवान् पुच्छात् प्रस्रावाद्द्राविडान्शकान् ।योनिदेशाच्च यवनान् शकृतः शवरान् बहून् ॥मूत्रतश्चासृजत् काञ्चीञ्छरभांश्चैव पार्श्वतःपौण्ड्रान् किरातान् यवनान् सिंहलान् वर्व्वरान्खशान् ॥चियुकांश्च पुलिन्दांश्च चीनान् हूनान् सके-रलान् ।ससर्ज्ज फेनतः सा गौर्म्लेच्छान् बहुविधानपि ॥"सा वशिष्ठस्य धेनुः । इति महाभारते । १ । १७६ ।३५ -- ३७ ॥) अन्यच्च । "शकजवनकाम्बोज-पारदपह्लवा हन्यमानास्तत्कुलगुरुं वशिष्ठंशरणं ययुः । अथैतान् वशिष्ठो जीवन्मृतकान्कृत्वा सगरमाह । वत्स वत्सालमेभिर्जीवन्मृत-कैरनुसृतैः । एते च मयैव त्वत्प्रतिज्ञापालनायनिजधर्म्मद्विजसङ्गपरित्यागं कारिताः । सतथेति तद्गुरुवचनमभिनन्द्य तेषां वेशान्य-त्वमकारयत् । जवनान्मुण्डितशिरसोऽर्द्धमुण्डान्शकान् प्रलम्बकेशान् पारदान् पह्लवांश्च श्मश्रु-धरान्निःस्वाध्यायवषट्कारानेतानन्यांश्च क्षत्त्रि-यांश्चकार । ते चात्मधर्म्मपरित्यागाद्ब्राह्मणैश्चपरित्यक्ता म्लेच्छतां ययुः ।" इति विष्णुपुराणे । ४ ।३ । १८ -- २१ ॥ * ॥ प्रकारान्तरेण तस्योत्-पत्तिर्यया, --सूत उवाच ।"वंशे स्वायम्भुवस्यासीदङ्गो नाम प्रजापतिः ।मृत्योस्तु दुहिता तेन परिणीतातिदुर्मुखी ॥सुतीर्था नाम तस्यास्तु वेनो नाम सुतःपुरा ।अधर्म्मनिरतः कामी बलवान् वसुधाधिपः ।लोकेऽप्यधर्म्मकृज्जातः परभार्य्यापहारकः ॥धर्म्मचारप्रसिद्ध्यर्थं जगतोऽस्य महर्षिभिः ।अनुनीतोऽपि न ददावनुक्षां स यदा ततः ॥शापेन मारयित्वैनमराजकभयार्द्दिताः ।ममन्थुर्ब्राह्मणास्तस्य बलाद्देहमकल्मषाः ॥तत्कायान्मथ्यमानात्तु निपेतुर्म्लेच्छजातयः ।शरीरे मातुरंशेन कृष्णाञ्जनसमप्रभाः ॥"इति मत्स्यपुराणे । १० । ३ -- ८ ॥ * ॥म्लेच्छभाषाभ्यासनिषेधो यथा, --"न सातयेदिष्टकाभिः फलानि वै फलेन तु ।न म्लेच्छभाषां शिक्षेत नाकर्षेच्च पदासनम् ॥"इति कौर्म्म्ये उपविभागे १५ अध्यायः ॥ * ॥तस्य मध्यमा तामसी गतिर्यथा, मानवे ।१२ । ४३ ।"हस्तिनश्च तुरङ्गाश्च शूद्रा म्लेच्छाश्च गर्हिताः ।सिंहा व्याघ्रा वराहाश्च मध्यमा तामसीगतिः ॥"(मन्त्रणाकाले म्लेच्छापसारणमुक्तं यथा, मनु-संहितायाम् । ७ । १४९ ।"जडमूकान्धवधिरांस्तैर्य्यग्योनान् वयोऽति-गान् ।स्त्रीम्लेच्छव्याधितव्यङ्गान् मन्त्रकालेऽपसार-येत् ॥""अथवा एवंविधा मन्त्रिणो न कर्त्तव्याः । बुद्धि-विभ्रमसम्भवात् ।" इति तद्भाष्ये मेधातिथिः ॥म्लेच्छानां पशुधर्म्मित्वम् । यथा, महाभारते । १ ।८४ । १५ ।"गुरुदारप्रसक्तेषु तिर्य्यग्योनिगतेषु च ।पशुधर्म्मिषु पापेषु म्लेच्छेषु त्वं भविष्यसि ॥")

म्लेच्छकन्दः, पुं, (म्लेच्छप्रियः कन्द इति मध्यपदलोपी कर्म्मधारयः ।) लशुनम् । इति राज-निर्घण्टः ॥ (तस्य पर्य्यायो यथा, --"लशुनस्तु रसोनः स्यादुग्रगन्धो महौषधम् ।अरिष्टो म्लेच्छकन्दश्च पवनेष्टो रसोनकः ॥"इति भावप्रकाशस्य पूर्ब्बखण्डे प्रथमे भागे ॥)

म्लेच्छजातिः, स्त्री, (म्लेच्छस्य जातिरिति षष्ठी-तत्पुरुषः म्लेच्छरूपा जातिरिति कर्म्मधारयोवा ।) गोमांसखादकबहुविरुद्धभाषकसर्व्वा-चारविहीनवर्णः । यथा, --"गोमांसखादको यस्तु विरुद्धं बहु भाषते ।सर्व्वाचारविहीनश्च म्लेच्छ इत्यभिधीयते ॥"इति प्रायश्चित्ततत्त्वधृतबौधायनवचनम् ॥अपि च ।"भेदाः किरातशवरपुलिन्दा म्लेच्छजातयः ॥"इत्यमरः । २ । ४० । २० ॥अन्यच्च ।"पौण्ड्रकाश्चौड्रद्रविडाः काम्बोजा शवनाःशकाः ।पारदाः पह्लवाश्चीनाः किराताः दरदाःखशाः ॥मुखबाहूरुपज्जानां या लोके जातयो बहिः ।म्लेच्छवाचश्चार्य्यवाचः सर्व्वे ते दस्यवः स्मृताः ॥"इति मानवे १० अध्यायः ॥

म्लेच्छदेशः, पुं, (म्लेच्छानां देशः म्लेच्छप्रधानोदेशो वा ।) चातुर्व्वर्ण्यव्यवस्थादिरहित-स्थानम् । तत्पर्य्यायः । प्रत्यन्तः २ । इत्यमरः ।२ । १ । ७ ॥ भारतवर्षस्यान्तं प्रतिगःप्रत्यन्तः । म्लेच्छति शिष्टाचारहीनो भवत्यत्रम्लेच्छः अल् । स चासौ देशश्चेति म्लेच्छदेशः ।किंवा म्लेच्छयन्ति असंस्कृतं वदन्ति शिष्टा-चारहीना भवन्तीति वा पचाद्यचि म्लेच्छानीचजातयः तेषां देशो म्लेच्छदेशः । भारतवर्ष-स्यान्तः शिष्टाचाररहितः कामरूपवङ्गादिः ।उक्तञ्च ।चातुर्व्वर्ण्यव्यवस्थानं यस्मिन् देशे न विद्यते ।

म्लेच्छदेशः स विज्ञेय आर्य्यावर्त्तस्ततः परमिति ॥" इति भरतः ॥(अपि च, मनुः । २ । २३ ।"कृष्णसारस्तु चरति मृगो यत्र स्वभावतः ।स ज्ञेयो यज्ञियो देशो म्लेच्छदेशस्ततःपरम् ॥")

म्लेच्छभोजनं, क्ली, (भुज्यते यदिति । भुज् + कर्म्मणिल्युट् । ततो म्लेच्छानां भोजनम् ।) यावकः ।इति शब्दरत्नावली ॥

म्लेच्छभोजनः, पुं, (भुज्यतेऽसौ इति । भुज् +ल्युट् । म्लेच्छानां भोजनः । (गोधूमः । इतित्रिकाण्डशेषः ॥

म्लेच्छमण्डलं, क्ली, (म्लेच्छानां मण्डलं समूहोऽत्र ।)

म्लेच्छदेशः । इति हेमचन्द्रः ॥

म्लेच्छमुखं, क्ली, (म्लेच्छे म्लेच्छदेशे मुखमुत्पत्ति-रस्य । इत्यमरटीकायां रघुनाथः ।) ताम्रम् ।इत्यमरः । २ । ९ । ९७ ॥ (तथास्य पर्य्यायः ।"ताम्रमौदुम्बरं शुल्वमुदुम्बरमपि स्मृतम् ।रविप्रियं म्लेच्छमुखं सूर्य्यपर्य्यायनामकम् ॥"इति भावप्रकाशस्य पूर्ब्बखण्डे प्रथमे भागे ॥"ताम्रमौडुम्बरं शूल्वं विद्यात् म्लेच्छमुख-न्तथा ॥"इति गारुडे २०८ अध्याये ॥)

म्लेच्छाशः, पुं, (म्लेच्छैरश्यते इति । अश् + कर्म्मणि+ घञ् ।) म्लेच्छभोजनः । गोधूमः । इतिकेचित् ॥

म्लेच्छास्यं, क्ली, (म्लेच्छे म्लेच्छदेशे आस्यमुत्पत्ति-रस्य ।) ताम्रम् । इति हारावली ॥

म्लेच्छितं, क्ली, (म्लेछ् देश्योक्तौ + क्तः ।) म्लेच्छ-भाषा । अपशब्दः । तत्पर्य्यायः । परभाषा २ ।इति हारावली ॥

https://sa.wikisource.org/wiki/शब्दकल्पद्रुमः/

म्लिष्ट न० म्लेच्छ--क्त नि० । १ अविस्पष्टवाक्ये २ तद्वाक्ययुक्ते ३ म्लाने च त्रि० मेदि०

म्लेच्छ अपशब्दे वा चु० उभ० पक्षे भ्वा० पर० अक०सेट् । म्लेच्छयति ते म्लेच्छति अमम्लेच्छत् त अम्लेच्छीत् **म्लेच्छ** पु० म्लेच्छ--घञ् । १ अपशब्दे "म्लेच्छोह वा नामयदप्रशब्द" इति श्रुतिः । कर्त्तरि अच् । २ पामरजातौ,३ नीचजातौ च पुंस्त्री० स्त्रियां ङीष् "गोमांसखादकोयस्तु विरुद्धं बहु भाषते । सर्च्चाचारविहीनश्च म्लेच्छइत्यभिधीयते" बौधायनः । ४ पापरते त्रि० मेदि० ।५ हिङ्गुले न० राजनि० ।

म्लेच्छकन्द पु० म्लेच्छप्रियः कन्दः शा० त० । लशुने राजनि०

म्लेच्छजाति स्त्री म्लेच्छाभिधा जातिः । गोमांसादिभक्षकेकिरातादिजातिभेदे अमरः ।

म्लेच्छदेश पु० म्लेच्छाधारो देशः । चातुर्वर्ण्याचाररहितेदेशे अमरः । "चातुर्वर्ण्यव्यवस्थानं यस्मिन् देशे नविद्यते । म्लेच्छदेशः स विज्ञेय आर्य्यावर्त्तस्ततःपरम्" ।

म्लेच्छभोजन न० म्लेच्छैर्भुज्यते भुज--कर्मणि ल्युट् ।१ यावके अन्नभेदे शब्दर० । २ गोधूमे पु० त्रिका० ।

म्लेच्छमण्डल न० ६ त० । म्लेच्छदेशे हेमच० ।

म्लेच्छमुख न० म्लेच्छानां मुखमिव रक्तत्वात् । ताम्रे अमरः ।म्लेच्छास्यमप्यत्र हारा० ।

म्लेच्छित न० म्लेच्छ--क्त । अपशब्दे असंस्कृतशब्दे हारा० ।

https://sa.wikisource.org/wiki/वाचस्पत्यम्

पञ्चजनः, पुं, (पञ्चभिर्भूतैर्जन्यतेऽसौ । पञ्च +जन् + कर्म्मणि घञ् । "जनिवध्योश्च" । ७ । ३ । ३५ ।इति न वृद्धिः ।) पुरुषः । इत्यमरः । २ । ६ । १ ॥(यथा, राजतरङ्गिण्याम् । ३ । ३५५ ।"सद्भावश्र्यादिका देव्यस्तेन श्रीशब्दलाञ्छिताः ।पञ्च पञ्चजनेन्द्रेण पुरे तस्मिन् निवेशिताः ॥")दैत्यविशेषः । स च संह्रादस्य कृतौ पत्न्यां जातः ।(यथा, श्रीभागवते । ६ । १८ । १४ ।"संह्रादस्य कृतिर्भार्य्याऽसूत पञ्चजनं ततः ॥"अपरो दैत्यभेदः । श्रीकृष्णः एनं हत्वा सान्दी-पनिमुनये अस्य मृतं पुत्त्रं गुरुदक्षिणास्वरूपं ददौ । यथा, भागवते । ३ । ३ । २ । "सान्दीपनेः सकृत् प्रोक्तं ब्रह्माधीत्य सविस्तरम् ।तस्मै प्रादात् वरं पुत्त्रं मृतं पञ्चजनोदरात् ॥")अस्यास्थ्ना पाञ्चजन्यनामा शङ्खो जातः । सचश्रीकृष्णस्य । यथा, --"पाञ्चजन्यं हृषीकेशो देवदत्तं धनञ्जयः ॥"इत्यादि श्रीभगवद्गीतायाम् । १ । १५ ॥ (प्रजा-पतिः । यथा, भागवते । ६ । ४ । ५१ ।"एषा पञ्चजनस्याङ्ग ! दुहिता वै प्रजापतेः ।असिक्नी नाम पत्नीत्वे प्रजेश ! प्रतिगृह्यताम् ॥"सगरराजपुत्त्रः । यथा, हरिवंशे । १५ । ६ ।"केशिन्यसूत सगरादसमञ्जसमात्मजम् ।राजा पञ्चजनो नाम बभूव स महाबलः ॥"गन्धर्ब्बाः पितरो देवा असुरा रक्षांसि पञ्चजन-पदवाच्यानि भवन्ति । इति चिन्तामणिः ॥)

पञ्चजनी, स्त्री, पञ्चानां जनानां समाहारः ततोङीप् । इति व्याकरणम् ॥ (विश्वरूपकन्या ।यथा, भागवते । ५ । ७ । १ । "तदनुशासनपरःपञ्चजनीं विश्वरूपदुहितरमुपयेमे ॥")**पञ्चजनीनः**, पुं, (पञ्चसु जनेषु व्यापृतः । दिक्-संख्ये संज्ञायामिति समासः । पञ्चजने हितम् ।पञ्चजनादुपसङ्ख्यानमितिखः । पां । ५ । १ । ९ ।वार्त्तिके ।) भण्डः । इति हलायुधः ॥ पञ्च-जनसम्बन्धिनि पञ्चजन्याः प्रभौ च त्रि ॥

https://sa.wikisource.org/wiki/शब्दकल्पद्रुमः/

पञ्चजन पु० पञ्चभिः भूतैर्जन्यते जन--घञ्, न वृद्धिः । १ मनुष्ये अमरः । २ मनुष्यसम्बन्धिनि प्राणादौ ३ मनुष्यतुल्ये देवादौ४ मनुष्यभेदे ब्राह्मणादौ च शा० सू० भाष्यम् तथाहि "यस्मिन् पञ्च पञ्च जना अ काशश्च प्रतिष्ठितः" तद्धृतवेदमन्त्रे "प्रञ्च पञ्च शब्दशब्देन पञ्चविंशतितत्त्वान्युच्यन्ते" इति सांख्यमतनिराकरणेन पञ्चजनशब्दस्य प्राणादिपरत्वं समर्थितं दिङ्मात्रमत्रोच्यते "कथं तर्हि पञ्च पञ्चजना इति, उच्यते, "दिक्सङ्ख्ये संज्ञायामिति" पा० विशेषस्मरणात् मंज्ञायामेव पञ्चशब्दस्य जनशब्देन समासः, ततश्च रूढ़त्वाभिप्रायेणैव केचित् पञ्चजना नाम विवक्ष्यन्ते, न साङ्ख्यतत्त्वाभिप्रायेण, ते कतीत्यस्यामाकाङ्क्षायां पुनः पञ्चेति प्रयुज्यते, पञ्चजना नाम केचित्, ते च पञ्चेत्यर्थः सप्तर्षयः सप्तेति यथा । के पुनस्ते पञ्चजना नामेति तदुच्यते । "प्राणादयो वाक्यशेषात्" शा० सू० "यस्मिन् पञ्च पञ्चजना इत्यत उत्तर- स्मिन्मन्त्रे ब्रह्मस्वरूपनिरूपणाय प्राणादयः पञ्च निर्दिष्टाः "प्राणस्य प्राणमुत चक्षुषश्चक्षुरुत श्चोत्रस्य श्रोत्रमन्नस्यान्नं षनसो ये मनो विदुः" इत्यत्र वाक्यशेषगताः सन्निधानात् पन्चजना विवक्ष्यन्ते । कथं पुनः प्राणादिषु पञ्चजन-शब्दप्रयोगः तत्त्वेषु वा कथं जनशब्दप्रयोगः समानेतु प्रसिद्ध्यतिक्रमे वाक्यशेषवशात् प्राणादय एव ग्रही- तव्या भवन्ति जनसम्बन्धाच्च प्राणादयो जनशब्दभाजोभवन्ति । जनवचनश्च पुरुषशब्दः प्राणेषु प्रयुक्तः, "ते वाएते पञ्च ब्रह्म पुरुषाः" इति, अत्र "प्राणो ह पिताप्राणो ह माता" इत्यादि च ब्राह्मणम् । समासबलाच्चसमुदायस्य रूढ़त्वमविरुद्धम् । कथं पुनरसति प्रथम-प्रयोगे रूढ़िः शक्याश्रयितुं शक्या, उद्भिदादिवदित्याह ।प्रसिद्धार्यसन्निधानेन ह्यप्रसिद्धार्थः शब्दः प्रयुज्यमानःसमभिव्याहारात् तद्विषयो नियम्यते यथोद्भिदा यजेतुयूपं चिनत्ति, वेदिं करोतीति । तथाऽयमपि पञ्चजन-शब्दः समासान्वाख्यानादवगतसंज्ञाभावः सेज्याकाङ्क्षीवाक्यशेषसमभिव्याहृतेषु प्राणादिषु वर्त्तिष्यते । कैश्चित्तुदेवाः पितरो गन्धर्वा असुरा रक्षांसि च पञ्च पञ्च जनाव्याख्याताः । अन्यैश्चत्वारो वर्णा निषादपञ्चमाःपरिगृहीताः" । तत्परिग्रहेऽपीह न कश्चिद्विरोधः ।आचार्यस्तु न पञ्चविंशतेस्तत्त्वानामिह प्रतीतिरस्तीत्येवं-परतया प्राणादयो वाक्यशेषादिति जगाद" ।२ दैत्यभेदे च यस्यास्थ्ना कृष्णस्य पाञ्चजन्यः शङ्खोजातः ।पाञ्चजन्यशब्दे दृश्यम् । ३ सृञ्जयनृपपुत्रे हरिवं० ३२ अ०अंशुमतः पितरि मगधनृपस्य ४ पुत्रभेदे १५० अ० ।५ प्रजापतिभेदे भाग० ६ । ४ । ४६ विश्वावसुदुहितरि भरतस्य६ पत्न्यां स्त्री ङीप् । भाग० ५ । ७ । १ **पञ्चजनीन** पञ्चसु जनेषु व्यापृतः ख । भण्डे हला० । https://sa.wikisource.org/wiki/वाचस्पत्यम्

நாலு nālu, *n.* < நால் +. 1. [Tu. *nālu.*] The number four; நான்கு. 2. Many, manifold; பல. நாலுவிஷயமும் தெரிந்தவன். 3. A few; சில. நாலு வார்த்தைதான் பேசினான். 4. See நாலடியார். ஆலும் வேலும் பல்லுக்குறுதி, நாலுமிரண்டுஞ் சொல்லுக் குறுதி.
நால்² nāl, n. [T. nāllu, K. M. nāl.] 1. The number four; நான்கு. நாலிரு வழக்கிற் றாபதப் பக்கமும் (தொல். பொ. 75).

पुरूरवा ह पुरा ऐडो राजा कल्याण आस तꣳ होर्वश्यप्सराभिदध्यौ। तꣳ संवत्सरं कामयमानानुचचारैवꣳ ह स्म वै पूर्वे ऽभिश्राम्यन्ति । तद्धातिचिरं मेने तस्य ह धावत: पुरो रथं कर्त दर्शयामास । तꣳ ह दृष्ट्वा राजावतस्थौ तꣳ हावस्थाय न ददर्शाथो ह पुनरातस्थौ तꣳ हास्थायैव ददर्श । स ह सारथिं पप्रच्छ सारथे किं पश्यसीति । त्वां भगव इति होवाच रथमश्वान्पन्थानमिति । स हेक्षां चक्रे दृप्यामि वै किलेति । तꣳ ह वागभ्यु-वाच न वै दृप्यस्यहं वै त्वामेतं कर्तमदीदृशमिति । अथ कस्त्वमित्यह-मुर्वश्यप्सरेति होवाच । सा त्वा संवत्सरं कामयमानान्वचारिषं तां मा जायां विन्दस्वेति । दुरुपचारा ह वै भवति देवा इति होवाच । का त उपचर्येति । शतं ममोपसद: । शतꣳशतं मा सर्पिष्कुम्भा अहरह-रागच्छेयुस्तदाशना स्यां न त्वा नग्नं पश्येयमिति । सर्वमेवैतद्भगवति सुकरमिति होवाच । कथा त्वपि जाया पतिं नग्नं न पश्यतीत्यन्तर्वासं वसीथा इति होवाचानग्नो भवेति । तया सहोवासान्तर्वासं वसान: सा ह स्म जाताञ्जातानेव पुत्रानपविध्यति । ताꣳ ह राजोवाच पुत्रकामा उ वै भगवति वयं मनुष्या: स्मो जाताञ्जातानु त्वमपविध्यसीति । सा होवाच पर्यवेतरात्रयो भवन्ति क्षीणायुषो ऽन्ये भूय: प्रियं करवावहा इति । सायुं चामावसुं च जनयां चकार । सा होवाचेमौ बिभृतेमौ सर्वमायुरेष्यत इति। प्राङायु: प्रवव्राज । तस्यैते कुरुपञ्चाला: काशिविदेहा इति । एतदायवं प्रव्राजं प्रत्यङमावसुस्तस्यैते गान्धारय स्पर्शवो ऽराट्टा इत्येतदामावसवम् ॥ ४४ ॥

XVIII.44

ŚADAUPAŚADAU

Purūravas, son of Iḍa, was a benevolent king. Apsaras Ūrvaśī became attached to him. Desiring him, she wandered for a year. Former people toiled in this manner. She felt the lingering too much. While he was traversing, she produced a pit in front of his chariot. Gazing at it, the king climbed down. Having climbed down, he did not perceive the pit. He again ascended (the chariot). After having ascended, he saw the pit. He asked the charioteer, "O charioteer, what do you see?" "You my lord" he said. "(I also see) the chariot, the horses and the path." He thought, "I have gone mad." A voice in the air uttered, "Thou hast not gone mad; I have made thee gaze the pit." "Who art thou?" "I am Apsaras Ūrvaśī. Desiring thee, I have been wandering for a year. Make me thy bride." "O lady, gods are difficult to approach" said he. "How shall I approach thee ?" "Mine shall be a hundred attendants. Every day I shall require a hundred pitchers of ghee. That will be my food. I shall not look at thee in a naked condition." "All this is possible, my lady," said he. "But how is it that a bride would not see her husband in a naked condition ?" "Do thou wear an inner garment so that thou wilt not be naked," said she. Wearing an inner garment, he lived with her, she threw away all the sons as soon as each one was born. The king said to her, "We, human beings are fond of sons, O lady. Thou art throwing away the sons as soon as each one is born." She said, "They have turned the night around; others were short-lived. We shall again enjoy." She generated two sons-Āyu and Amāvasu. She said, "Do you rear them; they shall live the full life." Āyu moved towards the east. Kuru-Pāñcāla and Kāśī-Videha were his regions. This is the realm of Āyu. Amāvasu proceeded towards the west. The Gāndhāris, Sparśus and Arāṭṭas were his regions. This is the realm of Amāvasu.

http://www.ignca.nic.in./eBooks/The_Baudhayana_Srauta-Sutra_Vol._III.pdf Baudhāyana Śrauta - *sutra* Edited & Translated by: C. G. KASHIKAR (2003)

1 http://arabian-archaeology.com/images/ic-007.jpg
2 George Coedes, *Histoire ancienne des Etats hindouises d'Extreme-Orient*,1944 (*Ancient History of the Hinduised States of Ancient Far East*)
3 https://www.dropbox.com/s/ykm93xf4unhordu/IndianLexicon.pdf?dl=0
4 Dennys Frenez and Massimo Vidale, 2012, *Harappa Chimaeras as 'Symbolic Hypertexts'. Some Thoughts on Plato, Chimaera and the Indus Civilization*
http://a.harappa.com/content/harappan-chimaeras
5 http://huntingtonarchive.osu.edu/Makara%20Site/makara
6 https://www.dropbox.com/s/ykm93xf4unhordu/IndianLexicon.pdf?dl=0
7 Three Gold pendants: Jewelry Marshall 1931: 521, pl. CLI, B3
8 Also called Indus Script. Decipherment demonstrates that an apt expression will be *Bharata Hieroglyphs.*
9 The total number of Harappa script inscriptions now total over 7,000.

10 S. Kalyanaraman, 2016, *Harappa Script & Language*, Amazon
11 *Indian Lexicon* – a comparative dictionary of 25+ languages of Bharata
https://www.dropbox.com/s/ykm93xf4unhordu/IndianLexicon.pdf?dl=0
12 Possehl, Gregory L. (1996). *Indus Age: The Writing System*. University of Pennsylvania Press.
13 A phrase used by linguist MB Emeneau
14 http://tinyurl.com/zdqqnds
15 Robert S. Wicks, 1992, Money, markets and trade in early Southeast Asia: the development of indigenous monetary systems to AD 1400, SEAP Publications, Cornell, Ithaca, NY, p.245; loc. Cit. Mulavarman's First Yūpa Inscription; FH van Naerssen, and RC longh, *The economic and administrative history of early Indonesia,* Leiden: Brill, 1977), p. 20; J. Ph. Vogel, 'The Yūpa inscription of Mulavarman from Koelei (East Borneo)', Bijdragen tot de Taal-, *Land-en Volkenkunde* 79, 1918: 213; Mulavarman's Second Yūpa Inscription, Vogel, 'Yūpa Inscriptionś, p. 214; Mulavarman's Third Yūpa Inscription; Vogel, '*Yūpa Inscriptions*', p. 215.
16 https://en.wikipedia.org/wiki/Tuisto
17 https://en.wikipedia.org/wiki/List_of_countries_by_largest_historical_GDP Pace Angus Maddison's work for OECD.

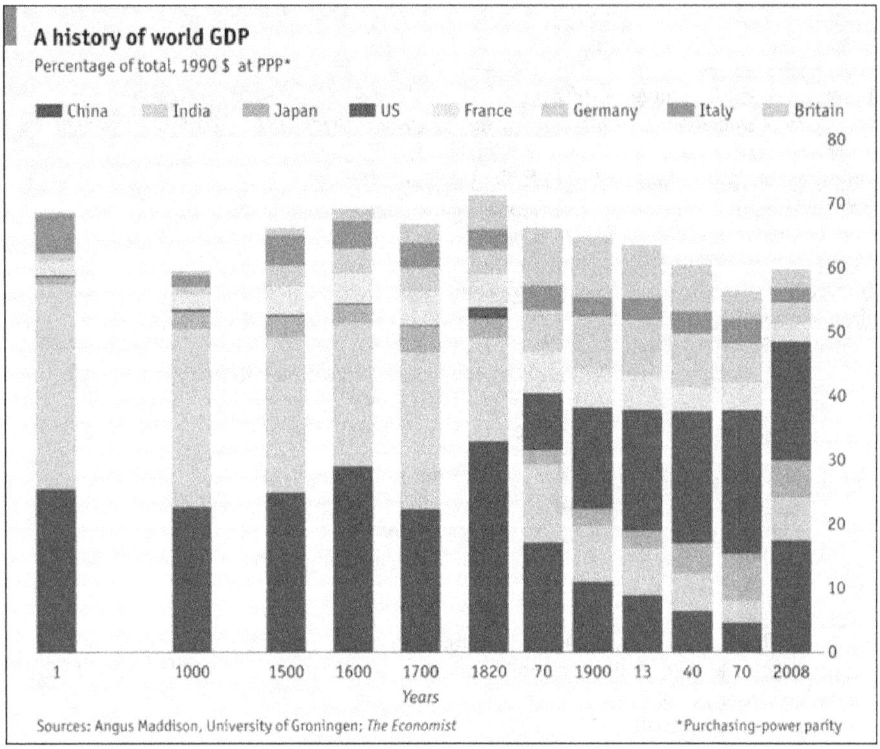

18 IndianLexicon.pdf (file://HP-PC/Users/HP/Google%20Drive/IndianLexicon.pdf)
19 http://huntingtonarchive.osu.edu/Makara%20Site/makara

www.ingramcontent.com/pod-product-compliance
Lightning Source LLC
Chambersburg PA
CBHW081111180526
45170CB00008B/2799
* 9 7 8 1 5 4 1 2 7 4 9 2 1 *